JUDGMENT AND DECISION MAKING

Baruch Fischhoff is a long-time leader in behavioral decision research. This book presents his most important work in the field, offering the reader both an understanding of how decisions are made and lessons on how to improve decision making. Relevant across a range of professional sectors and academic disciplines, this book is essential reading for those who need a breadth and depth of understanding of this field and how it can help us in the important decisions we face.

This book includes sections on hindsight bias, judgment under uncertainty, value elicitation, and decision making. Applied topics include adolescent risk taking, sexual assault, insurance, environment, medicine, technology, national security, and natural disasters. Introductory sections highlight important underlying themes: that understanding decision making requires research that integrates formal analysis (of how decisions should be made), empirical studies (of how decisions are actually made), and interventions (on how decision making can be improved). If science can clarify people's decision-making needs and abilities, while society provides the information and opportunities for them to steer their own lives, Fischhoff's view offers a feeling of guarded optimism regarding the opportunities for helping people to make sound choices. Thus, the readings provide scientific input to the ongoing debate on the proper roles of laypeople and technical experts in charting society's course.

Baruch Fischhoff is Howard Heinz University Professor, Carnegie Mellon University, USA, Department of Social and Decision Sciences, Department of Engineering and Public Policy.

EARTHSCAN RISK IN SOCIETY SERIES

Series editor: Ragnar E. Löfstedt

1. **Calculating Political Risk**
   Catherine Althaus

2. **The Citizens at Risk**
   From Urban Sanitation to Sustainable Cities
   Gordon McGranahan, Pedro Jacobi, Jacob Songsor, Charles Surjadi and Marianne Kjellen

3. **The Earthscan Reader on Risk**
   Edited by Ragnar E. Löfstedt and Åsa Boholm

4. **The Ethics of Technological Risk**
   Edited by Lotte Asveld and Sabine Roeser

5. **Facility Siting**
   Risk, Power and Identity in Land-Use Planning
   Edited by Åsa Boholm and Ragnar E. Löfstedt

6. **The Feeling of Risk**
   New Perspectives on Risk Perception
   Paul Slovic

7. **Foresight in Action**
   The Practice of Dealing with Uncertainty in Public Policy
   Marjolein B. A. van Asselt, Susan A. van 't Klooster, Philip W. F. van Notten and Livia A. Smits

8. **Global Environmental Risk**
   Jeanne X. Kasperson and Roger E. Kasperson

9. **Hazards, Vulnerability and Environmental Justice**
   Susan L. Cutter

10. **The Perception of Risk**
    Paul Slovic

11. **Public Safety and Risk Assessment**
    Improving Decision Making
    David J. Ball and Laurence Ball-King

12. **Risk Governance**
    Coping with Uncertainty in a Complex World
    Ortwin Renn

13. **Risk Management in Post-Trust Societies**
    Ragnar E. Löfstedt

14. **Risk, Media and Stigma**
    Understanding Public Challenges to Modern Science and Technology
    Edited by James Flynn, Paul Slovic and Howard Kunreuther

15. **Risk, Uncertainty and Rational Action**
    Carlo C. Jaeger, Ortwin Renn, Eugene A. Rosa and Thomas Webler

16. **The Social Contours of Risk (Volumes 1 & 2)**
    Jeanne X. Kasperson and Roger E. Kasperson

17. **Social Trust and the Management of Risk**
    Edited by George Cvetkovich and Ragnar E. Löfstedt

18. **The Spatial Dimension of Risk**
    How Geography Shapes the Emergence of Riskscapes
    Detlef Müller-Mahn

19. **The Tolerability of Risk**
    A New Framework for Risk Management
    Edited by Frédéric Bouder, David Slavin and Ragnar E. Löfstedt

20. **Transboundary Risk Governance**
    Rolf Lidskog, Linda Soneryd and Ylva Uggla

21. **Transboundary Risk Management**
    Edited by Joanne Linnerooth-Bayer, Ragnar E. Löfstedt and Gunnar Sjöstedt

22. **Trust in Cooperative Risk Management**
    Uncertainty and Scepticism in the Public Mind
    Michael Siegrist, Timothy C. Earle and Heinz Gutscher

23. **Trust in Risk Management**
    Uncertainty and Scepticism in the Public Mind
    Edited by Michael Siegrist, Timothy C. Earle and Heinz Gutscher

24. **Uncertainty and Risk**
    Multidisciplinary Perspectives
    Edited by Gabriele Bammer and Michael Smithson

25. **Judgment and Decision Making**
    Baruch Fischhoff

26. **Risk Analysis and Human Behavior**
    Baruch Fischhoff

JUDGMENT AND DECISION MAKING

Baruch Fischhoff

earthscan
publishing for a sustainable future
LONDON AND NEW YORK

First published 2012
by Earthscan
2 Park Square, Milton Park, Abingdon, Oxon OX14 4RN

Simultaneously published in the USA and Canada
by Earthscan
711 Third Avenue, New York, NY 10017

Earthscan is an imprint of the Taylor & Francis Group, an informa business

British Library Cataloguing in Publication Data
A catalogue record for this book is available from the British Library

Library of Congress Cataloging in Publication Data
Fischhoff, Baruch, 1946-
Judgment and decision making / Baruch Fischhoff.
p. cm. -- (Earthscan risk in society series)
Includes bibliographical references and index.
1. Judgment. 2. Decision making. 3. Risk-taking (Psychology) I. Title.
BF447.F57 2012
153.4'6--dc23
2011027664

ISBN: 978-1-84971-444-0 (hbk)
ISBN: 978-1-84971-445-7 (pbk)
ISBN: 978-0-203-14193-9 (ebk)

Typeset in Bembo
by GreenGate Publishing Services, Tonbridge, Kent

CONTENTS

PART V DECIDING 247

FIGURES

TABLES

ACKNOWLEDGMENTS

I am grateful to Mikey Jones, Alison Kuznets, Ragnar Löfstedt, Charlotte Russell and Rosa Stipanovic for their help in making this volume happen; to my teachers and mentors in the Detroit Public Schools, Wayne State University, and the Hebrew University of Jerusalem, for providing my education; to the governments and others that supported those institutions and my research; to my family for their love and support; and to all for helping me to see research as a form of public service.

ON THE COVER

In *Overgrown* the elk as ruminant encourages the viewer to ruminate across and on the image. As we traverse the branching nexus of antlers, entirely new and magical worlds open up. The experience is that of excited electrons surging through perpetually bifurcating pathways, forks in roadways that will never intersect, though trails extend toward each other. Endless rhizomatic possibility is promised. Yes, these are the branching decision trees, the individual trains of thought and the social networks that combine or proliferate to become the collective experience. The image conveys both the implications and consequences of individual decisions, if considering the elk head as undivided consciousness, and the tethering of myriad positions to a sovereign no longer central control, if looking through a political lens. The piece pivots on seeing a system so grand, a security system so exhaustive that it crashes under its own weight. It also conveys personal pain, anguish, turmoil, psychic trauma and physical paralysis. Considering the implications and risks of a potential decision or trying to retrace the (mis)steps back to a point of origin can be as damning as it can be enlightening.

Andrew Ellis Johnson
Overgrown, 1997

Oil on Belgian linen, 78" × 62"

http://artscool.cfa.cmu.edu/~johnson/

INTRODUCTION

From early in my career, I have been fortunate to have colleagues who valued research that was informed by both analytical and empirical sciences. Indeed, before I even had a career, Francine Wehmer and Ira Firestone, two psychology professors at Wayne State University, matched me with Samuel Komorita, as a research assistant for his pioneering work in what is now called behavioral game theory. At the time, I was too much of a novice to appreciate Sam's innovativeness in asking how people actually responded to prisoner's dilemma games—although I did realize that it was a suitable match for a math-psych major.

I had even less appreciation of the uniqueness of the guidance that I received from David Jonah, my advisor in the mathematics department, about the role of empirical knowledge in the evolution of that analytic discipline. As personal gifts, he gave me Otto Neugebauer's *The exact sciences in antiquity*, Carl Friedrich Gauss's *Disquisitiones Arithmeticae*, and G.H. Hardy's *A mathematician's apology*. I finished only the last, but from their gist (and from his confidence in me), I inferred that academic fields had no natural bounds, and could be informed by both thought and experience. My family prized education, but had little familiarity with academia. (Our father, an industrial chemist, was the first to go to high school, also at Wayne, along with my mother, sister, and brother.) The time being the Sixties also supported my feeling that anything was possible in academia—without being too specific about what that might be.

After graduating Wayne and three years of kibbutz living, I had the extraordinary good fortune to have Amos Tversky accept me as a student, bringing me into a world where anything was actually possible—for someone with his intellect and independence of mind. Soon afterward, he began his collaboration with Danny Kahneman, marked by their joint willingness to challenge any orthodoxy. At that time, the Hebrew University of Jerusalem was a remarkable place, in part because the university community included people who had grown up together, hence could reach across fields in ways that would have been impossible without that long-standing trust. Philosophers, statisticians, historians, engineers, and others dropped in on Amos and Danny's weekly seminars on judgment and decision making.

For me, this liberating atmosphere was amplified by two unrelated seminars. One was Danny's course in "Applications of Psychology," where each week's assignment required addressing an applied problem (e.g., designing hard-to-counterfeit bills, resolving conflicts in a cooperative farming village), using whatever psychology had to offer. The seminar taught me that psychology's validity depends on its usefulness, that no single theory or method could solve any serious problem, and that the search for solutions required both analysis and evidence. In Alan Baddeley's insightful formulation, psychology progressed through a combination of applied basic research, seeing how well basic results hold up in applied settings, and basic applied research, extracting new fundamental problems from applications.

The second unrelated seminar was an informal gathering of graduate students from different departments, convened by Reuven Kaminer, associate dean for foreign students (and a fellow Detroiter). Our main text was *The Sociological Imagination*, by C. Wright Mills (1959). Following Mills, we took our "personal worlds" (here, our chosen disciplines) as points of departure for examining our social worlds (here, the pressures shaping research questions). Although I remember few details (beyond the critique of Talcott Parsons as mind-numbingly dull), I absorbed something of Mills's turn of mind (or "imagination"), in which the scientific enterprise depends on continuously challenging how research problems are formulated.

The nascent state of decision-making research made it easier to adopt a challenging frame of mind. Economists, psychologists, statisticians, philosophers, and others had not yet sorted out their respective roles—sometimes even collaborating (as seen in Edwards, Lindman and Savage, 1963, and Edwards and Tversky, 1967). Sarah Lichtenstein's comprehensive "List" of researchers committed to collecting behavioral data framed in analytical terms had fewer than 100 members worldwide. Even that number included several in Eastern Europe whose opportunities were limited by their use of such suspect terms as "utility" and "subjective" probability. (As Stalin said, "We don't forecast, we plan.")

A further spur to my pursuing new research topics was the poor quality of scientific communication in those days. Far off in the Middle East, it seemed best to work on problems that no one else was likely to touch. My somewhat unusual preference for topics with both analytical and empirical elements made that somewhat easier. Compared to joining the fray on a hot problem, this research strategy has the benefit of getting to structure messy problems, along with the risk of working on marginal ones. Having stumbled across this strategy more or less as a matter of necessity, I have followed it more or less as a matter of taste. My next stop, working with Robyn Dawes, Lita Furby, Sarah Lichtenstein, and Paul Slovic, in Eugene, Oregon, afforded me superb help in cultivating these tastes. At Carnegie Mellon University, they are a norm.

The University views its strategic advantage as letting faculty do whatever they want, as long as they can eventually convince some serious audience of the value of their work.

The articles collected in this volume reflect my efforts to take advantage of the extraordinary mentoring that I've received and the freedom that academic life affords us. A companion volume collects basic applied research regarding a wide variety of risks ranging from waterborne diseases and nuclear power to sexual assault and adolescent sexuality. Given the diversity of these topics and the lack of boundaries on "real world" problems, my work has taken me into the fundamental topics around which this collection is organized: judgment, preference, and decision making. Along the way, I have had the further good fortune of wonderful collaborators. I hope that others have similar opportunities.

References

Baddeley, A. 1979. Applied cognitive and cognitive applied research. In L. G. Nilsson (Ed.), *Perspectives on memory research*. Hillsdale, NJ: Erlbaum.

Edwards, W., Lindman, H., and Savage, L.J. 1963. Bayesian statistical inference for psychologists. *Psychological Review, 70*, 193–242.

Edwards, W. and Tversky, A. (Eds.). 1967. *Decision making*. Wandsworth: Penguin.

Gauss, C.F. 1801, 1966. *Disquisitiones Arithmeticae* (translated by A.A.Clarke). New Haven: Yale University Press.

Hardy, G.H. 1940. *A mathematician's apology*. Cambridge: Cambridge University Press.

Mills, C.W. 1959. *The sociological imagination*. Oxford: Oxford University Press.

Neugebauer, O. 1957. *The exact sciences in antiquity*. Mineola, NY: Dover.

PART I

Background

The field currently called behavioral decision research was first identified by Ward Edwards (1954), in "A theory of decision making." Writing but a decade after von Neumann and Morgenstern's classic (1944) exposition of utility theory (see also Savage, 1954), Ward showed the essential role of behavioral research in translating utility theory from a brilliant abstraction to a practical way to understand and improve actual behavior. The drift in the field's name, through the intermediate behavioral decision making, reflects the growing realization that no single theory could cover the diversity of decision making. The field has, however, retained its unifying commitment to examining choices from both analytical and empirical perspectives, with utility theory as a point of departure for determining how decisions should be made. Decision analysis has always been the term for approaches trying to help people see their decisions in utility theory terms (Clemen and Reilly, 2003; Hammond, Keene and Raiffa, 1999; Raiffa, 1968).

"Judgment and decision making" (Chapter 1) offers a historical overview of the field. In the spirit of Ward's vision, it emphasizes the interplay between normative analysis of decisions' formal properties, descriptive research into how people actually behave, and prescriptive interventions for bridging the gap between the normative ideal and the descriptive reality. When interventions fail, any of the three stages could be at fault. The normative analysis might not have identified the issues that matter to the decision makers. The descriptive research might not have captured their intuitive perspectives faithfully. The intervention might not have addressed their problems well. The chapter traces research in confidence assessment, preference formation, clinical judgment, and other topics, often following the seminal formulations from the University of Michigan's mathematical psychology group, which included

Edwards, Clyde Coombs, Robyn Dawes, David Krantz, and Anatol Rapoport, among others. The chapter concludes by analyzing alternative default assumptions about where problems lie when decisions seem wrong.

"Amos Tversky" (Chapter 2) describes some of the path-breaking research of one of my graduate advisors, Amos Tversky, often in collaboration with my other advisor, Daniel Kahneman. That research exemplifies formulating problems in clear analytical terms, broadly informed by empirical research. The brief summaries here provide context for the chapters that follow, along with a small measure of gratitude for the advising that I received.

References

Clemen, R.T., and Reilly, T. (2003). *Making hard decisions.* Boston: Duxbury.

Edwards, W. (1954). A theory of decision making. *Psychological Bulletin, 54,* 380–397.

Edwards, W., Lindman, H.R., Savage, L.J. (1963). Bayesian statistical inference for psychological research. *Psychological Review, 70,* 193–242.

Hammond, J.S., Keeney, R.L., and Raiffa, H. (199). *Smart choices.* New York: Broadway.

Raiffa, J. (1968). *Decision analysis.* Belmont, MA: Addison-Wesley.

Savage, L.J. (1954). *The foundations of statistics.* Wiley: New York.

von Neumann, J., and Morgenstern, O. (1944). *Theory of games and economic behavior.* Princeton: Princeton University Press.

1

JUDGMENT AND DECISION MAKING

Baruch Fischhoff

Abstract

The study of judgment and decision making entails three interrelated forms of research: (1) *normative analysis*, identifying the best courses of action, given decision makers' values; (2) *descriptive studies*, examining actual behavior in terms comparable to the normative analyses; and (3) *prescriptive interventions*, helping individuals to make better choices, bridging the gap between the normative ideal and the descriptive reality. The research is grounded in analytical foundations shared by economics, psychology, philosophy, and management science. Those foundations provide a framework for accommodating affective and social factors that shape and complement the cognitive processes of decision making. The decision sciences have grown through applications requiring collaboration with subject matter experts, familiar with the substance of the choices and the opportunities for interventions. Over the past half century, the field has shifted its emphasis from predicting choices, which can be successful without theoretical insight, to understanding the processes shaping them. Those processes are often revealed through biases that suggest non-normative processes. The practical importance of these biases depends on the sensitivity of specific decisions and the support that individuals have in making them. As a result, the field offers no simple summary of individuals' competence as decision makers, but a suite of theories and methods suited to capturing these sensitivities.

Introduction

Decisions are easy when decision makers know what they want and what they will get, making choices from a set of well-defined options. Such decisions could be equally easy, but reach different conclusions, for people who see the facts similarly, but have different goals, or for people who have the same values but see the facts differently, or for people who disagree about both facts and values.

Decision making can become more difficult when there is uncertainty about either what will happen or what one wants to happen. Some decisions are so sensitive to estimates of fact or value that it pays to invest in learning, before acting. Other decisions will work out just as well, for any plausible estimates.

Thus, any account of decision-making processes must consider both the decisions and the individuals making them. The field of *behavioral decision research* provides such accounts. It entails three forms of research: (1) *normative*, identifying the best possible choice, given the state of the world and decision makers' values; (2) *descriptive*, characterizing how individuals make decisions, in terms comparable to the normative standard; and (3) *prescriptive*, attempting to close the gap between the normative ideal and the descriptive reality.

Although they can be described as an orderly progression, these three forms of research are deeply interrelated. Descriptive research is needed to reveal the facts and values that normative analysis must consider. Prescriptive interventions are needed to assess whether descriptive accounts provide the insight needed to improve decision making. Normative analyses are needed to understand the facts that decision makers must grasp and the practical implications of holding different values. Thus, understanding choices requires an iterative process, cycling through the three stages. This chapter follows the evolution of theory and method for seeking that understanding.

Behavioral decision research

Behavioral decision research emerged from normative models of decision making developed by philosophers, mathematicians, and economists.[1,2] These models describe how to determine the best possible course of action, given what individuals believe about the world and what they want from it. Individuals who follow these rules are said to be *rational*. Their choices are *optimal*, if they are well informed about the world and about their own values. Although normative models take strong positions on *how* decisions should be made, they are mute regarding *what* options, facts, and values should be considered. As a result, they require the empirical content provided by descriptive and prescriptive research to be anything but formalisms.

Comparability with normative analysis imposes important constraints on descriptive and prescriptive research. They cannot begin without first examining the world from decision makers' perspective. They cannot criticize choices without asking whether they might be rational, given what people want and believe. They cannot assess the importance of imperfections in decision-making processes, without embedding them in normative analyses, showing their practical implications. Imperfections can be theoretically informative without mattering much. Indeed, nonrational processes may survive because they have too little practical significance to provide the sharp negative feedback sometimes needed to change behavior.

Psychology progresses, in part, by applying what Berkeley and Humphreys[3] call the "bias heuristic," identifying departures from normative standards.[4] However, unless those standards are well defined, vaguely similar biases may proliferate. Different biases might share a common name (e.g., confirmation bias); the same bias might have different names (e.g., saliency, availability), impeding scientific progress.[5] Indeed, as discussed next, a major advance in early behavioral decision research was discovering that seemingly different theories were often indistinguishable.

Clinical judgment

World War II was a turning point for psychology, which showed its ability to assess efficiently the skills and problems of masses of individuals. After the war, attention turned to the effectiveness of those efficient assessments. These studies of *clinical judgment* quickly spread to topics as diverse as how psychologists decide whether clients are "neurotic" or "psychotic," radiologists sort ulcer X-rays into "benign" or "malignant," bank officers classify loans as "nonperforming," and brokers weigh stocks' prospects.[6–8]

Conducting studies of clinical judgment is straightforward: collect many judgments of cases described on a common set of possibly relevant *cues*. Use statistical methods (e.g., multiple regression) to predict those judgments from the cues. For example, Dawes[9] studied University of Oregon Psychology Department graduate admission committee evaluations of 384 applicants. Although applicants' files had many cues (e.g., letters of recommendation, full transcripts), the committee's ratings could be predicted well from just three: Graduate Record Examination (GRE) score, undergraduate grade point average (GPA), and quality of undergraduate institution (QI):

$$0.0032\ \text{GRE} + 1.02\ \text{GPA} + 0.0791\ \text{QI} \qquad (1)$$

This study illustrates four frequently replicated patterns:[9,10] (1) A simple model predicts a seemingly complex process. (2) Judges describe using very different strategies than that "captured" in the model. For example, committee members claimed that they considered more cues and used these three in more nuanced ways than just weighting and adding. (3) Even simpler models, replacing regression weights with unit weights on normalized variables, predict equally well. (4) Simple models predict the actual criterion (graduate school success) well.

There are at least three reasons why simple models predict surprisingly well. One is that people have difficulty introspecting into their own decision making.[11,12] A second is that people have difficulty executing complex strategies reliably, so that only simple patterns appear consistently. The third is that simple linear models can predict well without capturing the underlying processes,[9,13] as long as they use reliably measured correlates of the variables that actually affect decision making.

This good news for predictive research is bad news for explanatory research. Models using different variables, implying different processes, often predict equally well. As a result, regression weights need not capture how decisions are made. In many applications, good prediction suffices. For example, the health belief model[14,15] provides a structured way to identify variables correlated with health-related choices. Its application would, however, be misguided, if the weights on those variables were taken as reflecting how individuals think.[16]

For analogous reasons, behavioral decision researchers typically avoid the *revealed preference* analyses that are a staple of economics research.[17,18] For goods traded in efficient markets, prices show rational decision makers' values. If goods are characterized on common attributes, regression weights show those attributes' usefulness as predictors. For example, house prices might be predicted from their size, condition, age, school district, commuting distance, construction, and so on. Unfortunately, when predictors are correlated, regression weights can be unstable, complicating their interpretation as measures of importance.

One strategy for undoing these confounds is generating stimuli with uncorrelated cues. For example, one might create hypothetical graduate school candidates, using all possible combinations of GRE, GPA, and QI. A drawback to this *ANOVA design* is violating behavioral decision research's commitment to *probabilistic functionalism*,[4,19] the view that behavior is shaped by naturally occurring correlations among uncertain cues. Stimuli that violate these relationships lack ecological validity and require unnatural behavior, such as evaluations of implausible cue combinations (e.g., low GRE, high QI). An ANOVA design also gives equal weight to all cue combinations, however common or possible. Moreover, as with any design that presents many stimuli with a transparent cue structure, respondents may either lose focus (producing unreliable judgments) or improvise a mechanical response strategy (producing reliable, but unnatural judgments).

How people respond to novel tasks (e.g., grad candidates with low GRE and high QI) can be revealing. However, because *importance* is inherently context dependent, artificial contexts produce artificial importance weights. For example, although money is generally relevant to consumer decision making, other factors may dominate choices among similarly priced options. One possible reason why Eq. (1) did not include the variable "strength of letters of recommendation' is that candidates had similarly strong letters, written by faculty advisors who sell their students similarly. (QI should capture the reputations of those letter writers.)

As it discovered these limits to the explanatory value of predictive models, behavioral decision research shifted its focus from *what* choices people make to *how* they make them. As a result, studies describe decision-making processes that *can* come into play, as revealed by tasks with which it is relatively clear how a process would express itself. Applied researchers must then determine which of the possible processes are evoked by a specific decision.

Subjective expected utility

The normative analysis underlying behavioral decision research is founded on *expected utility theory*, classically codified by von Neumann and Morgenstern.[20] Its basic logic is straightforward: List the possible action options. For each option, enumerate its possible outcomes. For each such outcome, assess the value, or *utility*, of it happening. Assess the probability of its occurrence should each option be selected. Compute the *expected utility* of each option by multiplying the utility and probability of each outcome, should it be undertaken, then summing across outcomes. Choose the action with the greatest expected utility. When the probabilities reflect decision makers' beliefs, rather than scientific knowledge, the calculation produces *subjective expected utility*.[21] (As discussed below, some scholars view all probabilities as subjective.)

Descriptive research can look at how people undertake each element of this process: assessing the probabilities of possible outcomes, evaluating their utility (should they occur), and combining probabilities and utilities to identify the best option. The decisions can range from completely described and static to incompletely described and dynamic. Normative analyses exist for many kinds of decision.[22]

Individuals' performance on these tasks can be evaluated by *correspondence* or *coherence* tests. Correspondence tests ask how accurate their answers are. For example, how well can they predict whether they will graduate college or enjoy their major? Coherence tests ask how consistent responses are. For example, are probability judgments for an event at least as large as those for a subset ($p[A] \geq p[A \cap B]$)? Are outcomes equally valued, when described in formally equivalent ways (e.g., succeeding vs. not failing).

Predicting outcomes

Studies of how well people predict uncertain events have produced seemingly contradictory results. Sometimes, people do quite well; sometimes, quite poorly. To a first approximation, the difference depends on whether the task requires counting or inferences. With counting studies, the evidence is all of one type; with inference studies, the evidence is of different types. A counting study might display stimuli drawn randomly from a hidden population, then elicit estimates of a summary statistic (e.g., mean, range).[23] An inference study might require integrating *base-rate* evidence about what usually happens, with *individual information* about a specific case.

Counting tasks take advantage of individuals' ability to estimate the relative frequency of events that they observe—even without preparing to do so. For example, after producing rhymes for a set of words, people can estimate the number beginning with different letters.[24] Indeed, encoding frequencies has been called an automatic cognitive function, with research focusing on whether it relies on *tokens*, records of individual observations, or on *types*, category representatives reinforced with each observation.[25]

Assuming that individuals trust their frequency-encoding ability, Tversky and Kahneman[26] proposed the *availability heuristic*, whereby individuals estimate an event's probability by their ability to retrieve instances (tokens) or imagine them (types). Reliance on availability produces biased judgments when the observed events are an unrepresentative sample—and individuals cannot correct for the sampling bias. Researchers have identified many other possible biases, arising from reliance on judgmental heuristics. The strength of any claim of bias depends on the strength of the normative analysis.[27,28] The usefulness of any heuristic depends on how well its application can be predicted (e.g., how memory is searched for examples).

Inference studies tap individuals' lack of intuition and training for combining different kinds of evidence. Here, the normative standard has been the Bayesian approach to hypothesis evaluation.[5,29] Bayes theorem is an uncontroversial part of probability theory. Bayesian inference is more controversial, because it treats probabilities as subjective, thereby allowing inferences that combine diverse kinds of evidence.[30] *Frequentistic* probabilities require evidence of a single kind (e.g., coin flips, weather records). Subjective judgments are only probabilities if they pass coherence tests. Thus, probabilities are not just any assertion of belief.

A widely studied inferential bias is the "base-rate fallacy". Attributed to reliance on the representativeness heuristic,[26] it involves allowing even weak information about specific cases to outweigh knowledge of what generally happens (the base rate). Inadequately regressing judgments is the same bias with continuous variables. Absent strong information about specific cases, one should predict the mean of a distribution.

To avoid artifactual sources of bias,[31] behavioral decision research draws on the century-plus of psychophysics research into factors affecting quantitative judgments.[32–34] For example, because people avoid decimals, they are more likely to overestimate small risks in studies eliciting percentages (e.g., 0.1%) than in studies eliciting odds (e.g., 1 in 1000).[33] Knowing that, researchers can choose the method best suited to their question and reduce measurement artifacts.

Figure 1.1 depicts six such design features, critical to eliciting numbers. For example, Figure 1.1A shows that stimuli [S1, S2] elicit less of the response range when embedded in a larger range [L1, L2]. Figure 1.1C shows how values assigned to larger stimuli are cramped if the initial (standard) stimulus is large, relative to others in the set. Such effects occur because respondents must translate their perceptions into the investigators' terms. Where those terms are unnatural, respondents rely on *response preferences*.[35] For example, they try to use the entire response scale; they look for patterns in randomly ordered stimuli; they deduce an expected level of precision, such as what trade-off to make between speed and accuracy.

Ignoring response preferences leads to misinterpreting judgments. For example, subjects produced much higher estimates of annual US death toll, from[41] causes,

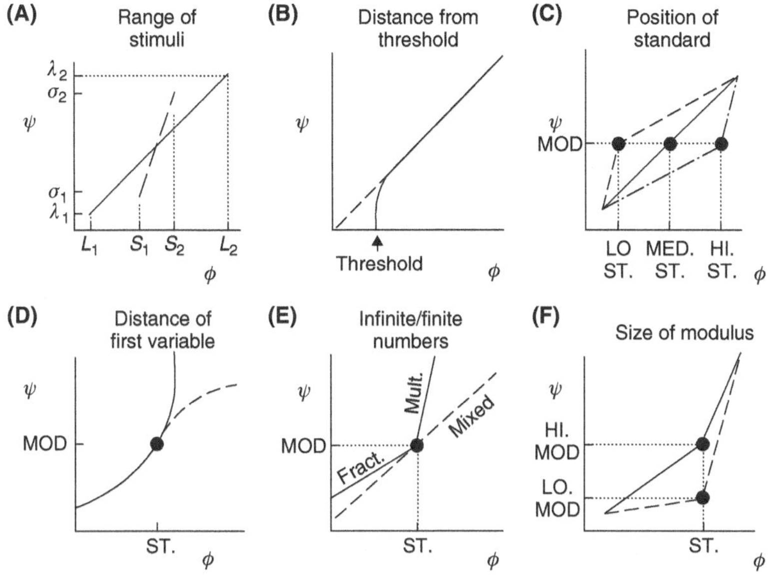

FIGURE 1.1 Six "laws of the new psychophysics," depicting the influence of experimental design on the numerical response used to describe the psychological state (Ψ) equivalent to a physical stimulus (ϕ). (A) A narrower stimulus range (*S*1, *S*2) will use a proportionately larger portion of the response range than would the same stimuli, when embedded in a larger response range (*L*1, *L*2). (B) The effects of assumptions regarding the treatment of stimuli below the threshold of perception or evaluation. (C) The effects of where a standard stimulus falls in the response range, after it has been assigned a numerical valuation (or modulus). (D) The effects of where the first judged stimulus is relative to the standard. (E) The effects of using fractional or integer response values, for stimuli smaller than the standard. (F) The reverse effects where a modulus value, for a given standard stimulus, falls within the response range (Reprinted with permission from Ref. 17 Copyright 2005 Elsevier).

when they received a high *anchor* (50,000 motor vehicles deaths), rather than a low anchor (1000 accidental electrocutions). Low frequencies were greatly overestimated with the high anchor, much less so with the low one. Estimates of relative frequency were similar, however the question was asked, suggesting robust risk perceptions, whose translation into numerical judgments was method dependent. The estimates were also biased in ways consistent with relying on the availability heuristic (e.g., homicides were overestimated relative to, less reported, suicides).[36]

Eliciting values

There are two streams of research into how people form preferences.[17] One follows psychophysics, treating the intensity of preferences like the intensity of physical experiences. The second follows the precepts of *decision analysis*, a consulting process designed to help individuals follow decision theory's normative model.[21,22]

Research in the psychophysical stream has individuals report their feelings directly, perhaps with a rating scale or a judgment of willingness-to-pay for a good. Attitude research is the archetype of this paradigm.

The correspondence test for psychophysical research asks how well elicited values predict behavior. Some attitude researchers hold that a fair test must elicit attitudes that are directly comparable to the target behavior.[37] For example, many behaviors could follow endorsement of "my faith is very important to me." Stronger predictions follow from "daily prayer with like-minded worshipers is very important to me." Even stronger predictions follow from specifying the form of worship. At the extreme, these judgments become statements of intention, rather than attitudes, representing general values. As such, their validity depends on how well people can predict their own experiences.[38]

The coherence standard for psychophysical judgments is *construct validity*. Expressed values should be sensitive to relevant changes in questions and insensitive to irrelevant ones. Applying this standard requires independently assessing relevance. For example, assuming that more is better, *scope tests* ask whether people put higher values on larger quantities of a good. Scope insensitive judgments represent incoherent preferences—except for individuals who feel that there can be too much of a good thing (e.g., rich food, conspicuous consumption). An "inside view" on individuals' basic values is needed to evaluate the coherence of their preferences.

Research in the decision analysis stream assumes that people cannot know what they want, in all possible situations. Rather, they must *construct* specific preferences from more basic values. In making these inferences, people may seek cues in a world that might be helpful, indifferent, or manipulative. The better people understand the factors shaping their inferences, the better chance they have of figuring out what they want.[39,40] Decision analysis structures that process. Its measurement is *reactive*, in the sense of changing people in the process of trying to help them discover their preferences. If successful, it deepens individuals' understanding of themselves.

Correspondence tests for constructed preferences compare elicited values with those that emerge from similar real-world processes. Thus, an intensive electoral campaign might be the standard for a study eliciting candidate preferences.[41,42] Intensive medical consultation might be the standard for preferences elicited with a medical decision aid.[43,44] Coherence tests for constructed preferences ask whether the elicitation session has included all perspectives that individuals might want to consider, while avoiding ones that would apply irrelevant influences.

Identifying the factors influencing behavior is, of course, psychology's central challenge. To study theoretically relevant factors, researchers must control irrelevant ones. Understanding these processes is an ongoing

TABLE 1.1 From artifact to main effect

Liability in judgment due to	*Led to*
Organism	
Inattention, laziness, fatigue, habituation, learning, maturation, physiological limitations, natural rhythms, experience with related tasks	Repeated measures Professional subjects Stochastic response models Psychophysiology Proactive and retroactive inhibition research
Stimulus presentation	
Homogeneity of alternatives, similarity of successive alternatives (especially first and second), speed of presentation, amount of information, range of alternatives, place in range of first alternative, distance from threshold, order of presentation, areal extent, ascending or descending series	Classic psychophysical methods The new psychophysics Attention research Range-frequency theory Order-effects research Regression effects Anticipation
Response mode	
Stimulus-response compatibility, naturalness of response, set, number of categories, halo effects, anchoring, very small numbers, response category labeling, use of end points	Ergonomics research Set research Attitude measurement Assessment techniques Contrasts of between-and within-subject design Response-bias research Use of blank trials
"Irrelevant" context effects	
Perceptual defenses, experimenter cues, social pressures, presuppositions, implicit payoffs, social desirability, confusing instructions response norms, response priming, stereotypic responses, second-guessing	New look in perception Verbal conditioning Experimenter demand Signal-detection theory Social pressure, comparison, and facilitation research

enterprise, which McGuire[45] depicted as turning "artifacts into main effects," worthy of independent investigation. Table 1.1 assembles parts of this history, in terms of the four essential elements of any behavior: the organism, the stimulus being evaluated, the response mode for expressing preferences, and potentially distracting contexts.[46] In terms of correspondence tests, these are all factors that could undermine the match between the conditions in which values are measured by researchers and expressed in life. In terms of coherence tests, these are all factors whose effects on expressed values could be compared with independent assessments of their relevance. That is, do changes in these factors affect valuations when, and only when, they should make a difference? Given the sheer number of potentially relevant factors, value elicitation requires broad understanding of behavioral science.

Making decisions

Non-normative theories

Knowing the limits to the theoretical insights possible with predictive models, applied in complex settings, behavioral decision researchers have focused on processes observed most clearly under experimental conditions. The robustness of observations in the lab is tested by varying those conditions (e.g., increasing economic incentives for good performance, changing information displays) and by identifying real-world analogs, in which a theoretically interesting process might play a practical role.

Foremost among these models is Kahneman and Tversky's[47] *prospect theory*. Its initial formulation identified several utility theory assumptions that were implausible psychologically. One is that people evaluate expected outcomes in terms of changes in their *net asset position*, namely, everything they have in the world. However, people are actually highly sensitive to changes and tend to forget the big picture—as witnessed in reminders to "count your blessings."[48] A second psychologically implausible assumption is that numerically equivalent changes in probabilities are equally important. However, the psychophysics of probability weighting places a premium on changes that lead to certain outcomes (e.g., from 90% to 100%) compared to mid-range changes (e.g., from 30% to 40%). A third such assumption is that people get increasingly averse as losses mount up, whereas psychology finds them increasingly apathetic.

One widely studied corollary of these principles is the *status quo bias*. It reflects how easily *reference points* can be shifted, varying how changes are viewed. For example, organ donation rates are much higher when drivers must opt out, when getting their drivers licenses, compared to when they must opt in.[49,50] Opting in makes surrendering organs seem like a loss, hence aversive. That formulation also suggests a social norm of organ donation and perhaps even a weaker right to refusal.

Any behaviorally realistic approach to decision making must accommodate the limits to cognitive computational capacity. Prospect theory accepts utility theory's cognitively implausible calculation of expected values. However, it uses more intuitively plausible elements and, as a linear model, is relatively robust to misestimating its parameters. Applying the theory requires identifying its elements with real-world equivalents, such as the reference points that decision makers use when assessing changes.[51] Fuzzy-trace theory[52] studies the processes by which individuals master the gist of recurrent decisions. Approaches building on the classic work of Herbert Simon[53] have examined individuals' ability to match simple decision-making heuristics to choices that would, otherwise, be unduly complex.[54] Query theory,[55] support theory,[56] and others [57] formalize the notion of weighting retrieved beliefs, embodied in the availability heuristic.

Emotions

Normative analyses can accommodate emotions as valued outcomes, such as the utility of being happy or the disutility of being fearful. For example, there are formal methods for incorporating such "psychological" outcomes, in analyses of risk decisions.[17,31] Descriptive research can accommodate emotions in terms of their effects on each element of decision making (defining options, predicting events, assessing personal values, integrating beliefs and values). For example, *cognitive appraisal theory*[58] predicts that anger increases the perceived probability of overcoming problems. In a field test with a nationally representative US sample, Lerner et al.[59] found that respondents were about 5% more optimistic, regarding their vulnerability to terror-related events, after an anger induction than after a fear induction. Prescriptive research can accommodate emotions by helping people to getting the right mix for particular choices.[60,61] For example, formal analyses might be used cautiously when they "anaesthetize" moral feeling[62]; decision aids for adolescents have focused on controlling emotions.[63]

The importance of emotion effects depends on their size. A 5% shift might tip a close decision, but not a clear-cut one. von Winterfeldt and Edwards[21] showed, mathematically, that decisions with continuous options (e.g., invest $X) are often insensitive to changes in input variables (i.e., probabilities, values). Thorngate[64] used simulations to examine the sensitivity of stylized decisions to errors due to imperfect heuristics, an approach that others have pursued extensively.[65, 66]

Decision-making competence (DMC)

The fundamental premise of experimental decision research is that people who master the skills that it studies make better real-world decisions.[67] Table 1.2 presents results from a study evaluating the external validity of seven experimental tasks, chosen to span the space of cognitive decision-making competencies.[68]

TABLE 1.2 Correlations between decision-making competence (DMC) and other variables

DMC correlated with	*Pearson r*	*Semi-partial correlation, controlling for*		
		Vocabulary	*ECF*	*Vocabulary and ECF*
Cognitive ability				
Vocabulary	.50	—	.28	—
ECF	.48	.26	—	—
Overall	p < .0001	p = .0009	p = .0008	—
Cognitive style				
Polarized thinking	–.34	–.20	–.24	–.19
Self-consciousness	.20	.14[b]	.05	.11
Self-monitoring	.24	.29[b]	.30[b]	.32
Behavioral coping	.32	.27[a]	.28[a]	.26
Overall	p < .0001	p < .0001	p < .0001	p < .0001
Risk behavior				
Antisocial disorders	–.19	–.18[b]	–.05	–.09
Externalizing behavior	–.32	–.28[b]	–.18	–.20
Delinquency	–.29	–.28[b]	–.18	–.21
ln(lifetime # of drinks)	–.18	–.22[b]	–.15	–.18
ln(lifetime marijuana use)	–.25	–.30[b]	–.20	–.25
ln(# times had sex)	–.24	–.30[b]	–.21	–.27
ln(# sexual partners)	–.30	–.33[b]	–.29[a]	–.31
Overall	p = .0004	p = .0002	p = .009	p = .002
Social and family influences				
Risk status (HAR = 1; LAR = 0)	–.35	–.27	–.23	–.21
SES	.35	.20	.21	.15
Social support	–.30	–.21	–.23	–.19
Positive peer environment	.33	.35[b]	.32[a]	.35
Overall	p = .0002	p = .002	p = .006	p = .007

ECF = executive cognitive function; HAR = high risk family; LAR = low risk family; SES = socio-economic status.

[a] Test A rejects the one-mediator null hypothesis.

[b] Test B rejects the one-mediator null hypothesis.

Reprinted with permission from Ref. 69 Copyright 2005 John Wiley and Sons, where the tasks are described more fully.

Respondents were 110 18- to 19-year-old males in a longitudinal study involving extensive assessments beginning at age 10. DMC scores, extracted from a factor analysis of performance on the seven tasks, showed good test–retest reliability, as did scores on an adult version.[69]

The first section shows positive correlations between DMC and standard measures of verbal and fluid intelligence (Vocabulary and ECF, respectively). The second section shows positive correlations between DMC and four "constructive" cognitive styles. The third section shows negative correlations between DMC and several important risk behaviors. The fourth section shows that DMC is higher for teens coming from low-risk (LAR) families, higher socioeconomic status (SES) families, and more positive peer environments. (The negative correlation with social support may reflect low DMC teens' greater gang membership.) Most correlations remained statistically significant after partialing out the two intelligence measures.

These results support the construct validity of DMC as a measure of decision-making skills that both cause and reflect important aspects of teens' lives. For example, teens with higher DMC come from families that might both model and reward good decision making. Bruine de Bruin et al.[69] found similar correlations between adult DMC and scores on a psychometrically validated Decision Outcome Inventory, eliciting self-reports of outcomes suggesting poor decisions, varying in severity (threw out food, bought clothes that were never worn, missed a train or bus, had a mortgage foreclosed, had a driver's license revoked, had an unplanned pregnancy) and inversely weighted by their frequency.

Prejudices about biases—and the rhetoric of competence

Over the past 40 years, the study of judgment and decision making has spread widely, first to social psychology,[70] then to application areas like accounting, health, and finance, finally penetrating mainstream economics under the banner of behavioral economics. That success owes something to the power of the approach, which liberated researchers previously bound by rational-actor models for describing behavior. It also owes something to the fascination of results that address a central aspect of the human condition, individuals' competence to manage their own affairs.[50,66,67] Very different social institutions may suit rational actors (e.g., free markets, civic engagement) and irrational ones (e.g., strong regulatory protection, deference to paternalistic experts).

Those seeking to extract general messages from this complex research literature have adopted several archetypal rhetorical stances. Familiarity with these stances can help in seeing the research through the stances. Table 1.3 summarizes several common themes, formulated in terms of their advocates' interpretation of the demonstrations of bias that tend to dominate the field.

It is not true

Examining research for possible flaws is central to any science. However, as seen in Table 1.1, the set of features that might conceivably change a research

TABLE 1.3 Debiasing methods according to underlying assumptions

Assumption	*Strategies*
Faculty tasks	
Unfair tasks	Raise stakes
	Clarify instructions
	Dispel doubts
	Use better response modes
	Discourage second guessing
	Ask fewer questions
Misunderstood tasks	Demonstrate alternative goal
	Demonstrate semantic disagreement
	Demonstrate impossibility of task
	Demonstrate overlooked distinction
Faulty judges	
Perfectible individuals	Warn of problems
	Describe problem
	Provide personalized feedback
	Train extensively
Incorrigible individuals	Replace them
	Recalibrate their responses
	Plan on error
Mismatch between judges and task	
Restructuring	Make knowledge explicit
	Search for discrepant information
	Decompose problem
	Consider alternative situations
	Offer alternative formulations
Re-education	Rely on experts
	Educate from childhood

Reprinted with permission from Ref. 71 Copyright 1982 Cambridge University Press.

result is very large, allowing endless criticisms by those who dislike a result ("the bias would disappear had you just changed …"). Such radical skepticism may be met by radical counter-skepticism ("you can't test for every conceivable confound"). A compromise asks whether confounds have general effects. The "unfair tasks" section of Table 1.3 lists common methodological criticisms (e.g., biases would vanish with higher stakes or clearer instructions). An early review of all studies studying these factors found no effect on hindsight bias or on overconfidence in beliefs. A more recent review found that financial incentives had mixed effects, sometimes improving performance, sometimes degrading it, but most often making no difference.[13]

It is true, but you should not say so

Demonstrations of bias allow researchers, who claim to know the answers, to fault others, who do not. Charging others with incompetence undermines their right to make decisions. As a result, researchers should avoid sweeping statements about human competence and stick to the details of domain-specific studies. They should convey both the "figure" of biases and the "background" of the heuristics producing them. They should recall that optical illusions reveal important properties of vision without hindering most activities.[21] They should resist those who promote their research because it serves their political ends.

People are doing something quite different—and doing it quite well

Describing decisions as suboptimal presumes a normative analysis, informed by knowledge of what people know and want. Without that analysis, evaluations can be unduly harsh (e.g., charging overconfidence when people have strategically overstated their beliefs) or lenient (e.g., excusing mistakes as attempts to learn by trial and error). Table 1.3's "misunderstood tasks" section lists some ways that actors and observers can interpret decisions differently. In experiments, *manipulation checks* can assess whether subjects understand tasks as intended. In the world, observers are typically left guessing. For example, there is an unresolved controversy over whether some Americans increased their travel risk, by driving rather than flying, right after the 9/11 attacks. However, the interpretation of their decisions requires knowing how they saw the costs, risks, and hassles of flying and driving. Without evidence on these beliefs, any evaluation of their choices is speculative.

But look at how well people do other things

Claims of bias seem strikingly at odds with the complex tasks that people routinely accomplish (including driving and flying). Perhaps, the biases are just laboratory curiosities, theoretically informative, but of limited practical importance. Or, perhaps the research denies people support that life typically affords them. The "restructuring tasks" section in Table 1.3 lists manipulations that have improved performance under lab conditions. For example, when prompted, people can generate reasons why they might be wrong (reducing overconfidence), ways that events might have turned out otherwise (reducing hindsight bias), or estimates of what normally happens (reducing base-rate neglect). If life provides similar cues, then these "debiasing" studies are most relevant for extrapolation to actual behavior.

Facing the problems

Arguably, by mid-adolescence, most people have the cognitive ability to acquire most of the skills needed to make better decisions.[52,67,69,72] Whether they do depends on the help that they get. Unfortunately, people often receive little training, feedback, and help in making decisions. Indeed, they often face marketers, politicians, and others trying to manipulate their choices.[44,73] Table 1.3's "perfectible individuals" section lists strategies that seem able to enhance individuals' decision-making abilities—recognizing that their success, in any specific setting, is an empirical question.[74,75] The "incorrigible individuals" section lists ways to live with fallibility. A historical example of recalibration was doubling engineers' chronic underestimates of the repair time for power plants.[76] A currently popular compromise is "nudging" people toward better decisions, by choosing better default choices (e.g., being an organ donor, contributing to pension plans).[50]

Conclusions

Judgment and decision making research both requires and allows an unusual degree of collaboration among scientists with diverse expertise. The core discipline of behavioral decision research entails familiarity with normative analyses, descriptive studies, and prescriptive interventions. Its execution involves input from experts in the subject matter of specific decisions, the other (social and affective) pressures on them, and the opportunities for change.[52] For example, Downs et al.[61] helped young women make better sex-related decisions, with an interactive DVD whose content reflected medical research (about sexually transmitted infections), behavioral decision research (about risk perceptions), and social psychology (about self-efficacy).

Behavioral decision research also provides a research platform where theoretical and practical research is mutually reinforcing. In the study of clinical judgment, such interactions showed the predictive power of simple models, a result that was invisible to researchers immersed in domain-specific research. In the study of judgment under uncertainty, these interactions revealed suboptimal strategies that survive because they are good enough to avoid major problems. In the study of value elicitation, they revealed the constructive nature of preference formation, as individuals infer what they want in the novel situations created by life and researchers. In the study of choice, they revealed the positive and negative interplay of cognition and affect. The field's future may exemplify Allan Baddeley's[77] call for the integrated pursuit of *applied basic* research, testing theory by its application, and *basic applied* research, creating theory from new phenomena observed through those tests.

Notes

1 Coombs CH, Dawes R, Tversky A. 1970. *Mathematical Psychology*. Englewood Cliffs, NJ: Prentice Hall.
2 Edwards W. 1954. A theory of decision making. *Psychol Bull* 54:380–397.
3 Berkeley D, Humphreys PC. 1982. Structuring decision problems and the "bias heuristic." *Acta Psychol* 5:201–252.
4 Hammond KR. 1966. *The Psychology of Egon Brunswik*. New York: Holt, Rinehart and Winston.
5 Fischhoff B, Beyth-Marom R. 1983. Hypothesis evaluation from a Bayesian perspective. *Psychol Rev* 90:239–260.
6 Goldberg LR. 1968. Simple models or simple processes? Some research on clinical judgments. *Am Psychol* 23:483–496.
7 Kelly EL, Fiske DW. 1951. *The Prediction of Performance in Clinical Psychology*. Ann Arbor: University of Michigan Press.
8 Meehl PE. 1954. *Clinical Versus Statistical Prediction: A Theoretical Analysis and A Review of the Evidence*. Minneapolis: University of Minnesota Press.
9 Dawes RM. 1979. The robust beauty of improper linear models in decision making. *Am Psychol* 34:571–582.
10 Dawes RM, Faust D, Meehl P. 1989. Clinical versus actuarial judgment. *Science* 243:1668–1674.
11 Ericsson A, Simon HA. 1994. *Verbal Reports as Data*. Cambridge, MA: MIT Press.
12 Nisbett RE, Wilson TD. 1977. Telling more than we know: Verbal reports on mental processes. *Psychol Rev* 84:231–259.
13 Camerer CF, Hogarth RM. 1999. The effects of financial incentives in experiments: A review and capital-labor-production framework. *J Risk Uncertainty* 19:7–42.
14 Becker MH. 1974. The health belief model and personal health behavior. *Health Educ Monogr* 2(4):324–473.
15 Hochbaum G. 1956. Why people seek diagnostic x-rays. *Publ Health Rep* 71:377–380.
16 Ogden J. 2003. Some problems with social cognition models: A pragmatic and conceptual analysis. *Health Psychol* 22:424–428.
17 Fischhoff B. 2005. Cognitive processes in stated preference methods. In: Mäler K-G and Vincent J, eds. *Handbook of Environmental Economics*. Amsterdam: Elsevier 937–968.
18 Viscusi WK. 1983. *Risk by Choice* Cambridge, MA: Harvard University Press.
19 Slovic P, Lichtenstein S. 1971. Comparison of Bayesian and regression approaches to the study of information processing in judgment. *Organ Behav Hum Perform* 6:649–744.
20 von Neumann J, Morgenstern O. 1944. *Theory of Games and Economic Behavior*. Princeton: Princeton University Press.
21 von Winterfeldt D, Edwards W. 1986. *Decision Analysis and Behavioral Research*. New York: Cambridge University Press.
22 Clemen RT. 2003. *Making Hard Decisions: An Introduction to Decision Analysis*. Belmont, CA: Duxbury.
23 Peterson CR, Beach LR. 1967. Man as an intuitive statistician. *Psychol Bull* 68:29–46.
24 Jonides J, Naveh-Benjamin M. 1987. Estimating frequency of occurrence. *J Exp Psychol Hum Learn Mem* 13(2):230–240.
25 Hasher L, Zacks RT. 1981. Automatic processing of fundamental information. *Am Psychol* 39:1372–1386.
26 Tversky A, Kahneman D. 1974. Judgment under uncertainty: Heuristics and biases. *Science* 185:1124–1131.
27 Gilovich T, Griffin D, Kahneman D, eds. 2002. *Heuristics and Biases: The Psychology of Intuitive Judgment*. New York: Cambridge University Press.

28 Kahneman D, Slovic P, Tversky A, eds. 1982. *Judgment Under Uncertainty: Heuristics and Biases*. New York: Cambridge University Press.
29 Edwards W, Lindman HR, Savage LJ. 1963. Bayesian statistical inference for psychological research. *Psychol Rev* 70:193–242.
30 Phillips LD. 1973. *Bayesian Statistics for Social Science* London: Nelson.
31 Fischhoff B. 2009. Risk perception and communication. In: Detels R, Beaglehole R, Lansang MA, Gulliford M, eds. *Oxford Textbook of Public Health*. 5th ed. Oxford: Oxford University Press; 940–952.
32 Poulton EC. 1968. The new psychophysics: Six models for magnitude estimation. *Psychol Bull* 69:1–19.
33 Poulton EC. 1994. *Behavioral Decision Making* Hillsdale, NJ: Lawrence Erlbaum.
34 Stevens SS. 1975. *Psychophysics: Introduction to Its Perceptual, Neural and Social Prospects*. New York, NY: John Wiley & Sons.
35 Tune GS. 1964. Response preferences: A review of some relevant literature. *Psychol Bull* 61:286–302.
36 Lichtenstein S, Slovic P, Fischhoff B, Layman M, Combs B. 1978. Judged frequency of lethal events. *J Exp Psychol Hum Learn Mem* 4:551–578.
37 Ajzen I, Fishbein M. 1980. *Understanding Attitudes and Predicting Social Behavior* Englewood Cliffs, NJ: Prentice-Hall.
38 Gilbert DT. 2006. *Stumbling on Happiness*. New York: Knopf.
39 Fischhoff B. 1991. Value elicitation: Is there anything in there? *Am Psychol* 46:835–847.
40 Lichtenstein S, Slovic P, eds. 2006. *Construction of Preferences*. New York: Cambridge University Press.
41 Lupia A. 1994. Shortcuts versus encyclopedias—information and voting behavior in California insurance reform elections. *Am Polit Sci Rev* 88:63–76.
42 Schläpfer F. 2008. Contingent valuation: a new perspective. *Ecol Econ* 64:729–740.
43 Politi MC, Han PKJ, Col N. 2007 Communicating the uncertainty of harms and benefits of medical procedures. *Med Decis Making* 27:681–695.
44 Schwartz LM, Woloshin S, Welch HCG. 2009. Using a drug facts box to communicate drug benefits and harms. *Annals of Internal Medicine* 150:516–527.
45 McGuire W. 1969. Suspiciousness of experimenter's intent. In: Rosenthal R, Rosnow RL, eds. *Artifact in Behavioral Research*. New York: Academic Press.
46 Fischhoff B, Slovic P, Lichtenstein S. 1980. Knowing what you want: Measuring labile values. In: Wallsten T, ed. *Cognitive Processes in Choice and Decision Behavior*. Hillsdale, NJ: Erlbaum; 117–141.
47 Kahneman D, Tversky A. 1979. Prospect theory: An analysis of decision under risk. *Econometrica* 47:263–281.
48 Helson H. 1964. *Adaptation Level Theory: An Experimental and Systematic Approach to Behavior* New York: Harper and Row.
49 Johnson EJ, Goldstein D. 2003. Do defaults save lives?. *Science* 302:1338–1339.
50 Thaler RH, Sunstein CR. 2008. *Nudge* New Haven: Yale University Press.
51 Fischhoff B. 1983. Predicting frames. *J Exp Psychol Learn Mem Cogn* 9:113–116.
52 Reyna VF, Farley F. 2006. Risk and rationality in adolescent decision making: Implications for theory, practice, and public policy. *Psychol Sci Publ Interest* 7:1–44.
53 Simon HA. 1957. *Models of Man*. New York: John Wiley & Sons.
54 Gigerenzer G, Selten R. 2001. *Bounded Rationality: The Adaptive Toolbox*. Cambridge, MA: MIT Press.
55 Johnson EJ, Haubl G, Keinan A. 2007. Aspects of endowment: A Query Theory of value construction. *J Exp Psychol* 33:461–474.
56 Tversky A, Koehler DJ. 1994. Support theory: A nonextensional representation of subjective probability. *Psychol Rev* 101:547–567.
57 Weber EU, Johnson EJ. 2009. Mindful judgment and decision making. *Annu Rev Psychol* 60:53–85.

58 Lerner JS, Keltner D. 2000. Beyond valence: Toward a model of emotion-specific influences on judgment and choice. *Cognit Emot* 14:473–493.
59 Lerner JS, Small DA, Fischhoff B. 2003. Effects of fear and anger on perceived risks of terrorism: A national field experiment. *Psychol Sci* 14:144–150.
60 Loewenstein G, Weber E, Hsee C, Welch N. 2001. Risk as feelings. *Psychol Bull* 67:267–286.
61 Slovic P, Peters E, Finucane ML, MacGregor D. 2005. Affect, risk and decision making. *Health Psychol* 24:S35–S40.
62 Tribe L. 1974. Ways not to think about plastic trees. *Yale Law J* 83:1315–1346.
63 Downs JS, Murray PJ, Bruine de Bruin W, White JP, Palmgren C. et al. 2004. An interactive video program to reduce adolescent females' STD risk: A randomized controlled trial. *Soc Sci Med* 59:1561–1572.
64 Thorngate W. 1980. Efficient decision heuristics. *Behav Sci* 25:219–225.
65 Ben-Haim, Y. 2006. *Info-gap Decision Theory: Decisions Under Severe Uncertainty*. 2nd ed. London: Academic Press.
66 Todd PM, Gigerenzer G. 2009. *Ecological Rationality: Intelligence in the World*. New York: Oxford University Press.
67 Stanovich KE. 2009. *Decision Making and Rationality in the Modern World* New York: Oxford University Press.
68 Parker A, Fischhoff B. 2005. Decision-making competence: External validity through an individual-differences approach. *J Behav Decis Making* 18:1–27.
69 Bruine de Bruin W, Parker A, Fischhoff B. 2007. Individual differences in adult decision-making competence (A-DMC). *J Pers Soc Psychol* 92:938–956.
70 Nisbett RE, Ross L. 1980. *Human Inference: Strategies and Shortcomings of Social Judgment* Englewood cliffs, NJ: Prentice-Hall.
71 Fischhoff B. 1982. Debiasing. In: Kahneman D, Slovic P, Tversky A, eds. *Judgment Under Uncertainty: Heuristics and Biases*. New York: Cambridge University Press; 422–444.
72 Fischhoff B. 2008. Assessing adolescent decision-making competence. *Dev Rev* 28:12–28.
73 Schwartz LM, Woloshin S, Welch HCG. 2008. *Know Your Chances* Berkeley, CA: University of California Press.
74 Baron J, Brown RV, eds. 1990. *Teaching Decision Making to Adolescents*. New Jersey: Erlbaum; 19–60.
75 Larrick RP, Morgan JN, Nisbett RE. 1990. Teaching the use of cost-benefit reasoning in everyday life. *Psychol Sci* 1:362–370.
76 Kidd JB. 1970. The utilization of subjective probabilities in production planning. *Acta Psychol* 34:338–347.
77 Baddeley AD. 1979. Applied cognitive and cognitive applied research. In: Nilsson LG, ed. *Perspectives on Memory Research*. Hillsdale, NJ: Lawrence Erlbaum.

Further reading

Fischhoff B, Kadvany J. 2011. *Risk: A Very Short Introduction*. London: Oxford University Press.

Fischhoff B, Lichtenstein S, Slovic P, Derby SL, and Keeney RL. 1981. *Acceptable Risk*. New York: Cambridge University Press.

Gilovich T, Griffin D, Kahneman D, eds. 2002. *Heuristics and Biases: The Psychology of Intuitive Judgment*. New York: Cambridge University Press.

Hastie R, Dawes RM. 2002. *Rational Choice in an Uncertain World*. San Diego: Russell Sage.

Kahneman D, Slovic P, and Tversky A, eds. 1982. *Judgment Under Uncertainty: Heuristics and Biases*. New York: Cambridge University Press.

Kahneman D, Tversky A, eds. 2000. *Choices, Values, and Frames*. New York: Cambridge University Press.

Lichtenstein S, Slovic P, eds. 2006. *Construction of Preferences*. New York: Cambridge University Press.

von Winterfeldt D, Edwards W. 1986. *Decision Analysis and Behavioral Research*. New York: Cambridge University Press.

Yates JF. 1990. *Judgment and Decision Making*. New York: John Wiley & Sons.

2

TVERSKY, AMOS (1937–96)

1. Amos Tversky

B. Fischhoff

Amos Tversky was born in Haifa, Israel, on March 16, 1937. His father was a veterinarian, his mother a member of Israel's Parliament, the Knesset. He fought in Israel's 1956, 1967, and 1973 wars, and received its highest honor for bravery. He received his BA in psychology at the Hebrew University of Jerusalem, in 1961. In 1965, he received his Ph.D from the University of Michigan's mathematical psychology program. He worked there with Clyde Coombs (his doctoral advisor), Ward Edwards, and David Krantz, among others. He returned to Jerusalem as a faculty member in 1967, moving to Stanford in 1978. A year at the Center for Advanced Study in the Behavioral Sciences provided Tversky and Daniel Kahneman with concentrated time to develop their approach to judgment under uncertainty.

Tversky died of melanoma, on June 6, 1996, in Stanford, California. At his death, he was Professor of Psychology at Stanford University, where he was also a Principal of its Center on Conflict and Negotiation. He held positions at Tel Aviv University, as Visiting Professor of Economics and Psychology and Permanent Fellow of the Sackler Institute of Advanced Studies. His wife, Barbara, is Professor of Psychology at Stanford, studying perception. He had three children, Oren, Tal, and Dona. He was known for his great energy, joy of life, and sense of humor.

Tversky made fundamental contributions to the understanding of human thought processes, and the mathematical foundations of the behavioral sciences. His work was distinguished by elegant formal models, tested in simple, illustrative experiments. Throughout much of his career, he worked together with Daniel Kahneman, on studies of judgment and decision making under conditions of risk and uncertainty. On these topics and others, he also collaborated with other scholars from psychology, economics, political science, law, statistics, philosophy, and mathematics. These projects, and the personal interactions surrounding them, helped to integrate these disciplines, at a time of increasing specialization. Among his many honors were the Distinguished Scientific Contribution Award of the American Psychological Association, a MacArthur Fellowship, and honorary doctorates at Yale, Chicago, Göteborg (Sweden), and New York at Buffalo. He was a member of the US National Academy of Sciences. It served as one venue for pursuing his life-long commitment to encouraging peace and understanding among diverse people.

2. Axiomatic theories of choice

An early paper (Tversky, 1969) typifies his approach. Normative theories of choice provide rules that individuals should follow, in order to make effective decisions. The most prominent of these is utility theory, founded on a set of intuitively appealing axioms. It describes options in terms of a set of attributes, or features, that an individual might like or dislike about them. When buying a car, the attributes might be (price, size, styling, reliability, comfort). According to utility theory, an individual should consider all relevant attributes, then rank the options according to their attractiveness (or "utility"). That ordering should have the property of *transitivity*. Someone who prefers Car A to Car B and Car B to Car C should also prefer Car A to Car C.

Transitivity is a decision-making norm that most people would endorse. If it accurately described their behavior, they would be better off. So would scientists and policy makers. The former would find it relatively easy to predict behavior, knowing that there is a stable overall ranking, which can be identified without examining every possible paired comparison. The latter would be able to create policy options (e.g., parks, housing, insurance schemes, transportation systems), knowing that the public's response will be relatively predictable and consistent.

Tversky showed systematic violations of transitivity, with experimental tasks, similar to choices encountered in everyday life. Moreover, he went beyond merely showing that utility theory described behavior imperfectly. Such a limited demonstration could have been easily attacked as a destructive curiosity that does little to help scientists do their job of explaining and predicting behavior. However, Tversky showed that intransitivity arose from essential psychological processes, which he captured in a simple, insightful model. These processes reflect the limits to human information processing capacity. In situations without the opportunity for trial-and-error learning, people can at best be boundedly rational, working in an orderly way on a subset of the problem (Simon, 1957).

The coping process that Tversky proposed has people comparing options initially on the most important attribute (e.g., car price), then checking whether differences on the second attribute (e.g., reliability), and perhaps others, are so great as to change the choice. For example, imagine sequential comparisons between overlapping pairs of cars (A with B, B with C, C with D, and so on).

The sequence is arranged so that each new car is both cheaper and less reliable that the previous one. Looking primarily at the price would make each new option more attractive, while the reductions in reliability slowly mount up. At some point, though, if the first car were compared with the last car, a buyer might decide that reliability has gotten so low that the reduction in price is insufficient compensation. In that case, the first car, which was long ago rejected, would suddenly be preferred.

Such intransitivity could, in principle, turn people into "money pumps," willing to pay something to move from one option to the next, but eventually going back to the first option and starting over again. As Tversky noted, car salespeople may exploit this cognitive weakness by adding options (and costs) to a favored car. Unlike the experimenter, though, they hope to stop just short of the point where customers balk at the increased price and revert to the stripped-down model. Hanson and Kysar (1999) discuss the implications of such exploitation for marketplace regulation.

Two themes in this research recur in Tversky's work. One is that choices, especially novel ones, can be very sensitive to how the options are presented. That insight is central to prospect theory (see below), which also provides expression for results from a long tradition of psychological research into "context effects." The second theme is that descriptively valid formal theories of human behavior should be assembled from psychologically plausible assumptions, rather than derived from normatively postulated principles.

The latter insight is expressed in many of Tversky's projects, which articulate the formal structure of particular choice situations in ways that allow the expression of plausible behavioral responses (e.g., Tversky, 1967, 1972). In some cases, these theories adapted existing mathematical approaches. In other cases, they drove the mathematics. The fundamental mathematical research associated with this research is summarized in three landmark volumes (Krantz et al., 1971/1989/1990). The study of alternative choice rules for dealing with multiattribute options has become a cornerstone of a *conjoint measurement*, a family of procedures used widely in applied research (e.g., marketing multiattribute consumer products).

3. Prospect theory

These strands of research came together in Kahneman and Tversky's (1979) "prospect theory." Published in a central economics journal, it offered an alternative to utility theory, the core of that discipline. It claimed to achieve greater descriptive validity—but at the price of abandoning the normative aspirations of economics. It depicted individuals as thinking in an orderly, sensible way when faced with choices. However, the rules that people follow differ from those prescribed by the axioms of utility theory, in ways that produce systematic violations. People are sensitive to features of choice tasks that have no representation in utility theory, and insensitive to features that are in the theory.

Superficially, prospect theory resembles utility theory. "Prospects" (i.e., options) are represented as sets of possible consequences. The value (rather than "utility") assigned to a prospect is the sum of the values assigned to each consequence, weighted by the chances of it happening. However, the components of this evaluation differ from their utility theory analogs in ways designed to make the theory more realistic psychologically (Tversky and Kahneman, 1991):

(a) The evaluations are made relative to a *reference point*, rather than with respect to an individual's *net asset position* (or entire worth). That reference point is typically the status quo. However, it could be another salient value (e.g., evaluating a raise relative to what one expected to get or to what others got, rather than relative to one's current salary). This feature of prospect theory reflects people's widely observed sensitivity to changes.
(b) People care less about a given absolute difference in gains or losses as the overall change increases. The idea the people care less about a fixed gain as they have or win more is shared with utility theory. However, the idea that people care proportionately less as losses mount up is not. Utility theory explains insurance buying (among other things) in terms of people's special aversion to large losses. Prospect theory reflects the psychological principle that the just-noticeable difference between situations is greater when the overall change is larger—whatever the direction of change.
(c) Before being combined with value judgments, the probabilities of the consequences are subjected to a "probability weighting function." It reflects the observation that people pay particular attention to sure things, while being relatively indifferent to probability differences in the middle range (e.g., the difference between 0.35 and 0.45 matters less than that between 0.90 and 1.00).

As combined in the theory, these principles (along with some others) predict (among other things) *framing effects*. Changing the reference point can reverse choices, in ways that have no reflection in utility theory (which has no such point). For example, an emergency medical intervention that could save 200 of 600 possible lives is more attractive when described in terms of the 200 lives saved than the 400 lives lost. Prospect theory's challenge to utility theory evoked vigorous responses. One extreme involved radical methodological skepticism. Partisans attacked any aspect of prospect theory's research method that could, conceivably, have contributed artifactually to behaviors offered as evidence supporting the theory. Although some of these critiques were ad hoc claims, disconnected from the research literature, others raised fundamental questions with theoretical and methodological implications.

One such question concerns the conflicting norms of experimental economics and experimental psychology regarding the provision of financial incentives. Economists typically believe that what people say and do has little meaning, unless they are rewarded for their performance. In this view, even if they are systematic, responses to hypothetical questions could reflect no more than acquiescence to implicit task demands. In contrast, psychologists typically view all behavior as "real," with the attendant challenge of determining how individuals have construed their task. Even a hypothetical choice can induce high intrinsic and extrinsic motivation, as individuals try to impress themselves and those watching them. Camerer and Hogarth (1999) summarize the evidence, regarding the varied impacts of monetary rewards on decision-making performance. The other extreme involves economists who accept the validity of prospect theory's results and attempt to reshape their science in its light. Some have examined its implications for the axiomatic basis of utility theory, either by refining those axioms to fit the data better or by replacing them with ones expressing different normative principles. Others have examined the expression and implications of non-rational behavior in real-world circumstances, prompting new fields like behavioral finance, which questions widely accepted efficient market theories.

Still others have sought to help people reduce such inconsistencies in their preferences. These "constructivist" approaches assume that people's inconsistencies arise because evaluating options requires an inferential process, whereby people derive a specific valuation from more general, basic values. If that process is incomplete, then people will be unduly influenced by the initial problem presentation. Reducing that risk requires a more interactive process, explicitly considering alternative perspectives—trying to balance those suggestions, so that people come to know what they want. Fischhoff and Manski (1999) show the convergence among economists and psychologists concerned with these questions.

4. Judgment under uncertainty: heuristics and biases

For economists and psychologists concerned with decision making, values are a matter of personal preference. As a result, showing them to be suboptimal means demonstrating inconsistencies. With beliefs, suboptimality can be found by comparing judgments to one another (looking for inconsistency) and to an accepted external standard (looking for error). In their first collaboration, Tversky and Kahneman (1971) demonstrated errors in the judgments of mathematical psychologists and statisticians, when estimating the statistical power of research designs. These individuals knew how to perform the relevant calculations. However, when forced to rely on intuition, they systematically exaggerated the chances of observing anticipated results—acting as though the law of large numbers applied to samples with small numbers of observations.

Tversky and Kahneman (1974) proposed that these judgments arose from relying on the *representativeness* heuristic. According to this rule of thumb, an event is judged likely to the extent that it captures (or "represents") the salient features of the situation that might produce it. Conversely, reliance on representativeness should lead to neglecting factors lacking a salient role in such processes, even if they are normatively relevant. Sample size is such a factor. If a situation can produce an event, why should it matter how much one observes? A second normatively relevant factor with no role in representativeness is the relative frequency of an event in the past. Such *base-rate* (or *a priori*) information should guide predictions unless there is strongly diagnostic *individuating* information, showing why a specific case is different from the norm. (Medical students are sometimes told, "When you hear hoof beats, think horses, not zebras," at least in North America.) However, studies have found that even flimsy evidence can attract the lion's share of attention. It is natural to think about how well someone fits the image of an engineer (or how much a mammogram image looks like a tumor). It is not so natural (or easy) to combine that vision or theory with a statistical summary of how frequent engineers (or tumors) are in the population. There have been many subsequent demonstrations of underweighting base rates (including in the interpretation of mammograms), as well as vigorous debate (sometimes with data, sometimes not) regarding the conditions determining its extent (Gilovich et al., 2002, Kahneman et al., 1982).

Analogous reasoning led to predicting that reliance on representativeness would encourage underestimating regression to the mean and exaggerating the predictability of random sequences. The latter was demonstrated in studies showing an illusory belief in being able to identify streaks in basketball free-throw shooting. Both collegiate and professional players (and fans) interpreted patches of random sequences as reflecting transient "hot hands" (Gilovich and Tversky, 1985).

Tversky and Kahneman (1974) proposed the representativeness heuristic as one exemplar of a general strategy that people use when confronted with unfamiliar tasks (and no opportunity for systematic study or calculation). A second such heuristic, *availability*, judges the probability of an event by the ease with which examples come to mind. That rule has some validity (in many domains, things that one hasn't seen tend to be less likely than things that one has). However, it can predictably lead one astray, as when the news media disproportionately report one risk (e.g., homicide vs. asthma) (Slovic, 1987). Subsequently, Tversky and Koehler (1994) formalized the relationship between evaluations of evidence and probability judgments in "support theory." Like its predecessors, it predicts the conditions under which simple, intuitive thought leads to good and bad judgments. A third heuristic, also with historical roots in psychology, is *anchoring and adjustment*. Users begin an estimation process from a salient value (the "anchor"), then adjust from there, in response to other concerns that come to mind. Typically, that adjustment is insufficient. If the anchor is far from the correct value, then the resulting estimate will be as well.

One cross-cutting theme in Tyersky's judgment research (as with that on choice) is questioning the normative standards used to evaluate people's performance. Here, too, the result has been re-evaluating those standards and their domains of applicability (Shafer and Tversky, 1985). Another recurrent theme is the importance of comparison processes in judgment. Representativeness asks how well an event fits the process that would have to produce it. Availability asks how well an example fits the archetype of the category whose probability is being predicted. In a parallel research program, Tversky (1977) created a theory of similarity judgments, based on measure-theoretic concepts and, again, simple, elegant experiments.

Bibliography

Camerer C F, Hogarth R M 1999. The effects of financial incentives in experiment. *Journal of Risk and Uncertainly* 19: 7–42.

Fischhoff B, Manski C F (eds) 1999. Editor's introduction. Elicitation of preferences. *Journal of Risk and Uncertainty* 19: 1–3.

Gilovich T, Griffin D, Kahneman D (eds.) 2002. *The Psychology of Judgment: Heuristics and Biases*. Cambridge University Press, New York.

Gilovich T, Tversky A 1985. The hot hand in basketball. *Cognitive Psychology* 17: 295–314.

Hanson J D, Kysar D A 1999. Taking behavioralism seriously: Some evidence of market manipulation. *Harvard Law Review* 112: 1420–1572.

Kahneman D, Slovic P, Tversky A (eds.) 1982. *Judgment under Uncertainty: Heuristics and Biases*. Cambridge University Press, New York.

Kahneman D, Tversky A 1979. Prospect theory: An analysis of decision under risk. *Econometrica* 47: 263–81.

Krantz D, Luce R D, Suppes P, Tversky A 1971/1989/1990. *Foundations of Measurement*, Vols. 1–3. Academic Press, New York.

Shafer G, Tversky A 1985. Languages and designs for probability judgment *Cognitive Science* 9: 309–39.

Simon H A 1957. *Models of Man: Social and Rational*. Wiley, New York.

Slovic P 1987. Perception of risk. *Science* 236: 280–5.

Tversky A 1967. A general theory of polynomial conjoint measurement. *Journal of Mathematical Psychology* 4: 1–20.

Tversky A 1969. The intransitivity of preferences. *Psychological Review* 76: 31–48.

Tversky A 1972. Elimination by aspects: A theory of choice. *Psychological Review* 79: 281–99.

Tversky A 1977. Features of similarity. *Psychological Review* 84: 327–52.

Tversky A, Kahneman D 1971. Belief in 'law of small numbers". *Psychological Bulletin* 76: 105–10.

Tversky A, Kahneman D 1974. Judgment under uncertainty: Heuristics and biases. *Science* 185: 1124–31.

Tversky A, Kahneman D 1991. Advances in prospect theory. *Journal of Risk and Uncertainty* 5: 297–323.

Tversky A, Koehler D J 1994. Support theory: A nonextensional representation of subjective probability. *Psychological Review* 101: 547–67.

[illegible]

Bibliography

[illegible]

PART II

Hindsight bias

My first serious research project was my doctoral dissertation, on hindsight bias. It was inspired by Paul Meehl's (1973) "Why I do not attend case conferences" and strongly influenced by Roberta Wohlstetter's *Pearl Harbor: Warning and Decision* (1962), which meticulously dissected how the bias led to blaming Navy commander Kimmel (and President Roosevelt) for missing signals that were buried in noise before the attack. My contribution was devising experimental tasks that allowed clearly demonstrating the bias – as distinguished from the genuine learning that can come with knowing what happened – and evaluating alternative explanations.

"Hindsight ≠ foresight: The effect of outcome knowledge on judgment under uncertainty" (Chapter 3) summarizes three studies from my dissertation. The first finds that reporting an outcome makes it seem more likely than it did in foresight, even when it did not actually happen. Outcome knowledge also changes the apparent importance of different facts describing the situation, suggesting that the same facts can be interpreted to "explain" whatever is reported. The second study finds that people imagine that, had they been asked, they would have seen things that way in foresight. The third study finds that people believe that others should have been equally foresightful, making them unrealistically accountable for knowing things that they could not reasonably have predicted. The discussion distinguishes learning from hindsight bias, concluding that outcome knowledge can teach us things, if we don't imagine that we knew them all along.

"The early history of hindsight research" (Chapter 4) was my response to being asked to comment on papers at a 2004 conference dedicated to hindsight bias. I had to decline, partly because I had not followed the literature, having decided long ago that I wanted to test myself by working on different problems. The organizers graciously allowed me to reminisce, instead, on the circumstances leading to the research, namely, the confluence of my interest in historiography and the excitement of the Tversky-Kahneman seminar. In the chapter, I speculate why these studies had proven so (unpredictably) successful. One reason was the candid, supportive criticism of that seminar. A second may have been the habit that I developed while writing it, of mulling alternative wording before putting anything down. I imagined that style as reflecting my training in mathematics, trying to be sure about what one had established before moving ahead. It makes me a slow writer and sometimes difficult co-author.

"For those condemned to study the past: Reflections on historical judgment" (Chapter 5) bookmarks a research program that my dissertation was meant to start, but which I have yet to fulfill. Realizing that no one has thought as hard about historical judgment as historians, I read a good deal of historiography, took Shaul Friedlander's course in psychohistory, and even attended an informal group of graduate students in medieval history (courtesy of a common friend, Roger Hurwitz). I envisioned a synthesis that combined insights from history and behavioral decision research. I intended this article as a progress report. However, it is as far as I have gotten. A decade later, I reluctantly cancelled my subscription to History and Theory, glad to have had one article (Fischhoff, 1978) in a journal edited by the likes of Raymond Aron and Isaiah Berlin, but sad not to have written a book that was reviewed there.

References

Fischhoff, B. 1978. Intuitive use of formal models. A comment on Morrison's "Quantitative Models in History.' *History and Theory, 17*, 207-210.

Meehl, P.E. 1973. Why I do not attend case conferences. In *Psychodiagnosis: Selected papers* (pp. 225-302). New York: Norton.

Wohlstetter, R.A. 1962. *Pearl Harbor: Warning and Decisions*. Stanford University Press: Palo Alto, CA.

3

HINDSIGHT ≠ FORESIGHT

The effect of outcome knowledge on judgment under uncertainty

Baruch Fischhoff

HEBREW UNIVERSITY OF JERUSALEM, ISRAEL

Abstract

One major difference between historical and nonhistorical judgment is that the historical judge typically knows how things turned out. In Experiment 1, receipt of such outcome knowledge was found to increase the postdicted likelihood of reported events and change the perceived relevance of event-descriptive data, regardless of the likelihood of the outcome and the truth of the report. Judges were, however, largely unaware of the effect that outcome knowledge had on their perceptions. As a result, they overestimated what they would have known without outcome knowledge (Experiment 2), as well as what others (Experiment 3) actually did know without outcome knowledge. It is argued that this lack of awareness can seriously restrict one's ability to judge or learn from the past.

Introduction

Hindsight and foresight differ formally in the information available to the observer. The hindsightful judge possesses outcome knowledge, that is, he knows how things turned out. The foresightful judge does not. Although foresight usually implies looking at the future, in the absence of outcome knowledge, past and future events can be equally inscrutable.

The studies presented here ask two questions about the judgmental differences between hindsight and foresight: (a) How does receipt of outcome knowledge affect judgment? (b) How aware are people of the effects that outcome knowledge has on their perceptions? Answers to these questions will shed light on how people do learn and might better learn from history.

The two hypotheses explored are: (a) Reporting an outcome's occurrence increases its perceived probability of occurrence; and (b) people who have received outcome knowledge are largely unaware of its having changed their perceptions in the manner described in the first hypothesis. In combination, these two hypotheses indicate that reporting an outcome produces an unjustified increase in its perceived predictability, for it seems to have appeared more likely than it actually was.

Indirect support for the first hypothesis may be found in a variety of sources. For example, the historian Georges Florovsky (1969) notes:

> The tendency toward determinism is somehow implied in the method of retrospection itself. In retrospect, we seem to perceive the logic of the events which unfold themselves in a regular or linear fashion according to a recognizable pattern with an alleged inner necessity. So that we get the impression that it really could not have happened otherwise. (p. 369)

An apt name for this hypothesized tendency to perceive reported outcomes as having been relatively inevitable might be "creeping determinism"—in contrast with philosophical determinism, which is the conscious belief that whatever happens has to happen.

Phenomena resembling creeping determinism have been noted by psychologists as well as historians. One example is Tversky and Kahneman's (1971) "law of small numbers," the belief that data which were observed more or less had to be observed. A second example is the tendency to rework or reconstruct the biographies of deviants to show that their present diagnoses (labels) are inevitable products of their life histories (Lofland, 1969; Rosenhan, 1973; Schur, 1971). A third is the defensive attribution of responsibility for accidents, a process in which people carefully scrutinize the data describing accidents in order to uncover or impose a pattern that will increase their perceived predictability and avoidability (Walster, 1967).

All of this evidence for creeping determinism is, however, either indirect, imprecise, unsystematic (anecdotal), or confounded by motivational and emotional issues. Experiment 1 directly tested the validity of the creeping determinism hypothesis and explored some of the concomitant effects of outcome knowledge on judgment.

Experiment 1

Method

Design. The six subexperiments described in this section are identical except for the stimuli used. In each, subjects were randomly assigned to one of five experimental groups, one Before group and four After groups. In each subexperiment, the Before group read a brief (150 word) description of a historical or clinical event for which four possible outcomes were provided. The After groups read identical passages to which a final sentence presenting one of the possible outcomes as the "true" outcome had been added. As the possible outcomes were mutually exclusive, three of the four After groups received "true" outcomes that actually had not happened. Subjects in all groups were asked to (a) estimate the likelihood of occurrence of each of the four possible outcomes, and (b) evaluate the relevance of each datum in the event description. In two of the subexperiments subjects were also asked to indicate the relative extent to which they relied on the passage and on outside information.

Instructions. The cover sheet of each questionnaire read:

> In this questionnaire we are interested in knowing how people judge the likelihood of possible outcomes of social events. A passage describing an unfamiliar historical event appears below. We will ask you to evaluate the probability of occurrence of each of the four possible outcomes of the event [including that which actually happened—for After subjects] in the light of the information appearing in the passage.

A typical passage, as taken from Woodward's (1938) *The Age of Reform*, was:

> [1] For some years after the arrival of Hastings as governor-general of India, the consolidation of British power involved serious war. [2] The first of these wars took place on the northern frontier of Bengal where the British were faced by the plundering raids of the Gurkas of Nepal. [3] Attempts had been made to stop the raids by an exchange of lands, but the Gurkas would not give up their claims to country under British control, [4] and Hastings decided to deal with them once and for all. [5] The campaign began in November, 1814. It was not glorious. [6] The Gurkas were only some 12,000 strong; [7] but they were brave fighters, fighting in territory well-suited to their raiding tactics. [8] The older British commanders were used to war in the plains where the enemy ran away from a resolute attack. [9] In the mountains of Nepal it was not easy even to find the enemy. [10] The troops and transport animals suffered from the extremes of heat and cold, [11] and the officers learned caution only after sharp reverses. [12] Major-General Sir D. Octerlony was the one commander to escape from these minor defeats. (pp. 383–384)

The possible outcomes offered were: (a) British victory, (b) Gurka victory, (c) military stalemate with no peace settlement, and (d) military stalemate with a peace settlement. For After subjects, the appropriate outcome was appended to the passage in the form of an additional sentence, such as, "The two sides reached a military stalemate, but were unable to come to a peace settlement."

Following the passage, subjects were asked, "In the light of the information appearing in the passage, what was the probability of occurrence of each of the four possible outcomes listed below. (The probabilities should sum to 100%)."

On the following page, each datum appeared on a separate line followed by a seven-point scale on which subjects were asked to indicate "how relevant or important each datum in the event description was in determining the event's outcome." The numbers in the passage above indicate the division into datum units. They did not appear in the passage presented to subjects.

Stimulus selection. Four different events were used to achieve greater generality for the results obtained: Event A, the British-Gurka struggle cited above; Event B, the near-riot in Atlanta, Georgia in July 1967, as described in the *Kerner Commission Report on Civil Disorders* (National Advisory Commission, 1968, pp. 53–5); Events C and D, clinical cases reported by Albert Ellis (1966, pp. 262–264).[1] For Events C and D, the word "social" in the instructions was replaced by "individual" and the word "historical" was deleted.

Several methodological considerations guided the event selection process: (a) The event should be sufficiently familiar to permit intelligent responses and sufficiently unfamiliar to rule out the possibility of subjects knowing what really happened—especially those receiving false outcome reports. (b) Past events were used to allow provision of "true" outcomes to the After groups.[2] (c) The space of possible outcomes had to be readily partitionable. For Events B, C, and D, the set of outcomes was constructed to be mutually exclusive and exhaustive. Although this is not the case for Event A, pretests indicated that these four outcomes constituted an effective partition.

Subjects. Approximately equal numbers of subjects participated in each group in each subexperiment. Event A (Gurkas) was administered twice, once to a group of 100 English-speaking students recruited individually at The Hebrew University campus in Jerusalem and once to a class of 80 Hebrew-speaking subjects at the University of the Negev in Beer Sheba. Event B (riot) was administered to two separate classes at The Hebrew University, one containing 87 Hebrew-speaking psychology majors with at least one year's study of statistics and one of 100 Hebrew-speaking students with no knowledge of statistics. Event C (Mrs. Dewar) was administered to the 80 University of the Negev students; Event D (George) to the 100 Hebrew University students without statistics training.

Procedure. Questionnaires for the various experimental groups were distributed randomly. Subjects devoted 20–30 min to the completion of each questionnaire.

Results

Probability estimates. Table 3.1 presents the mean probability assigned to each outcome by subjects in each experimental group for each subexperiment.

Similar patterns of data emerged in the two subexperiments using Event A (differing in subjects' language) and in the two using Event B (differing in subjects' knowledge of statistics). For the sake of tabular brevity, only one subexperiment in each pair is presented.

TABLE 3.1 Mean probabilities assigned to each outcome, Experiment 1

			Outcome evaluated			
Experimental group	*n*	*Outcome provided*	*1*	*2*	*3*	*4*
Event A: British-Gurka struggle (English-speaking subjects)						
Before	20	none	33.8	21.3	32.3	12.3
After	20	1	57.2	14.3	15.3	13.4
	20	2	30.3	38.4	20.4	10.5
	20	3	25.7	17.0	48.0	9.9
	20	4	33.0	15.8	24.3	27.0
Event B: Near-riot in Atlanta (subjects with knowledge of statistics)						
Before	20	none	11.2	30.8	43.8	14.2
After	20	1	30.6	25.8	23.3	20.3
	20	2	5.5	51.8	24.3	18.5
	20	3	3.9	23.9	50.8	21.4
	20	4	16.7	31.9	23.4	27.9
Event C: Mrs. Dewar in therapy						
Before	19	none	26.6	15.8	23.4	34.4
After	13	1	43.1	13.9	17.3	25.8
	17	2	26.5	23.2	13.4	36.9
	16	3	30.6	14.1	34.1	21.3
	17	4	21.2	10.2	22.6	46.1
Event D: George in therapy						
Before	17	none	27.4	26.9	39.4	6.3
After	18	1	33.6	20.8	37.8	8.0
	18	2	22.4	41.8	28.9	7.1
	20	3	20.5	22.3	50.0	7.3
	17	4	30.6	19.5	37.7	12.3

Note. The actual outcomes are numbers 1, 1, 4, and 2 for Events A, B, C, and D, respectively. Outlined cells are those with After probabilities of reported outcomes.

The creeping determinism hypothesis predicts that After subjects told that a particular outcome has happened will assign it a higher probability than will Before subjects. Four outcomes reported to different groups in each of six subexperiments afford 24 opportunities to test the hypothesis. The critical comparisons are between the outlined diagonal cells (those indicating the mean probability assigned to an outcome by subjects for whom that outcome was reported to have happened) and the Before cell in the top row above them.

In each of the 24 cases, reporting an outcome increased its perceived likelihood of occurrence ($p < .001$; sign test). Twenty-two of these differences were individually significant ($p < .025$; median test). Thus the creeping determinism effect was obtained over all variations of subject population, event description, outcome reported, and truth of outcome reported. The differences between mean Before and After probabilities for reported outcomes ranged from 3.6% to 23.4%, with a mean of 10.8%. Slightly over 70% of After subjects assigned the reported outcome a higher probability than the mean assignment by the corresponding Before subjects.

No outcome was judged inevitable by any Before subject, whereas a small proportion (2.1%) of After subjects did assign 100% to reported outcomes. Evidently, most After subjects felt that in the light of the facts given in the description, other (unreported) outcomes were still possible (e.g., "The Gurkas had a 70% chance of winning, but the British still might have pulled it off"). Similarly, After subjects found a higher percentage of unreported outcomes to have been impossible (as indicated by a probability of 0%) than did Before subjects (11.5% versus 8.0%).

Another way to appraise the extent of creeping determinism is to translate mean Before probabilities into the form of a priori odds and the mean After probabilities for reported outcomes into a posteriori odds. The ratio of prior and posterior odds for outcome *i* provides a sort of average likelihood ratio for the impact of the datum "Outcome *i* did actually occur" (where the two hypotheses are "Outcome *i* occurs" and "Outcome *i* does not occur"). Over the 24 outcomes reported, these likelihood ratios varied from 1.2 to 3.5 (M = 1.96). Thus in the present sense, reporting an outcome's occurrence approximately doubles its perceived likelihood of occurrence.

Because the outcomes varied considerably in their mean Before probability (from 6.3% to 44.0%), reporting their occurrence may be seen as confirming (or disconfirming) subjects' expectations to varying degrees. There was a highly significant negative correlation ($t = .435$; $p < .001$) between the prior odds and likelihood ratios associated with reported outcomes (as computed in the preceding paragraph). Thus, the more unlikely an outcome report, the greater the impact it has.

Relevance judgments. Table 3.2 presents the mean relevance judgments for each datum in one subexperiment. Inspection reveals that the relevance attributed to any datum is highly dependent on which outcome, if any, subjects believe to be true. Some of these differences seem readily interpretable. For example, the fact that "the British officers learned caution only after sharp reverses" (Datum No. 11) was judged most relevant by subjects told of a British victory, and rather irrelevant by subjects told of a Gurka victory.

A less impressionistic analysis on the effects of outcome knowledge on relevance judgments proceeded in the following manner. For each subexperiment, a two-way (outcome reported × datum evaluated) fixed model analysis of variance (ANOVA) was performed on subjects' judgments of data relevance.

TABLE 3.2 Mean data relevance judgments for Event A, Experiment 1 (Hebrew-speaking subjects)

	Datum number											
Outcome reported	*1*	*2*	*3*	*4*	*5*	*6*	*7*	*8*	*9*	*10*	*11*	*12*
None	4.50	5.11	4.22	5.78	4.50	6.00	5.50	5.44	4.39	4.56	4.28	5.56
British victory	4.78	4.44	5.28	4.83	4.61	4.44	4.61	4.56	5.72	5.33	5.78	4.11
Gurka victory	3.66	4.83	3.55	4.44	5.89	5.11	4.11	4.61	3.72	5.22	4.11	4.78
Stalemate												
with peace treaty	4.50	4.72	4.55	5.89	5.50	4.17	4.22	5.00	4.22	5.22	4.89	4.94
without peace treaty	4.94	5.50	4.39	5.11	5.33	5.11	4.78	4.39	4.17	3.72	4.50	4.61

To accommodate the varying number of subjects in the experimental groups of subexperiments, the following procedure was adopted: The analysis was repeated three times to produce maximum, minimum, and middle solutions. For the maximum solution, subjects were randomly sampled from the smaller experimental groups and their responses duplicated, equating the size of all cells. For the minimum solution, subjects were randomly deleted from the larger groups until cell size was equated. For the middle solution, a combination of duplication and elimination was performed.

The same ANOVA was performed on the three sets of data. The results discussed hold for all three solutions. (a) In each subexperiment, there was a significant Outcome Reported × Datum Evaluated interaction reflecting the differential effect of outcome knowledge on relevance judgments discussed above and shown in Table 3.2. (b) Over the six subexperiments, only two weak outcome-reported effects emerged. Thus, there is no indication of an entire set of data having greater mean relevance for one outcome than another. (c) Datum-evaluated effects appear in all but one subexperiment. They reflect data perceived to be either relevant or irrelevant whatever happens (e.g., Hastings' decision to deal with the Gurkas "once and for all" [Datum No. 4] was universally perceived as relevant).

Reliance. Subjects in two of the subexperiments were asked to indicate with a number from 0% to 100% the extent to which they had relied on the material presented in the passage compared with general (outside) knowledge. In both cases, Before subjects indicated relying significantly ($p < .05$; median test) more on the passage than did After subjects.

Discussion

Reporting an outcome's occurrence consistently increases its perceived likelihood and alters the judged relevance of data describing the situation preceding the event. Hindsight-foresight differences in perceived data relevance have also been noted by historians observing the creeping determinism effect. Consider, for example, Tawney (1961): "Historians give an appearance of inevitability to an existing order by dragging into prominence the forces which have triumphed and thrusting into the background those which they have swallowed up" (p. 177). Or, consider Wohlstetter (1962): "It is much easier after the event to sort the relevant from the irrelevant signals. After the event, of course, a signal is always crystal clear. We can now see what disaster it was signaling since the disaster has occurred, but before the event it is obscure and pregnant with conflicting meanings" (p. 387).

How justified are the judgmental changes affected by receipt of outcome knowledge? It is hard to say, simply because there is no unequivocal way to assign probabilities or relevance weights to unique events like the British–Gurka struggle. If, for example, someone claims that there was no chance (or a 7% chance or a 98.6% chance) of a thermonuclear war during the 1960s, who can prove him wrong? Indeed, the only wrong estimate is that it was 100% likely.

Were such events well defined and re-occurring, the wisdom of increasing the postdicted probability of some reported outcomes would be readily apparent. Consider a judge who has previously encountered four British–Gurka-type struggles, each indistinguishable from the one used here, two of which were won by the Gurkas. Upon learning of another Gurka victory, he may properly update that outcome's predictive (Before) probability of 50% to a higher postdictive (After) probability. Hearing of a Gurka victory may also justify some change in relevance judgments by showing, for example, the true importance of British suffering from climatic extremes. It may also teach him something about the nature of 19th century colonialism—and thus change the sort of "laws" or reasons he uses in drawing inferences from the event description.

Thus, the judgmental changes which we have called creeping determinism could conceivably reflect what judges learn from outcome reports. The skepticism expressed in the anecdotal observations presented above (e.g., Florovsky, Schur, Tawney), however, suggests that this is not the case. In the light of these comments, it appears that what passes for the wisdom of hindsight often contains heady doses of sophistry—that the perceived inevitability of reported outcomes is imposed upon, rather than legitimately inferred from, the available evidence.

As described in these accounts, postdictive likelihood estimates are exaggerated through a largely unconscious process evoked by receipt of outcome knowledge. How aware people are of the effect that outcome knowledge has on their perceptions was examined in Experiment 2. Aside from helping to clarify the nature of creeping determinism, these results have considerable intrinsic interest. Awareness is clearly crucial to knowing what one has learned from the past (i.e., from outcome knowledge). It may be necessary for learning from the past at all.

Experiment 2

Method

Design. Subjects were presented stimulus materials identical to those used in the After groups of Experiment 1, with each event description accompanied by a "true" outcome. They were asked to respond "as they would have had they not known the outcome." For each of the four events there were four After(ignore) groups, one receiving each possible outcome as true. If subjects are aware of the effect of outcome knowledge on their judgments, the responses of all of the After(ignore) groups should resemble those of that Before group in Experiment 1 which dealt with the same event. If After(ignore) subjects are completely unable to ignore the effect of outcome knowledge, their responses should resemble those of the After group in Experiment 1 which received the same outcome as "true."

Instructions. The cover of each test booklet read:

> A number of short descriptions of real social and personal events appear below, each with a number of possible outcomes. On the basis of these data, we ask you to evaluate the likelihood of the outcomes listed. We thank you for your participation.

Each remaining page of the test booklet was identical to the corresponding page of the Experiment 1 booklet, except that each response section was preceded with the instruction to "answer as you would have had you not known what happened."

Subjects. Eighty members of an introductory statistics class at the University of the Negev participated.

Procedure. Questionnaires were randomly distributed to a single group of subjects. Each subject received one version of each of the four different events. In a test booklet, Events A, B, and C alternated systematically as the first three events, with Event D (the least interesting) always appearing last. Order was varied to reduce the chances that subjects sitting in adjoining seats either copied from one another or discovered the experimental deception. All materials were in Hebrew. Questionnaires were anonymous.

Results

Probability estimates. Table 3.3 presents mean probability assignments by subjects in each of the After(ignore) groups along with the responses of the corresponding Before groups from Experiment 1. (The Hebrew-speaking group is used for Event A, the pooled responses of both relevant subexperiments for Event B.) The entries in each row will be called a *profile.* They indicate the probabilities subjects believed they would have assigned to the outcomes had they not known "what really happened."

TABLE 3.3 Mean probabilities assigned by subjects responding "As if you did not know what happened," Experiment 2

Experimental group	*n*	*Outcome provided*	*Outcome evaluated* 1	2	3	4
		Event A: British–Gurka struggle				
Before	17	none	29.4	23.5	34.7	12.4
After(ignore)	20	1	29.8	27.4	24.9	18.4
	15	2	38.0	21.7	19.7	20.7
	18	3	22.1	31.8	31.9	14.3
	18	4	18.1	32.9	28.9	21.2
		Event B: Near-riot in Atlanta				
Before	39	none	11.3	29.0	43.9	16.3
After(ignore)	17	1	24.6	27.0	28.3	19.8
	21	2	9.0	41.5	36.4	13.1
	20	3	6.3	24.5	43.5	25.8
	20	4	13.3	20.3	36.5	24.0
		Event C: Mrs. Dewar in therapy				
Before	19	none	26.6	15.8	23.4	34.4
After(ignore)	19	1	36.4	10.2	16.1	37.4
	19	2	24.7	28.8	15.5	31.9
	15	3	25.1	13.7	34.9	26.4
	20	4	18.3	12.3	21.8	52.8
		Event D: George in therapy				
Before	17	none	26.4	26.9	39.4	6.3
After(ignore)	17	1	41.8	16.5	35.3	6.5
	18	2	24.6	35.9	32.4	7.0
	20	3	18.3	20.4	57.3	4.0
	18	4	21.0	21.1	38.4	19.6

Note. In each case the Before results are taken from the corresponding before(no outcome) group in Experiment 1 (subjects who actually responded not knowing what happened). Outlined cells are those with After(ignore) probabilities of reported outcomes.

These reconstructed probabilities indicate no more than marginal awareness of the effects of outcome knowledge. In 13 of 16 cases, the mean After(ignore) probability of the reported outcome was higher than the mean Before probability for the same event. For reported outcomes the mean Before–After(ignore) difference of 9.2% was slightly but not significantly less than the 10.8% mean Before–After difference in Experiment 1 ($p > .10$; Mann-Whitney *U* test).

The After(ignore) profiles closely resembled the corresponding After profiles. For 14 of 16 profiles, the mean absolute difference between corresponding cells was smaller for the After(ignore)–After comparison than for the relevant After(ignore)–Before comparison ($p < .002$; sign test). The median absolute difference between corresponding cells was 3.7% for After(ignore)–After, and 6.4% for After(ignore)–Before ($p < .001$; Mann–Whitney *U* test). There is no apparent reason, other than sampling error, for the weaker results obtained with Event A.

Relevance judgments. If After(ignore) subjects are able to ignore outcome knowledge, the outcome report they received should have no effect on their reconstructed relevance judgments. Instead, however, these relevance judgments clearly reflected the outcomes that After(ignore) subjects believed to have happened (but were instructed to ignore). For example, in Experiment 1 After subjects told of a British victory assigned substantially greater importance to the fact that "British officers learned caution only after sharp reverses" (Datum No. 11) than did Before subjects; those told of a Gurka victory assigned it slightly less importance. After(ignore) subjects in Experiment 2 who were asked to ignore a report of British victory believed that even without the report they would have perceived the relevance of Datum No. 11; those told to ignore a report of Gurka victory believed that they in foresight would have seen its irrelevance. When the relevance judgment ANOVA of Experiment 1 is repeated on the present data, this dependence is reflected in highly significant ($p < .0005$) Outcome Reported X Datum Evaluated interactions.

Interestingly, for 128 of the 184 individual datum units evaluated by subjects in the four outcome groups of the four events After and After(ignore) relevance judgments were either both higher or both lower than the corresponding Before judgments (as was the case in the example, Datum No. 11, given above) ($z = 5.23$; sign test). There was no tendency for After and After(ignore) relevance judgments to be consistently higher or lower than Before relevance judgments, which might in itself account for this result.

Discussion

Experiment 1 showed that receipt of outcome knowledge affects subjects' judgments in the direction predicted by the creeping determinism hypothesis. Experiment 2 has shown that subjects are either unaware of outcome knowledge having an effect on their perceptions or, if aware, they are unable to ignore or rescind that effect. Both the relevance and the probability judgments of After(ignore) subjects suggest that subjects fail to properly reconstruct foresightful (Before) judgments because they are "anchored" in the hindsightful state of mind created by receipt of outcome knowledge.

It might be asked whether this failure to empathize with ourselves in a more ignorant state is not paralleled by a failure to empathize with outcome-ignorant others. How well people manage to reconstruct the perceptions that others had before the occurrence of some event is a crucial question for historians, and indeed for all human understanding. The assumption that we clearly perceive how others viewed situations before receipt of outcome knowledge underlies most second-guessing of their decisions. Experiment 3 examined this question.

Experiment 3

Method

Design. Subjects were presented with stimulus materials identical to those used in Experiments 1 and 2. They were asked to respond as had other student judges who had not known the true outcome. Before(others) subjects were not provided with any outcome knowledge. After(others) subjects received versions of the stimulus events with one of the four possible outcomes presented as the true outcome (what had actually happened). After(others) subjects' task was essentially to ignore outcome knowledge in order to respond like Before(others) subjects.

Instructions. The cover of each test booklet read:

> Short descriptions of a number of real social and personal events appear below, each with several possible outcomes. These descriptions were presented to students of social science in other universities in Israel. (However, they were not told which of the possible outcomes actually happened.) We will ask you to guess the judgments of these students regarding the likelihood of possible outcomes. We thank you for your participation.

The section in parentheses only appeared in the instructions for After(others) subjects. Each page of the test booklets was identical to the corresponding page of the Experiment 1 test booklets, except for the addition of a reminder, "Answer as you think other students (who did not know what happened) answered" before each response section.

Subjects. Ninety-four members of an intermediate statistics class at the University of the Negev participated.

Results

Probability estimates. Table 3.4 presents mean probability assignments by subjects in each group. After(others) subjects' inability to ignore the effects of creeping determinism is clearly evident. For 14 of the 16 reported outcomes ($p < .002$; sign test), they attributed higher probabilities to outcome-ignorant others than did Before(others) subjects. As in Experiment 2, being told to ignore outcome knowledge slightly, but not significantly ($p > .10$; Mann–Whitney U test), reduced its impact. The mean Before(others)–After(others) difference was 8.7% compared with the mean Before–After difference of 10.8% in Experiment 1.

TABLE 3.4 Mean probabilities assigned by subjects responding "As did other students who did not know what happened," Experiment 3

Experimental group	*n*	*Outcome provided*	*Outcome evaluated* 1	2	3	4
		Event A: British–Gurka struggle				
Before	21	none	26.4	24.5	29.5	19.5
After(ignore)	17	1	39.4	22.4	20.3	18.8
	17	2	18.8	42.6	20.3	20.0
	22	3	31.1	21.2	26.6	20.0
	17	4	28.2	21.9	23.7	26.2
		Event B: Near-riot in Atlanta				
Before	20	none	11.0	24.0	41.8	23.2
After(ignore)	17	1	15.0	24.7	36.5	23.8
	18	2	13.2	36.0	35.2	14.6
	19	3	4.8	22.5	51.1	21.6
	16	4	12.3	26.4	38.4	22.8
		Event C: Mrs. Dewar in therapy				
Before	21	none	19.6	15.9	24.0	40.5
After(ignore)	18	1	20.3	20.0	28.3	31.4
	18	2	31.9	23.3	14.8	30.0
	16	3	30.6	12.5	26.9	30.1
	19	4	12.5	20.4	22.6	44.4
		Event D: George in therapy				
Before	19	none	30.7	22.4	39.2	7.8
After(ignore)	15	1	46.0	15.3	30.0	8.7
	16	2	22.5	36.6	34.1	6.9
	17	3	19.8	14.8	57.7	7.8
	16	4	23.5	18.3	40.3	17.8

Note. Outlined cells are those with After(ignore) probabilities of reported outcomes.

Relevance judgments. After(ignore) subjects who had received different outcome reports attributed markedly different relevance judgments to the outcome-ignorant others. The dependence of the relevance judgments that they attributed on the outcome knowledge they were to ignore produced significant ($p < .01$) Outcome Reported × Datum Evaluated interactions for each of the four events. Thus, After(ignore) subjects expected other subjects to have seen in foresight patterns of data relevance that they themselves only saw in hindsight.

Projection. Comparing Tables 3.1 and 3.4 and Tables 3.3 and 3.4, it is apparent that the entries in corresponding Before and Before(others) cells are quite similar, as are corresponding After(others) and After(ignore) cells. The mean absolute difference between entries in corresponding cells is 3.5% for the first comparison, 5.1% for the latter. This suggests that when asked to respond like similar others, subjects respond as they believe they themselves would have responded in similar circumstances (i.e., by projection). Both the probability and relevance judgments of After(others) subjects more closely resembled those of After(ignore) and After subjects than those of Before(others) subjects.

Reasons. Some 87% of the subjects provided reasons for their judgments. Although content analysis of these reasons proved intractable, one interesting finding is that After(others) subjects offered consistently more reasons than Before subjects (p <.05; median test). In Experiment 1, After subjects reported relying more on outside information (as compared with the text) than did Before subjects. Perhaps in both cases knowing what happened facilitates knowing where to look and what to accept as reasons.

General discussion

Finding out that an outcome has occurred increases its perceived likelihood. Judges are, however, unaware of the effect that outcome knowledge has on their perceptions. Thus, judges tend to believe that this relative inevitability was largely apparent in foresight, without the benefit of knowing what happened.

In a fourth study (Fischhoff and Beyth, 1975), subjects were asked on the eve of former President Nixon's trips to China and the USSR (in early 1972) to estimate the probability of various possible outcomes of the visits (e.g., Nixon's meeting Chairman Mao, visiting Lenin's tomb, or announcing that the trips were successful). From 2 wk to 6 mo after the trips' completion, these same subjects were asked to remember as best they could their own original predictions. They were also asked to indicate for each event whether or not they believed that it had actually happened.

Results showed that subjects remembered having given higher probabilities than they actually had to events believed to have occurred and lower probabilities to events that hadn't occurred. Their original predictions showed considerable overestimation of low probabilities; that is, too many events that they judged to be extremely unlikely or impossible did occur. The probability judgments that they remembered, however, consistently underestimated low probabilities. Indeed, almost no events to which they remembered assigning low probabilities were perceived to have occurred.

Thus, undiagnosed creeping determinism not only biases people's impressions of what they would have known without outcome knowledge, but also their impressions of what they themselves and others actually *did* know in foresight.

Explanations

The simplest hypothesis regarding the manner in which judges process outcome knowledge suffices to account for these results. Assume that upon receipt of outcome knowledge judges immediately assimilate it with what they already know about the event in question. In other words, the retrospective judge attempts to make sense, or a coherent whole, out of all that he knows about the event. The changes in relevance judgments could reflect such assimilative meaning adjustment.

Assimilation of this type would tend to induce creeping determinism for judges using any of the techniques for producing subjective probability estimates appearing in Tversky and Kahneman's (1974) compendium. Judges using the heuristic of "representativeness" perceive outcomes as likely when they match or represent the dominant features of the situation that produced them. Assimilation of outcome knowledge should certainly increase the perceived "fit" between the reported outcome and the situation that preceded it. A second heuristic leads judges to evaluate an outcome's likelihood by the relative "availability" of scenarios leading to its occurrence and nonoccurrence. The judge who knows what happened, and has adjusted his perceptions in the light of that knowledge, may well find it difficult to imagine how things could have turned out otherwise.

An alternative mode of explanation focuses on structural differences between the tasks of hindsight and foresight. Judges possessing outcome knowledge may, for example, tend to reverse their temporal perspective and produce scenarios that proceed backward in time, from the outcome to the preceding situation. Such scenario retrodiction may effectively obscure the ways in which events might not have taken place, much as solving a maze backward can obscure the ways in which one might have gotten lost entering from the beginning.

Receipt of outcome knowledge may also restructure the task of judges using the "anchoring and adjustment" heuristic (Tversky and Kahneman, 1974). Judges may estimate the likelihood of a reported outcome by initially assigning it 100%, the most salient possible value, and then looking for reasons to adjust downward from there. Adjustment from initial values is typically inadequate and would produce creeping determinism with this task.

Of these explanations, those based on assimilation most readily account for the underestimation of creeping determinism found in Experiments 2 and 3 and in the Fischhoff and Beyth (1975) study. Making sense out of what one is told about the past seems so natural and effortless a response that one may be unaware that outcome knowledge has had any effect at all on him. Judges who are aware that outcome knowledge has affected their perceptions still face the unenviable task of reconstructing their foresightful state of mind. "Undiagnosed creeping determinism" would characterize the responses of subjects who, in reconstruction, were unable to adequately unanchor themselves from the perspective of hindsight.

Implications

In the short run, failure to ignore outcome knowledge holds substantial benefits. It is quite flattering to believe, or lead others to believe, that we would have known all along what we could only know with outcome knowledge, that is, that we possess hindsightful foresight. In the long run, however, unperceived creeping determinism can seriously impair our ability to judge the past or learn from it.

Consider a decision maker who has been caught unprepared by some turn of events and who tries to see where he went wrong by recreating his pre-outcome knowledge state of mind. If, in retrospect, the event appears to have seemed relatively likely, he can do little more than berate himself for not taking the action which his knowledge seems to have dictated. He might be said to add the insult of regret to the injury inflicted by the event itself. When second-guessed by a hindsightful observer, his misfortune appears to have been incompetence, folly, or worse.

In situations where information is limited, and indeterminate, occasional surprises—and resulting failures—are inevitable. It is both unfair and self-defeating to castigate decision makers who have erred in fallible systems, without admitting to that fallibility and doing something to improve the system. According to historian Roberta Wohlstetter (1962), the lesson to be learned from American surprise at Pearl Harbor is that we must "accept the fact of uncertainty and learn to live with it. Since no magic will provide certainty, our plans must work without it" (p. 401).

When we attempt to understand past events, we implicitly test the hypotheses or rules we use to both interpret and anticipate the world around us. If, in hindsight, we systematically underestimate the surprises which the past held and holds for us, we are subjecting those hypotheses to inordinately weak tests and, presumably, finding little reason to change them. Thus, the very outcome knowledge which gives us the feeling that we understand what the past was all about may prevent us from learning anything from it.[3]

Acknowledgements

Preparation of this report was supported by the Advanced Research Projects Agency of the Department of Defense (ARPA Order 2449) and was monitored by the Office of Naval Research under Contract No. N00014-73-C-0438 (NR 197–026). The research reported constitutes part of a doctoral dissertation submitted to The Hebrew University of Jerusalem. I am deeply indebted to Amos Tversky, Daniel Kahneman, Paul Slovic, Ruth Beyth, and Sarah Lichtenstein for their contributions to this project. The detailed comments of two anonymous reviewers on a previous draft are gratefully acknowledged.

Notes

1 Copies of these stimuli with the four offered outcomes will be supplied with all requested reprints. The permission granted by the Oxford University Press and Ronald Press to use these copyrighted materials is gratefully acknowledged.

2 It might be wondered whether Before subjects might not behave like After subjects with reference to their predictions, which they know are actually postdictions. In a series of five experiments (Fischhoff, 1976), we found that manipulating the temporal setting of possible outcomes has no effect on their perceived likelihood.

3 Elaboration on this point as well as speculation on how hindsight can be improved may be found in Fischhoff (1974) and Fischhoff and Beyth (1974).

References

Ellis, A. 1966. Psychosexual and marital problems. In L. A. Berg and L. A. Pennington (Eds.), *An introduction to clinical psychology*. New York: Ronald.

Fischhoff, B. 1974. Hindsight: Thinking backward? *Oregon Research Institute Research Monograph*, *14*, No. 1.

Fischhoff, B. 1976. The effect of temporal setting on likelihood estimates. *Organizational Behavior and Human Performance*, 15, 180–194.

Fischhoff, B., and Beyth, R. 1974. Failure has many fathers (Review of *Victims of groupthink* by I. Janis). *Oregon Research Institute Research Bulletin*, *14*, No. 17.

Fischhoff, B., and Beyth, R. 1975. "I knew it would happen"—remembered probabilities of once-future things. *Organizational Behavior and Human Performance*, *13*, 1–16.

Florovsky, G. 1969. The study of the past. In R. H. Nash (Ed.), *Ideas of history* (Vol. 2). New York: Dutton.

Lofland, R. 1969. *Deviance and identity*. Englewood Cliffs, N.J.: Prentice-Hall.

National Advisory Commission. 1968. *Kerner Commission report on civil disorders*. New York: Bantam.

Rosenhan, D. 1973. On being sane in insane places. *Science*, *79*, 250–252.

Schur, E. 1971. *Labelling deviant behavior*. New York: Harper and Row.

Tawney, R. H. 1961. *The agrarian problem in the sixteenth century*. New York: Franklin.

Tversky, A., and Kahneman, D. 1971. Belief in the law of small numbers. *Psychological Bulletin*, *76*, 105–110.

Tversky, A., and Kahneman, D. 1974. Judgment under uncertainty: Heuristics and biases. *Science*, *185*, 1124–1131.

Walster, E. 1967. "Second guessing" important events. *Human Relations*, *20*, 239–250.

Wohlstetter, R. 1962. *Pearl Harbor: Warning and decision*. Stanford, Calif.: Stanford University Press.

Woodward, E. L. 1938. *Age of reform*. London: Oxford University Press.

4

FOR THOSE CONDEMNED TO STUDY THE PAST

Reflections on historical judgment

Baruch Fischhoff

Learning from the past looks easy, but there are a thousand pitfalls.

I often think it odd that history should be so dull, for a great deal of it must be invention.

Catherine Morlund

Benson (1972) has identified four reasons for studying the past: to entertain, to create a group (or national) identity, to reveal the extent of human possibility, and to develop systematic knowledge about our world, knowledge that may eventually improve our ability to predict and control. On a conscious level, at least, we behavioral scientists restrict ourselves to the last motive. In its pursuit, we do case studies, program evaluations, and literature reviews. We even conduct experiments, creating artificial histories upon which we can perform our post mortems.

Three basic questions seem to arise in our retrospections: (1) Are there patterns upon which we can capitalize so as to make ourselves wiser in the future? (2) Are there instances of folly in which we can identify mistakes to avoid? (3) Are we really condemned to repeat the past if we do not study it? That is, do we really learn anything by looking backward?

Whatever the question we are asking, it is generally assumed that the past will readily reveal the answers it holds. Of hindsight and foresight, the latter appears as the more troublesome perspective. One can explain and understand any old event if an appropriate effort is applied. Prediction, however, is acknowledged to be rather more tricky. The present essay investigates this presumption by taking a closer look at some archetypal attempts to tap the past. Perhaps its most general conclusion is that we should hold the past in a little more respect when we attempt to plumb its secrets. While the past entertains, ennobles, and expands quite readily, it enlightens only with delicate coaxing.

Looking for wisdom

Formal Modeling. While the past never repeats itself in detail, it is often viewed as having repetitive elements. People make the same kinds of decisions, face the same kinds of challenges, and suffer the same kinds of misfortune often enough for behavioral scientists to believe that they can detect recurrent patterns. Such faith prompts psychometricians to study the diagnostic secrets of ace clinicians, clinicians to look for correlates of aberrant behavior, brokers to hunt for harbingers of price increases, and dictators to ponder revolutionary situations. Their search usually has a logic paralleling that of multiple regression or correlation. A set of relevant cases is collected and each member is characterized on a variety of dimensions. The resulting matrix is scoured for significant relationships that might aid us in predicting the future.

The Daily Racing Form, for example, offers the earnest handicapper some one hundred pieces of information on each horse in any given race. The handicapper with a flair for data processing might commit to some computer's memory the contents of a bound volume of the *Form* and try to derive a formula predicting speed as a weighted sum of scores on various dimensions. For example:

$$\tilde{y} = b_1x_1 + b_2x_2 + b_3x_3 \qquad (1)$$

where $\tilde{y}$ is our best guess at a horse's speed, x_1 is its percentage of victories in previous races, x_2 is its jockey's percentage of winning races, and x_3 is the weight it will carry in the present race. When scores are standardized (by subtracting the mean and then dividing by the standard deviation), the b_i reflect the importance of the different factors. If $b_1 = 2b_2$, then a given change in the horse's percentage of wins affects our speed prediction twice as much as an equivalent change in jockey's percentage, because past performances have proved twice as sensitive to x_1 as to x_2.

Sounds easy, but there are a thousand pitfalls. One emerges when the predictors (x_i) are correlated, as might (and in fact does) happen were winning horses to draw winning jockeys or vice versa. In such cases of multicollinearity, each variable has some independent ability to explain past performance and the two have some shared ability. When the weights are determined, that shared explanatory capacity will somehow be split between the two. Typically, that split renders the weights (b_i) uninterpretable with any degree of precision. Thus the regression equation cannot be treated as a theory of horse racing, showing the importance of various factors.

A more modest theoretical goal would simply be to determine which factors are and which factors are not important, on the basis of how much each adds to our understanding of y. The logic here is that of stepwise regression; additional variables are added to the equation as long as they add something to its overall predictive (or explanatory) power. Yet even this minimalistic strategy can run afoul of multicollinearity. If many reflections of a particular factor (such as different aspects of breeding) are included, their shared explanatory ability may be divided up into such small pieces that no one aspect makes a "significant" contribution.

Of course, these nuances may be of relatively little interest to handicappers as long as the formula works well enough to help them somewhat in beating the odds. We scientist types, however, want wisdom as well as efficacy from our techniques. It is hard for us to give up interpreting weights. Regression procedures not only express, but also produce, understanding (or, at least, results) in a mechanical, repeatable fashion. Small wonder then that they have been pursued doggedly despite their limitations.

One of the best documented pursuits has been in the study of clinical judgment. Clinical judgment is exercised by a radiologist who sorts X-rays of ulcers into "benign" and "malignant," by a personnel officer who chooses the best applicants from a set of candidates, or by a crisis center counselor who decides which callers threatening suicide are serious. In each of these examples, the diagnosis involves making a decision on the basis of a set of cues or attributes. When, as in these examples, the decision is repetitive and all cases can be characterized by the same cue, it is possible to model the judge's decision-making policy statistically. One collects a set of cases for which the expert has made a summary judgment (benign, serious) and then derives a regression equation, like (1), whose weights show the importance the judge has assigned to each cue.

Two decades of such policy-capturing studies persistently produced a disturbing pair of conclusions: (1) simple linear models, using a weighted sum of the cues, did an excellent job of postdicting judges' decisions, although (2) the judges claimed that they were using much more complicated strategies (Goldberg, 1968, 1970; Slovic and Lichtenstein, 1971). A commonly asserted form of complexity is called "configural" judgment, in which the diagnostic meaning of one cue depends upon the meaning of other cues (for example, "that tone of voice makes me think 'not suicidal' unless the call comes in the early hours of the morning").

Two reasons for conflict between measured and reported judgment policies have emerged from subsequent research, each with negative implications for the usefulness of regression modeling for "capturing" the wisdom of past decisions. One was the growing realization that combining enormous amounts of information in one's head, as required by such formulae, overwhelms the computational capacity of anyone but an *idiot savant.* A judge trying to implement a complex strategy simply would not be able to do so with great consistency. Indeed, it is difficult to learn and use even a non-configural, weighted sum, decision rule when there are many cues or unusual relationships between the cues and predicted variable (Slovic, 1974).

The second realization that has emerged from clinical judgment research is that simple linear models are extraordinarily powerful predictors. As long as one can identify and measure the attributes relevant to an individual, one can mimic his or her decisions to a large degree with simple models bearing no resemblance to actual cognitive processes. That is, under very general conditions, one can misspecify weights and even combination rules and still do a pretty good job of predicting decisions (Dawes, 1979). Thus, whatever people are doing will look like the application of a simple linear model. In Hoffman's (1960) term, such models are *paramorphic* in that they reproduce the input–output relations of the phenomena they are meant to describe without any guarantee of fidelity to the underlying processes.

Empirically discovering an analytical result by Wilks (1938), Dawes and Corrigan (1974) showed that considerable predictive success is possible without almost any modeling at all. All one has to do is to identify the variables (or attributes) to which a decision maker attends and decide whether they are positively or negatively related to the decision criterion. If these variables are expressed in standard units, they can be given unit weights (+ 1 or – 1, as appropriate). Such a unit weighted model will, under very general conditions, predict decisions as well as a full-blown regression model does.

Thus, a simple substantive theory indicating what variables people care about when making decisions may be all one needs to make reasonably good predictions of teir behavior. If some signs encourage a diagnosis or decision and others discourage it, simply counting the number of encouraging and discouraging signs will provide a fair guess at the individual's behavior. The result, however, will be a more modest theory than one can derive by flashy regression modeling.

Obviously, some factors are more important than others. Therefore a theory using importance weights should be more faithful to reality than one using unit weights. However, any unreliability or misspecification of those weights due to poor procedure or multicollinearity reduces their usefulness very quickly. Indeed, models using poorly conceived or executed weighting schemes may succeed in spite of rather than because of their increased sophistication (Fischhoff, Goitein, and Shapira, 1982), Thus, while the past seems to be right out there to be understood, our standard statistical procedures do not always tell us what we want to know. If not used carefully, they may mislead us, leaving us less wise than when we started. It is tempting to embrace highly complicated theories in their entirety without realizing that their power comes from very simple underlying notions, rather than from having captured the essence of the past.

Looking for folly

Searching for wisdom in historic events requires an act of faith, a belief in the existence of recurrent patterns waiting to be discovered. Searching for wisdom in the behavior of historical characters requires a somewhat different act of faith, confidence that our predecessors knew things we do not know. The first of these faiths is grounded in philosophy; it distinguishes those who view history as a social science, not an ideographic study of unique events. The second of these faiths is grounded in charity and modesty. It distinguishes those who hope to see further by standing on the shoulders of those who came before and those satisfied with standing on their faces. Idioms like "those who do not study history are condemned to repeat it" suggest that the latter faith is relatively rare.

An active search for folly is, of course, not without merit. Not only do individuals for whom things do not go right often have a lot of explaining to do, but such explanations are crucial to learning from their experience. By seeing how things went wrong, we hope to make them go right in the future.

Assuming that we know what has happened, we are then in a position to exploit the wisdom of our own hindsight in explaining and evaluating the past behavior of others. On closer examination, however, the advantages of knowing how things turned out may be oversold (Fischhoff, 1975). In hindsight, people consistently exaggerate what could have been anticipated in foresight. They not only tend to view what has happened as being inevitable, but also to view it as having appeared "relatively inevitable" before it happened. People believe that others should have been able to anticipate events much better than was actually the case. They even misremember their own predictions so as to exaggerate in hindsight what they knew in foresight (Fischhoff and Beyth, 1975).

As described by historian Georges Florovsky (1969): "The tendency toward determinism is somehow implied in the method of retrospection itself. In retrospect, we seem to perceive the logic of the events which unfold themselves in a regular or linear fashion according to a recognizable pattern with an alleged inner necessity. So that we get the impression that it really could not have happened otherwise" (p. 369). An apt name for this tendency to view reported outcomes as having been relatively inevitable might be "creeping determinism" in contrast with philosophical determinism, the conscious belief that whatever happens has to happen.

One corollary tendency is to telescope the rate of historical processes, exaggerating the speed with which "inevitable" changes are consummated (Fischer, 1970). For example, people may be able to point to the moment when large landed estates (latifundia) were doomed, without realizing that they took two and a half centuries to disappear. Another is the tendency to remember people as having been much more like their current selves than was actually the case (Yarrow, Campbell, and Burton, 1970). A third may be seen in Barraclough's (1972) critique of the historiography of the ideological roots of Nazism. Looking back from the Third Reich, one can trace its roots to the writings of many authors from whose writings one could not have projected Nazism. A fourth is to imagine that the participants in a historical situation were fully aware of its eventual importance ("Dear Diary, The Hundred Years' War started today," Fischer, 1970). A fifth is the myth of the critical experiment, unequivocally resolving the conflict between two theories or establishing the validity of one. In fact, "the crucial experiment is seen as crucial only decades later. Theories do not just give up, since a few anomalies are always allowed. Indeed, it is very difficult to defeat a research programme supported by talented and imaginative scientists" (Lakatos, 1970, pp. 157–158).

In the short run, failure to ignore outcome knowledge holds substantial benefits. It is quite flattering to believe, or lead others to believe, that we would have known all along what we could only know with outcome knowledge, that is, that we possess hindsightful foresight. In the long run, however, undetected creeping determinism can seriously impair our ability to judge the past or learn from it.

Consider decision makers who have been caught unprepared by some turn of events and who try to see where they went wrong by recreating their pre-outcome knowledge state of mind. If, in retrospect, the event appears to have seemed relatively likely, they can do little more than berate themselves for not taking action that their knowledge seems to have dictated. They might be said to add the insult of regret to the injury inflicted by the event itself. When second-guessed by a hindsightful observer, their misfortune appears as incompetence, folly, or worse.

In situations where information is limited and indeterminate, occasional surprises and resulting failures are inevitable. It is both unfair and self-defeating to castigate decision makers who have erred in fallible systems, without admitting to that fallibility and doing something to improve the system. According to historian Roberta Wohlstetter (1962), the lesson to be learned from American surprise at Pearl Harbor is that we must "accept the fact of uncertainty and learn to live with it. Since no magic will provide certainty, our plans must work without it" (p. 401).

When we attempt to understand past events, we implicitly test the hypotheses or rules we use both to interpret and to anticipate the world around us. If, in hindsight, we systematically underestimate the surprises that the past held and holds for us, we are subjecting those hypotheses to inordinately weak tests and, presumably, finding little reason to change them. Thus the very outcome knowledge which gives us the feeling that we understand what the past was all about may prevent us from learning anything about it.

Protecting ourselves against this bias requires some understanding of the psychological processes involved in its creation. It appears that when we receive outcome knowledge, we immediately make sense out of it by integrating it into what we already know about the subject. Having made this reinterpretation, the reported outcome now seems a more or less inevitable outgrowth of the reinterpreted situation. "Making sense" out of what we are told about the past is, in turn, so natural that we may be unaware of outcome knowledge having had any effect on us. Even if we are aware of there having been an effect, we may still be unaware of exactly what it was. In trying to reconstruct our foresightful state of mind, we will remain anchored in our hindsightful perspective, leaving the reported outcome too likely looking.

As a result, merely warning people about the dangers of hindsight bias has little effect (Fischhoff, 1977). A more effective manipulation is to force oneself to argue against the inevitability of the reported outcome, that is, try to convince oneself that it might have turned out otherwise. Questioning the validity of the reasons recruited to explain its inevitability might be a good place to start (Koriat, Lichtenstein, and Fischhoff, in press; Slovic and Fischhoff, 1977). Since even this unusual step seems inadequate, one might further try to track down some of the uncertainty surrounding past events in their original form. Are there transcripts of the information reaching the Pearl Harbor Command prior to 7 A M on December 7? Is there a notebook showing the stocks you considered before settling on Waltham Industries? Are there diaries capturing Chamberlain's view of Hitler in 1939? An interesting variant was Douglas Freeman's determination not to know about any subsequent events when working on any given period in his definitive biography of Robert E. Lee (Commager, 1965). Although admirable, this strategy does require some naive assumptions about the prevalence of knowledge regarding who surrendered at Appomattox.

Looking at all

Why look? Study of the past is predicated on the belief that if we look, we will be able to discern some interpretable patterns. Considerable research suggests that this belief is well founded. People seem to have a remarkable ability to find some order or meaning in even randomly produced data. One of the most familiar examples is the gamblers' fallacy. Our feeling is that in flipping a fair coin, four successive "heads" will be followed by a "tail" (Lindman and Edwards, 1961). Thus, in our minds, even random processes are constrained to have orderly internal properties. Kahneman and Tversky (1972) have suggested that of the thirty-two possible sequences of six binary events only one actually looks "random."

Although the gamblers' fallacy is usually cited in the context of piquant but trivial examples, it can also be found in more serious attempts to explain historical events. For example, after cleverly showing that Supreme Court vacancies appear more or less at random (according to a Poisson process), with the probability of at least one vacancy in any given year being .39, Morrison (1977) claimed that:

> [President] Roosevelt announced his plan to pack the Court in February, 1937, shortly after the start of his fifth year in the White House. 1937 was also the year in which he made his first appointment to the Court. That he had this opportunity in 1937 should come as no surprise, because the probability that he would go five consecutive years without appointing one or more justices was but .08, or one chance in twelve. In other words, when Roosevelt decided to change the Court by creating additional seats, the odds were already eleven to one in his favor that he would be able to name one or more justices by traditional means that very year [pp. 143–144].

However, if vacancies do appear at random, then this reasoning is wrong. It assumes that the probabilistic process creating vacancies, like that governing coin flips, has a memory and a sense of justice, as if it knows that it is moving into the fifth year of the Roosevelt presidency and that it "owes" FDR a vacancy. However, on January 1, 1937, the past four years were history, and the probability of at least one vacancy in the coming year was still .39 (Fischhoff, 1978).

Feller (1968) offers the following anecdote involving even higher stakes: Londoners during the blitz devoted considerable effort to interpreting the pattern of German bombing, developing elaborate theories of where the Germans were aiming (and when to take cover). However, when London was divided up into small, contiguous geographic areas, the frequency distribution of bomb hits per area was almost a perfect approximation of the Poisson distribution. Natural disaster constitutes another category of consequential events where (threatened) lay people see order when experts see randomness (Kates, 1962).

One secret to maintaining such beliefs is failure to keep complete enough records to force ourselves to confront irregularities. Historians acknowledge the role of missing evidence in facilitating their explanations with comments like "the history of the Victorian Age will never be written. We know too much about it. For ignorance is the first requisite of the historian – ignorance which simplifies and clarifies, which selects and omits, with placid perfection unattainable by the highest art" (Strachey, 1918).

Even where records are available and unavoidable, we seem to have a remarkable ability to explain or provide a causal interpretation for whatever we see. When events are produced by probabilistic processes with intuitive properties, random variation may not even occur to us as a potential hypothesis. For example, the fact that athletes chastized for poor performance tend to do better the next time out fits our naive theories of reward and punishment. This handy explanation blinds us to the possibility that the improvement is due instead to regression to those players' mean performance (Furby, 1973; Kahneman and Tversky, 1973).

Fama (1965) has forcefully argued that the fluctuations of stockmarket prices are best understood as reflecting a random walk process. Random walks, however, have even more unintuitive properties than the binary processes to which they are formally related (Carlsson, 1972). As a result, we find that market analysts have an explanation for every change in price, whether purposeful or not. Some explanations are inconsistent: for example, when the market rises following good economic news, it is said to be responding to the news; if it falls, that is explained by saying that the good news had already been discounted. Other explanations seem to deny the possibility of any random factor – for example, that ultimate fudge factor, the "technical adjustment."

The pseudo-power of our explanations can be illustrated by analogy with regression analysis. Given a set of events and a sufficiently large or rich set of possible explanatory factors, one can always derive post-dictions or explanations to any desired degree of tightness. In regression terms, by expanding the set of independent variables one can always find a set of predictors with any desired correlation with the independent variable. The price one pays for overfitting is, of course, shrinkage, failure of the derived rule to work on a new sample of cases. The frequency and vehemence of methodological warnings against overfitting suggest that correlational overkill is a bias that is quite resistant to even extended professional training (for references, see Fischhoff and Slovic, 1980).

One way of thinking of an overfitted theory is like a suit tailored so precisely to one individual in one particular pose that it will not fit anyone else or even that same individual in the future or even in the present if new evidence about him comes to light (for instance, he lets out his breath to reveal a pot-belly). An historian who had built an airtight case accounting for all available evidence in explaining how the Bolsheviks won might be in a sad position were the USSR to release suppressed documents showing that the Mensheviks were more serious adversaries than had previously been thought. The price investment analysts pay for overfitting is their long-run failure to predict any better than market averages (Dreman, 1980—although the cynic might say that they actually make their living through the generation of hope (and commissions).

Overfitting works because of capitalization on chance fluctuations. If measurement is sufficiently fine, two cases differing on one variable will also differ on almost any other variable one chooses to name. As a result, one can calculate a non-zero (actually, in this case, perfect) correlation between the two variables and derive an "interesting" substantive theory. Processes analogous to this two-dimensional case work with any m observations in the n-space defined by our set of possible explanatory concepts.

In these examples, the data are fixed and undeniable, while the set of possible explanations is relatively unbounded; one hunts until one finds an explanation that fits. Another popular form of capitalization on chance leaves the set of explanations fixed (usually at one candidate) and sifts through data until supporting evidence is found. While the crasser forms of this procedure are well known, others are more subtle and even somewhat ambiguous in their characterization. For example, you run an experiment and fail to receive an anticipated result. Thinking about it, you note an element of your procedure that might have mitigated the effect of the manipulated variable. You correct that; again no result, but again a possible problem. Finally, you (or your subjects) get it right and the anticipated effect is obtained. Now, is it right to perform your statistical test on that n'th sample (for which it shows significance) or the whole lot of them? Had you done the right experiment first, the question would not even have arisen. Or, as a toxicologist, you are "certain" that exposure to Chemical X is bad for one's health, so you compare workers who do and do not work with it in a particular plant for bladder cancer, but still no effect. So you try intestinal cancer, emphysema, dizziness, …, until you finally get a significant difference in skin cancer. Is that difference meaningful? Of course, the way to test these explanations or theories is by replication on new samples. That step, unfortunately, is seldom taken and often not possible for technical or ethical reasons (Tukey, 1977).

Related complications can arise even with fixed theories and data sets. Diaconis (1978) notes the difficulty of evaluating the surprisingness of ESP results, even in the rare cases in which they have been obtained in moderately supervised settings, because the definition of the sought event keeps shifting. "A major key to B.D.'s success was that he did not specify in advance the result to be considered surprising. The odds against a coincidence *of some sort* are dramatically less than those against any prespecified *particular one* of them" (p. 132).

Tufte and Sun (1975) discovered that the existence or non-existence of bellwether precincts depends upon the creativity and flexibility allowed in defining the event (for what office, in what elections, how good is good, are precincts that miss consistently to be included?). They are commonly believed to exist because we have an uncommonly good ability to find a signal even in total noise.

Have we seen enough? Given that we are almost assured of finding something interpretable when we look at the past, our next question becomes "have we understood it?" The hindsight research described earlier suggests that we are not only quick to find order, but also poised to feel that we knew it all along in some way, or would have been able to predict the result had we been asked in time. Indeed, the ease with which we discount the informativeness of anything we are told makes it surprising that we ever ask the past, or any other source, many questions, This tendency is aggravated by tendencies (1) not to realize how little we know or are told, leaving us unaware of what questions we should be asking in search of surprising answers (Fischhoff, Slovic, and Lichtenstein, 1977, 1978) and (2) to draw far-reaching conclusions from even small amounts of unreliable data (Kahneman and Tversky, 1973; Tversky and Kahneman, 1971).

Any propensity to look no further is encouraged by the norm of reporting history as a good story, with all the relevant details neatly accounted for and the uncertainty surrounding the event prior to its consummation summarily buried, along with any confusion the author may have felt (Gallie, 1964; Nowell-Smith, 1970). Just one of the secrets to doing this is revealed by Tawney (1961). "Historians give an appearance of inevitability to an existing order by dragging into prominence the forces which have triumphed and thrusting into the background those which they have swallowed up" (p. 177).

Although an intuitively appealing goal, the construction of coherent narratives exposes the reader to some interesting biases. A completed narrative consists of a series of somewhat independent links, each fairly well established. The truth of the narrative depends upon the truth of the links. Generally, the more links there are and the more detail in each link, the less likely the story is to be correct in its entirety. However, Slovic, Fischhoff, and Lichtenstein (1976) have found that adding detail to an event description can increase its perceived probability of occurrence, evidently by increasing its thematic unity. Bar Hillel (1973) found that people consistently exaggerate the probability of the conjunction of a series of likely events. For example, her subjects generally preferred a situation in which they would receive a prize if seven independent events each with a probability of .90 were to occur to a situation in which they would get the same prize if a fair coin fell on "heads." The probability of the compound event is less than .50, whereas the probability of the single event is .50. In other words, uncertainty seems to accumulate at much too slow a rate.

What happens if the sequence includes one or a few weak or unlikely links? The probability of its weakest link should set an upper limit on the probability of an entire narrative. Coherent judgments, however, may be compensatory, with the coherence of strong links "evening out" the incoherence of weak links. This effect is exploited by attorneys who bury the weakest link in their arguments near the beginning of their summations and finish with a flurry of convincing, uncontestable arguments.

Coles (1973) presents a delicious example of the overall coherence of a story obscurring the unlikelihood of its links: Freud's most serious attempt at psychohistory was his biography of Leonardo da Vinci. For years, Freud had sought the secret to understanding Leonardo, whose childhood and youth were basically unknown. Finally, he discovered a reference by Leonardo to a recurrent memory of a vulture touching his lips while he was in the cradle. Noting the identity of the Egyptian hieroglyphs for "vulture" and "mother" and other circumstantial evidence, Freud went on to build an imposing and coherent analysis of Leonardo. While compiling the definitive edition of Freud's works, however, the editor discovered that the German translation of Leonardo's recollection (originally in Italian) which Freud had used was in error, and that it was a kite and not a vulture which had stroked his lips. Despite having the key to Freud's analysis destroyed, the editor decided that the remaining edifice was strong enough to stand alone. As Hexter (1971) observed, "Partly because writing bad history is pretty easy, writing very good history is rare" (p. 59).

Conclusion

What general lessons can we learn about the study of the past, beyond the fact that understanding is more elusive than may often be acknowledged?

Presentism. Inevitably, we are all captives of our present personal perspective. We know things that those living in the past did not. We use analytical categories (such as feudalism, Hundred Years War) that are meaningful only in retrospect (Brown, 1974). We have our own points to prove when interpreting a past which is never sufficiently unambiguous to avoid the imposition of our ideological perspective (Degler, 1976). Historians do "play new tricks on the dead in every generation" (Becker, 1935).

There is no proven antidote to presentism. Some partial remedies can be generalized from the discussion of how to avoid hindsight bias when second-guessing the past. Others appear in almost any text devoted to the training of historians. Perhaps the most general messages seem to be (1) knowing ourselves and the present as well as possible; "the historian who is most conscious of his own situation is also most capable of transcending it" (Croce, quoted in Carr, 1961, p. 44); and (2) being as charitable as possible to our predecessors; "the historian is not a judge, still less a hanging judge" (Knowles, quoted in Marwick, 1970, p. 101).

Methodism. In addition to the prison of our own time, we often further restrict our own perspective by voluntarily adopting the blinders that accompany strict adherence to a single scientific method. Even when used judiciously, no one method is adequate for answering many of the questions we put to the past. Each tells us something and misleads us somewhat. When we do not know how to get the right answer to a question, an alternative epistemology is needed: use as broad a range of techniques or perspectives as possible, each of which enables us to avoid certain kinds of mistakes. This means a sort of interdisciplinary cooperation and respect different from that encountered in most attempts to comingle two approaches. Matches or mismatches like psychohistory too often are attempted by advocates insensitive to the pitfalls in their adopted fields (Fischhoff, 1981). Hexter (1971) describes the historians involved in some such adventures as "rats jumping aboard intellectually sinking ships" (p. 10).

Learning. Returning to Benson, if we want the past to serve the future, we cannot treat it in isolation. The rules we use to explain the past must also be those we use to predict the future. We must cumulate our experience with a careful eye to all relevant tests of our hypotheses. One aspect of doing this is compiling records that can be subjected to systematic statistical analysis; a second is keeping track of the deliberations preceding our own decisions, realizing that the present will soon be past and that a well-preserved record is the best remedy to hindsight bias; a third is to make predictions which can be evaluated. One disturbing lesson from Three Mile Island is that it is not entirely clear what that ostensibly diagnostic event told us about the validity of the Reactor Safety Study (U.S. Nuclear Regulatory Commission, 1975) which attempted to assess the risks from nuclear power. A fourth is to get a better idea of the validity of our own feelings of confidence, insofar as confidence in present knowledge controls our pursuit of new information and interpretation (Fischhoff, Slovic, and Lichtenstein, 1977). Thus we want to structure our lives so as to facilitate learning.

Indeterminacy. In the end, though, there may be no answers to many of the questions we are posing. Some are ill-formed. Others just cannot be answered with existing or possible tools. As much as we would like to know "how the pros do it," there may be no way statistically to model experts' judgmental policies to the desired degree of precision with realistic stimuli. Our theories are often of "such complexity that no single quantitative work could even begin to test their validity" (O'Leary and others, 1974, p. 228). When groups we wish to compare on one variable also differ on another, there is no logically sound procedure for equating them on that nuisance variable (Meehl, 1970). When we have tried many possible explanations on a fixed set of data, there is no ironclad way of knowing just how many degrees of freedom we have used up, just how far we have capitalized on chance (Campbell, 1975). When we use multiple approaches, the knowledge they produce never converges neatly. In the end, we may have to adopt Trevelyan's philosophical perspective that "several imperfect readings of history are better than none at all" (cited in Marwick, 1970, p. 57).

Acknowledgements

My thanks to Lita Furby, Lewis Goldberg, Sarah Lichtenstein, and Paul Slovic for their perceptive comments on earlier drafts.

References

Bar-Hillel, M. 1973. "On the Subjective Probability of Compound Events." *Organizational Behavior and Human Performance, 9*, 396–406.

Barraclough, G. 1972. "Mandarins and Nazis." *New York Review of Books, 19*(6), 37–42.

Becker, C. 1935. "Everyman His Own Historian." *American Historical Review, 40*, 221–236.

Benson, L. 1972. *Toward the Scientific Study of History Selected Essays*. Philadelphia: Lippincott.

Brown, E. A. R. 1974. "The Tyranny of a Construct: Feudalism and Historians of Medieval Europe." *American Historical Review, 79*, 1063–1088.

Campbell, D. T. 1975. "'Degrees of Freedom' and the Case Study." *Comparative Political Studies, 8*, 178–193.

Carlsson, G. 1972. "Random Walk Effects in Behavioral Data." *Behavioral Science, 17*, 430–437.

Carr, E. H. 1961. *What Is History*? London, Macmillan.

Coles, R. 1973. "Shrinking History." *New York Review of Books*, Part I in Feb 22, 1973, *20*, pp. 15–21 and Part II in Mar. 8, *20*, pp. 25–29.

Commager, H. S. 1965. *The Nature and Study of History*. Columbus, Ohio: Merrill.

Dawes, R. M. 1979. "The Robust Beauty of Improper Linear Models in Decision Making." *American Psychologist, 34*, 571–582.

Dawes, R. M., and Corrigan, B. 1974. "Linear Models in Decision Making." *Psychological Bulletin, 81* (2), 95–106.

Degler, C. N. 1976. "Why Historians Change Their Minds." *Pacific Historical Review, 48*, 167–189.

Diaconis, P. 1978. "Statistical Problems in ESP Research." *Science, 201*, 131–136.

Dreman, D. 1980. *Contrarian Investment Strategy*. New York: Random House.

Fama, E. F. 1965. "Random Walks in Stock Market Prices." *Financial Analysts Journal, 21*, 55–60.

Feller, W. 1968. *An Introduction to Probability Theory and Its Applications*. (3rd ed.) Vol. 1. New York: Harper and Row.

Fischer, D. H. 1970. *Historian's Fallacies*. New York: Harper and Row.

Fischhoff, B. 1975. "Hindsight ≠ Foresight: The Effect of Outcome Knowledge on Judgment Under Uncertainty." *Journal of Experimental Psychology: Human Perception and Performance, 1*, 288–299.

Fischhoff, B. 1977. "Perceived Informativeness of Facts." *Journal of Experimental Psychology: Human Perception and Performance, 3*, 349–358.

Fischhoff, B. 1978. "Intuitive Use of Formal Models. A Comment on Morrison's 'Quantitative Models in History." *History and Theory, 17*, 207–210.

Fischhoff, B. 1981. "No Man Is a Discipline." In J. Harvey (Ed.), *Cognition, Social Behavior, and the Environment*. Hillsdale, N.J.: Erlbaum.

Fischhoff, B., and Beyth, R. 1975. "'I Knew It Would Happen'—Remembered Probabilities of Once-Future Things." *Organizational Behavior and Human Performance, 13*, 1–16.

Fischhoff, B., Goitein, B., and Shapira, Z. 1982. "The Experienced Utility of Expected Utility Approaches." In N. Feather (Ed.), *Expectancy, Incentive, and Action*. Hillsdale, N.J.: Erlbaum.

Fischhoff, B., and Slovic, P. 1980. "A Little Learning ...: Confidence in Multicue Judgment." In R. Nickerson (Ed.), *Attention and performance, VIII* (pp. 779–800). Hillsdale, NJ: Erlbaum.

Fischhoff, B., Slovic, P., and Lichtenstein, S 1977. "Knowing with Certainty: The Appropriateness of Extreme Confidence." *Journal of Experimental Psychology: Human Perception and Performance, 3*, 552–564.

Fischhoff, B., Slovic, P., and Lichtenstein, S. 1978. "Fault Trees: Sensitivity of Estimated Failure Probabilities to Problem Representation." *Journal of Experimental Psychology: Human Perception and Performance, 4*, 330–344.

Florovsky, G. 1969. "The Study of the Past." In R. H. Nash (Ed.), *Ideas of History*. Vol. 2, New York: Dutton.

Furby, L. 1973. "Interpreting Regression Toward the Mean in Developmental Research." *Developmental Psychology, 8*, 172–179.

Gallie, W. B. 1964. *Philosophy and the Historical Understanding*. London: Chatto and Windus.

Goldberg, L. R. 1968. "Simple Models or Simple Processes? Some Research on Clinical Judgments." *American Psychologist, 23*, 483–496.

Goldberg, L. R. 1970. "Man Versus Model of Man: A Rationale, Plus Some Evidence, for a Method of Improving on Clinical Inference." *Psychological Bulletin, 73*, 422–432.

Hexter, J. H. 1971. *The History Primer*. New York: Basic Books.

Hoffman, P. J. 1960. "The Paramorphic Representation of Clinical Judgment." *Psychological Bulletin, 47*, 116–131.

Kahneman, D., and Tversky, A. 1972. "Subjective Probability: A Judgment of Representativeness." *Cognitive Psychology, 3*, 430–454.

Kahneman, D., and Tversky, A. 1973. "On the Psychology of Prediction." *Psychological Review, 80*, 237–251.

Kates, R. W. 1962. *Hazard and Choice Perception in Flood Plain Management*, Research Paper No. 78. Chicago: University of Chicago, Department of Geography.

Koriat, A., Lichtenstein, S., and Fischhoff, B. 1980. "Reasons for Confidence." *Journal of Experimental Psychology: Human Learning and Memory*, 6, 107–118.

Lakatos, I. 1970. "Falsification and Scientific Research Programmes." In I. Lakatos and A. Musgrave (Eds.), *Criticism and the Growth of Scientific Knowledge*, Cambridge: Cambridge University Press.

Lindman, H. G., and Edwards, W. 1961. "Supplementary Report: Unlearning the Gamblers' Fallacy" *Journal of Experimental Psychology, 62*, 630.

Marwick, A. 1970. *The Nature of History*. London: Macmillan.

Meehl, P. E. 1970. "Nuisance Variables and the Ex Post Facto Design." In M. Radner and S. Winokur (Eds), *Minnesota Studies in the Philosophy of Science*. Minneapolis: University of Minnesota Press.

Morrison, R. J. 1977. "Franklin D. Roosevelt and the Supreme Court: An Example of the Use of Probability Theory in Political History." *History and Theory, 16*, 137–146.

Nowell-Smith, P. H. 1970. "Historical Explanation." In H. E. Kiefer and M. K. Munitz (Eds), *Mind, Science, and History* Albany: State University of New York Press.

O'Leary, M. K., Coplin, W. D., Shapiro, H. B., and Dean, D. 1974. "The Quest for Relevance." *International Studies Quarterly, 18,* 211–237.

Slovic, P. 1974. "Hypothesis Testing in the Learning of Positive and Negative Linear Functions" *Organizational Behavior and Human Performance, 11,* 368–376.

Slovic, P., and Fischhoff, B. 1977. "On the Psychology of Experimental Surprises." *Journal of Experimental Psychology: Human Perception and Performance, 3,* 544–551.

Slovic, P., Fischhoff, B., and Lichtenstein, S. 1976. "Cognitive Processes and Societal Risk Taking." In J. S. Carroll and J. W. Payne (Eds.), *Cognition and Social Behavior,* Hillsdale, N.J.: Erlbaum.

Slovic, P., and Lichtenstein, S. 1971. "Comparison of Bayesian and Regression Approaches to the Study of Information Processing in Judgment." *Organizational Behavior and Human Performance, 6,* 649–744.

Strachey, L. 1918. *Eminent Victorians.* New York: Putnam's.

Tawney, R. H. 1961. *The Agrarian Problem in the Sixteenth Century,* New York: Franklin.

Tufte, E. R., and Sun, R. A. 1975. "Are There Bellwether Electoral Districts?" *The Public Opinion Quarterly, 39,* 1–18.

Tukey, J. W. 1977. "Some Thoughts on Clinical Trials, Especially Problems of Multiplicity." *Science, 198,* 679–690.

Tversky, A., and Kahneman, D. 1971. "The Belief in the 'Law of Small Numbers.'" *Psychological Bulletin, 76,* 105–110.

Wilks, S. S. 1938. "Weighting Systems for Linear Functions of Correlated Variables When There Is No Dependent Variable." *Psychometrika, 3,* 23–40.

Wohlstetter, R. 1962. *Pearl Harbor: Warnings and Decision.* Stanford, Calif: Stanford University Press.

Yarrow, M., Campbell, J. D., and Burton, R. V. 1970. "Recollections of Childhood: A Study of the Retrospective Method." *Monographs of the Society for Research in Child Development, 35* (5).

5

AN EARLY HISTORY OF HINDSIGHT RESEARCH

Baruch Fischhoff

CARNEGIE MELLON UNIVERSITY

The first studies of hindsight bias reflected the confluence of two desires. One arose from being part of the heady early days of Amos Tversky and Danny Kahneman's heuristics-and-biases research program. As the paradigm evolved, the challenge for the participating graduate students was to find a heuristic to call one's own or to find a way to elaborate one of the three "classics" (availability, representativeness, anchoring and adjustment). Maya Bar Hillel and Ruth Beyth-Marom chose the latter route. However, I was still struggling to reconcile the political motives that had brought me to Israel, planning to live in a kibbutz for the rest of my life, with academic life—and its more realistic view of the pace of change in human affairs.

For one meeting of the seminar, we read Paul Meehl's (1973) "Why I Do Not Attend Case Conferences." One of his many insights concerned clinicians' exaggerated feeling of having known all along how cases were going to turn out. To me, this sounded a lot like the exaggerated claims of understanding political processes that permeated the political discussions to which I had long subjected myself. Those discussions often left me wondering, "If we're so prescient, why aren't we running the world?" Psychological research provided an opportunity (and an obligation) to discipline such observations with systematically collected evidence, interpreted in the context of evolving theory. It occurred to me that I might have my bias, if I could provide the evidence and theory.

The evidence came first. The research group was very sensitive to the need for normative analysis, establishing the standard for evaluating performance. The task that provided the clearest demonstration of hindsight bias came in the initial study. President Nixon was about to leave for his historic trips to China and the Soviet Union. Ruth and I asked people to assess the probabilities of various possible outcomes (e.g., Pres. Nixon will meet Chairman Mao; Pres.

Nixon will announce that the trip was a success). After the trips were over, we asked subjects to recall their predictions. Fortunately, we had the foresight to ask them what they thought had happened, so that we could evaluate any bias in their memory, relative to their own beliefs (which often differed from press accounts of what had happened).

Assuming that subjects accepted our claim of testing their memory, rather than their knowledge of international relations, identifying bias was relatively straightforward. Did they remember having given higher probabilities than they actually had for events that they thought had happened (and lower probabilities for ones that had not)? Theoretically, though, the bias seemed multiply determined, fitting any account of revising beliefs that left prior representations less available, if not lost entirely. In the spirit of the times, we thought of hindsight bias as arising from the use of a sense-making heuristic, whereby people integrate all they know about a topic into a coherent mental model. Like other heuristics, it was good for some things (looking forward with a full set of beliefs), but bad for others (reconstructing previous perspectives).

However, the recall design has its limits. Having to find participants twice makes it hard to study hindsight for temporally distant events. (We recently did it over a year's time (Fischhoff, Gonzalez, Lerner, and Small, 2005).) Moreover, the design cannot be used with events that people did not judge in foresight. A task that can be used with any event is to ask people what they would have predicted in a situation, had they been asked, after being told what had happened. Another friend, Aron Hirt-Mannheimer, and I searched history books for the right kind of event. Aron found the British–Gurkha story. Once we had prepared it as a stimulus, producing others was straightforward. (About 20 years later, someone asked for the other stimuli. I was stunned to discover how inappropriate one had become. Taken from Ellis (1966), it described a "troubled" homosexual, in terms that fit now-antiquated mores and theories. Although I sent it, I argued that it should not be used. In addition to being offensive, it had changed its meaning so much that using it would not represent a replication.)

Fischhoff (1975) describes other attempts to deal with possible demand characteristics with "what would you have said?" instructions (as do Slovic and Fischhoff, 1977). A nagging normative question was how much people had actually learned from observing an event. At one seminar meeting, Danny Kahneman suggested an analogy with drawing colored balls from an urn (a popular research task in those days; Slovic and Lichtenstein, 1971). In historical terms, the parallel involves a class of "four British–Gurkha-type struggles, each indistinguishable from the one used [in the study], two of which were won by the Gurkhas. Upon learning of another Gurkha victory, [one might] properly update that outcome's predictive (Before) probability of 50% to a higher postdictive (After) probability" (p. 293).

A less formal and direct form of normative analysis involved asking what experts in historical judgment advise. Fischhoff (1975) offers some connections to relevant sources in historiography; Fischhoff (1982) has more. The best guidance, though, came from a philosopher of science. Despite not believing in the social sciences, Imre Lakatos (1970) offered an account of scientific thinking that made good cognitive sense. His "myth of the critical experiment" showed how scientists exaggerated how fully the implications of experiments (like Michelson–Morley) were understood when they first appeared (see also Motterlini, 1999).

The 1973 War interrupted the research (and many other things). The recriminations over how surprised Israel had been seemed to have an element of hindsight bias, adding insult to the injury of the war. Despairing of getting something published in the general press, Ruth and I managed to place a book review (Fischhoff and Beyth, 1974) of Irving Janis's (1972) *Victims of Groupthink*, which had become part of the popular debate. While conceding Janis's insights into group processes, we wondered how much his selection and interpretation of good and bad cases (Cuban Missile Crisis, Bay of Pigs) had been influenced by knowing how they had turned out. I only had one opportunity to meet Prof. Janis, about ten years later. Our review was the first thing that he wanted to discuss.

It is challenging to reconstruct what were realistic expectations for hindsight research, at the time of these studies. Recently, Roediger (2004) wrote a presidential column for the *APS Observer* on most-cited dissertations. Mine had not come to his attention. Out of curiosity, I looked to see how it rated. To my amazement, it would have made the top ten. The two main articles from it, Fischhoff (1975) and Fischhoff and Beyth (1975) had received 533 and 207 citations, respectively. Given the base rate of citations for scientific publications, no evidence available at that time should have been diagnostic enough to predict anything like that. I suppose that the work deserves some credit. So does the creativity of those who have found it a useful platform for studying a variety of processes.

(The third article from the dissertation, Fischhoff (1976) considered whether being set in the past per se changed an event's apparent probability. It had been cited 19 times. I was surprised that anyone had read it at all, given that it concluded by accepting the null hypothesis of no effect. Its two reviewers were divided: One thought the result was obvious; the other was sure that it wasn't true. That contradiction and the fact that I was halfway around the world led the editor to accept it without changes.)

The clearest prediction at that time came from the woman running the office that processed dissertations. She took one look at my stack of six copies and said, "You won't have any trouble." I asked how she knew, given that she had not even seen the title. "The thin ones go right through." Seeing my puzzlement, she took me into the next room and pointed to one that would have trouble. Each two-volume copy was thicker than my full set. We looked to see what the topic was. Something in French literature. I have often wondered how its author fared.

References

Ellis, A. 1966. Psychosexual and marital problems. In L.A. Berg and L.A. Pennington (Eds.), *An introduction to clinical psychology*. New York: Ronald.

Fischhoff, B. 1975. Hindsight ≠ foresight: The effect of outcome knowledge on judgment under uncertainty. *Journal of Experimental Psychology: Human Perception and Performance, 104*, 288–299.

Fischhoff, B. 1976. The effect of temporal setting on likelihood estimates. *Organizational Behavior and Human Performance, 15*, 180–194.

Fischhoff, B. 1982. For those condemned to study the past: Reflections on historical judgment. In D. Kahneman, P. Slovic, and A. Tversky (Eds.), *Judgment under uncertainty: Heuristics and biases*. New York: Cambridge University Press.

Fischhoff, B., and Beyth, R. 1974. Failure has many fathers. Review of I. Janis, *Victims of Groupthink. Megamot, 28*, 325–328 (in Hebrew). Reprinted in *Policy Sciences*, 7, 388–393.

Fischhoff, B., and Beyth, R. 1975. "I knew it would happen": Remembered probabilities of once–future things. *Organizational Behavior and Human Performance, 13*, 1–16.

Fischhoff, B., Gonzalez, R.M., Lerner, J.S., and Small, D.A. 2005. Evolving judgments of terrorism risks: Foresight, hindsight, and emotion. *Journal of Experimental Psychology: Applied, 11*, 124–139.

Janis, I. 1972. *Victims of groupthink*. Boston: Houghton Mifflin.

Lakatos, I. 1970. Falsification and the scientific research programmes. In I. Lakatos and A. Musgrave (Eds.), *Criticism and the growth of scientific knowledge*. Cambridge, UK: Cambridge University Press.

Meehl, P.E. 1973. Why I do not attend case conferences. In *Psychodiagnosis: Selected papers* (pp. 225–302). New York: Norton.

Motterlini, M. 1999. *For and against method*. Chicago: University of Chicago Press.

Roediger, H.L., III. 2004 Great dissertations: Mark I. *Observer, 17*(10).

Slovic, P., and Fischhoff, B. 1977. On the psychology of experimental surprises. *Journal of Experimental Psychology: Human Perception and Performance, 3*, 544–551.

Slovic, P., and Lichtenstein, S. 1971. Comparison of Bayesian and regression approaches to the study of information processing in judgment. *Organizational Behavior and Human Performance, 6*, 649–744.

PART III

Judgment

Bayesian inference is a general approach to making orderly inferences about uncertain events. It assumes that people have prior beliefs about any question – if only a state of ignorance. When they encounter new data, they interpret it in light of their general knowledge about such situations, leading to updated (or posterior) beliefs. Bayesian inference treats all beliefs as subjective, even if there is wide agreement about them (e.g., the chances of winning a fair roulette play). However, it insists that beliefs be consistent with one another and with gambles based on them, before calling them probabilities. As a result, subjective does not mean arbitrary. Bayesian inference provides the foundation for the normative analyses underlying much behavioral decision research.

"Hypothesis evaluation from a Bayesian perspective" (Chapter 6) revisits Edwards, Lindman and Savage (1963), the landmark introduction of Bayesian inference to psychologists. Ruth Beyth-Marom and I wrote our article because we feared that the original was inaccessible to readers without its authors' mathematical training, just as Larry Phillips's masterful (1974) text had not changed the statistical practices of most social scientists. We hoped that a more psychological exposition, informed by the ensuing 20 years of research would help researchers to tell when judgments are biased—and to distinguish one bias from another. Well-characterized biases are a vital research tool, suggesting reasons why people would do things wrong, when motivated to do them right.

"Knowing with certainty: The appropriateness of extreme confidence" (Chapter 7) examines one key element of Bayesian inference: knowing how much confidence to place in one's beliefs. Many studies have found that people have an imperfect feeling for how much they know (Lichtenstein et al., 1982). Generally speaking, they are too confident when their knowledge is limited, not confident enough when they know a lot. We wondered to what extent those results reflected difficulty using numeric probabilities to express confidence (e.g., "there is a 75% chance that my answer is correct"). We thought that people should find it relatively easy to express extreme confidence (e.g., 1000 to 1 odds). However, we still found overconfidence, even when we asked people to bet on their confidence, tried not to raise unrealistic expectations, and avoided capitalizing on trick questions (which people get wrong, while being certain that they are right). Subsequent studies showed how hard it is to make tasks seem really difficult—as though once people understand a task, they feel like they know something about it (Fischhoff and Slovic, 1980).

"Fault trees: Sensitivity of estimated failure probabilities to problem representation" (Chapter 8) studies one possible reason for overconfidence: people exaggerate how thoroughly they have analyzed a problem. We modeled our experiments on fault-tree analysis, a method for assessing the chances of complex systems (e.g., nuclear power plants, space missions) going awry. The trees in our studies used the everyday risk of a car not starting. We found that problems that are out of sight are also out of mind, so that people do not realize when they are overlooking even familiar problems (e.g., running out of gas). Such oversight is not easily remedied by asking people to think hard about "all other problems" or by showing them details about one problem (e.g., the battery system), hoping that they could imagine similar detail about other problems. Russo and Kolzow (1994) later found how sensitive such judgments can be to how problems are described when people lack common terms (e.g., does the "ignition system" include the battery?).

"Assessing uncertainty in physical constants" (Chapter 9) examines the uncertain judgments of world experts. Max Henrion observed that physicists provide something like Bayesian confidence intervals, when estimating physical constants, such as the rest mass of the electron or the speed of light (before it was defined). That is, they say something like "We are 90% confident that the value is between X and Y." Treating their judgments in that spirit, we found overconfidence like that observed with lay people for seemingly similar reasons (e.g., trusting poor data and discarding sound data based on preconceptions). Although physicists do not follow formal procedures for eliciting judgments (Morgan and Henrion, 1990), they do take their confidence intervals seriously enough that unexpected observations can lead to doubting the laws of the universe, rather than their own judgment. For a while, it looked like we might join the Particle Data Group in one of its periodic assessments of recommended values. Kammen et al. (1994) extended our results to other realms, as did Morgan and Keith (2008).

"Fifty/fifty = 50?" (Chapter 10) deals with an anomaly in the middle of the probability range, rather than the extremes. As part of a project studying judgments of breast cancer risks (Byram et al., 2008), our research group noticed a peculiar bimodal distribution, with most people giving probabilities around 20%, but a minority saying 50%. Wändi Bruine de Bruin observed that teens participating in semi-structured interviews sometimes used "50" in the sense of 50–50. Searching our archive of transcripts, we found that such instances seemed most common with unfamiliar or threatening events. Needing to provide a number, but not knowing what to say, teens said "50," but without meaning a numeric probability. These studies led to Wändi's dissertation, which included finding excess 50s even in expert judgments (Bruine de Bruin et al., 2002).

References

Bruine de Bruin, W., Fischbeck, P.S., Stiber, N.A., and Fischhoff, B. 2002. What number is "fifty-fifty"?: Distributing excessive 50% responses in elicited probabilities. *Risk Analysis, 22*, 725–735.

Byram, S., Schwartz, L., Woloshin, S., and Fischhoff, B. 2008. Women's perceptions of breast cancer risk factors: A mental models approach. In J. Krueger (ed.) *Rationality and social responsibility: Essays in honor of Robyn Mason Dawes*. (pp. 245–274) Mahwah, NJ: Lawrence Erlbaum Associates.

Edwards, W., Lindman, H.R., Savage, L.J. 1963. Bayesian statistical inference for psychological research. *Psychological Review, 70*, 193–242.

Fischhoff, B. and Slovic, P. 1980. A little learning...: Confidence in multicue judgment. In R. Nickerson (Ed.), *Attention and performance, VIII* (pp. 779–800). Hillsdale, NJ: Erlbaum.

Kammen, D.M., Shlyakhter, A. I., Broido, C., and Wilson, R. 1994. Quantifying credibility of energy projections from trends in past data: the U.S. energy sector. *Energy Policy, 22*, 119–131.

Lichtenstein, S., Fischhoff, B., and Phillips, L.D. 1982. Calibration of probabilities: State of the art to 1980. In D. Kahneman, P. Slovic, and A. Tversky (Eds.), *Judgment under uncertainty: Heuristics and biases* (pp. 306–334). New York: Cambridge University Press.

Morgan, M.G., and Henrion, M. 1990. *Uncertainty*. New York: Cambridge University Press.

Morgan, M.G., and Keith, D. 2008. Improving the way we think about projecting future energy use and emissions of carbon dioxide. *Climatic Change, 90*, 189–215.

Phillips, L.D. 1974. *Bayesian statistics for social scientists*. New York: Thomas Y. Crowell.

Russo, J.E. and Kolzow, K.J 1994. Where is the fault in the fault trees? *Journal of Experimental Psychology: Human Perception and Performance, 20*(1), 17–32.

6

HYPOTHESIS EVALUATION FROM A BAYESIAN PERSPECTIVE

Baruch Fischhoff

DECISION RESEARCH, PERCEPTRONICS, INC. EUGENE, OREGON AND MEDICAL RESEARCH COUNCIL APPLIED PSYCHOLOGY UNIT CAMBRIDGE, ENGLAND

Ruth Beyth-Marom

DECISION RESEARCH, PERCEPTRONICS, INC. EUGENE, OREGON

Abstract

Bayesian inference provides a general framework for evaluating hypotheses. It is a normative method in the sense of prescribing how hypotheses should be evaluated. However, it may also be used descriptively by characterizing people's actual hypothesis-evaluation behavior in terms of its consistency with or departures from the model. Such a characterization may facilitate the development of psychological accounts of how that behavior is produced. This article explores the potential of Bayesian inference as a theoretical framework for describing how people evaluate hypotheses. First, it identifies a set of logically possible forms of non Bayesian behavior. Second, it reviews existing research in a variety of areas to see whether these possibilities are ever realized. The analysis shows that in some situations several apparently distinct phenomena are usefully viewed as special cases of the same kind of behavior, whereas in other situations previous investigations have conferred a common label (e.g., *confirmation bias*) to several distinct phenomena. It also calls into question a number of attributions of judgmental bias, suggesting that in some cases the bias is different than what has previously been claimed, whereas in others there may be no bias at all.

Introduction

Hypothesis evaluation is a crucial intellectual activity. Not surprisingly, it is also a focus of psychological research. A variety of methods have been applied to understand how people gather and interpret information in order to evaluate hypotheses. Either implicitly or explicitly, some theory of how people *should* evaluate hypotheses provides the conceptual framework for studies of how they *do* evaluate them. Such a prescriptive theory provides a set of articulated terms for describing tasks and a definition of appropriate behavior against which actual performance can be compared. The theory might even be descriptively valid at a certain level if people are found to follow its dictates, either due to natural predilections or because they have been trained to do so. Even when behavior is suboptimal, some psychological insight may be obtained by asking whether that behavior may be described as a systematic deviation from the theory. Reference to a normative theory can also identify performance deficits that need to be understood and rectified.

One general and popular normative scheme is Bayesian inference, a set of procedures based upon Bayes' theorem and the subjectivist interpretation of probability. These procedures show how to (a) identify the data sources that are most useful for discriminating between competing hypotheses, (b) assess the implications of an observed datum vis-à-vis the truth of competing hypotheses, (c) aggregate the implications of different data into an overall appraisal of the relative likelihood of those hypotheses being correct, and (d) use that appraisal to select the course of action that seems best in light of available evidence. Excellent detailed expositions of the scheme may be found in Edwards, Lindman, and Savage (1963), Lindley (1965), Novick and Jackson (1974), and Phillips (1973).

The present article presents a simple version of Bayesian inference. From this scheme, it derives a taxonomy of logically possible deviations. This taxonomy is then used to characterize several published studies reporting biased hypothesis evaluation. In some cases, the result is to reiterate the claims of the original investigators; in other cases, those claims are countered by alternative interpretations that suggest other biases that may be involved or ways in which observed behavior might be construed as being properly Bayesian. In still other cases, research conducted in other traditions is cast in Bayesian terms in the hope of profiting from others' experience and drawing different fields together.

This review is not meant to be exhaustive; rather, it is meant to illustrate how the Bayesian perspective can be used to illuminate a variety of tasks. As a result, it emphasizes new interpretations and, by reference to the taxonomy, it identifies potential biases for which positive evidence is lacking—raising the question of whether it was an opportunity to observe suboptimal behavior that researchers missed or an opportunity to exhibit suboptimal behavior that subjects "missed." The theory is complete in the sense that it treats all of the basic issues that arise in Bayesian inference and that must be faced when one assesses the optimality of behavior. It is incomplete in that it does not show how the theory can be adapted to model all possible situations. For example, all of the hypotheses we consider here are discrete, both for simplicity's sake and because the studies we cite have used discrete hypotheses. The treatment of continuous hypotheses (e.g., "Today's mean temperature is *x*") is considered by Edwards et al. (1963) and others.

Theory

Definition of probability

From the Bayesian perspective, knowledge is represented in terms of statements or hypotheses, H_i, each of which is characterized by a subjective probability, $P(H_i)$, representing one's confidence in its truth (DeFinetti, 1976). For example, one might be .75 confident that it will snow tomorrow. Such probabilities are subjective in the sense that individuals with different knowledge (or beliefs) may legitimately assess $P(H_i)$ quite differently. The term *assess* is used rather than *estimate* to emphasize that a probability expresses one's own feelings rather than an appraisal of a property of the physical world. Thus, there is no "right" probability value for a particular statement. Even if a very low probability proves to be associated with a true statement, one cannot be sure that it was not an accurate reflection of the assessor's (apparently erroneous) store of knowledge.

The constraints on subjective probabilities emerge when one considers sets of assessments. Formally speaking, a set of probabilities should be orderly or *coherent* in the sense of following the probability axioms (Kyburg and Smokler, 1980; Savage, 1954).[1] For example, $P(H) + P(\bar{H})$ should total 1.0.

Updating

Additional derivations from the probability axioms lead to Bayes' theorem, which governs the way in which one's beliefs in hypotheses should be updated in the light of new information. In its simplest form, the theorem treats the implications of an observation that produces the datum (D) for determining whether a hypothesis (H) is true, relative to its complement, $\bar{H}$. In such cases, Bayes' theorem states that

$$\frac{P(H/D)}{P(\bar{H}/D)} = \frac{P(D/H)}{P(D/\bar{H})} \cdot \frac{P(H)}{P(\bar{H})} \qquad (1)$$

Reading from the right, the three terms in this formula are (a) the prior odds that H (and not $\bar{H}$) is true in the light of all that is known before the receipt of D, (b) the likelihood ratio, representing the information value of D with respect to the truth of H, and (c) the posterior odds that H is true in the light of all that is known after the receipt of D. Equation 1 could also be applied to any pair of competing hypotheses, A and B (with A replacing H and B replacing $\bar{H}$).

Although the subjectivist interpretation of probability is controversial, Bayes' theorem generally is not. The axioms from which it is derived are common to most interpretations of probability. Bayesians have more frequent recourse to the theorem because the subjectivist position enables them to incorporate prior beliefs explicitly in their inferential processes.[2]

Likelihood ratios

If the probability of observing *D* given that *H* is true is different from the probability of observing *D* when *H* is not true, then the likelihood ratio is different from 1 and the posterior odds are different from the prior odds. That is, the odds favoring *H* have become smaller or greater as a result of having observed *D*. Such a datum is considered to be informative or *diagnostic*. Its degree of diagnosticity can be expressed in terms of how different the likelihood ratio is from 1. Clearly, diagnosticity depends upon the hypotheses being tested. A datum that distinguishes one hypothesis from its complement may be completely uninformative about another pair of hypotheses. Data do not answer all questions equally well.

The value of the likelihood ratio is independent of the value of the prior odds. One could in principle observe a datum strongly supporting a hypothesis that is initially very unlikely. If that happened, one's posterior odds favoring *H* might still be very low, although not as low as they were before. There is also no necessary relationship between the values of the numerator and the denominator of the likelihood ratio. A datum might strongly favor *H* even if it is very unlikely given the truth of $\bar{H}$. Similarly, observing a datum that is a necessary concomitant of *H*, that is, $P(D/H) \approx 1$, may be uninformative if it is also a necessary concomitant of $\bar{H}$.

Action

The apparatus of Bayesian inference also provides tools for converting one's beliefs in hypotheses into guides to action. In simplest terms, these tools translate the cost associated with erroneously acting as though a hypothesis is true and the cost of erroneously acting as though a hypothesis is false into a *critical ratio*. If the posterior odds favoring *H* are above this value, then one is better off acting as though *H* is true; if they are below the value, then one is better off acting as though *H* is false.

The value of the critical ratio depends, of course, on the particular kinds of action that are contemplated and the outcomes that are associated with them. Where one's posterior odds stand vis-à-vis the critical ratio depends on both one's prior odds and the evidence that has subsequently been received. Two individuals who had agreed on the costs of the different errors and on the meaning of the different data might still act differently, if they had different prior odds. On the other hand, if the cumulative weight of the new evidence was sufficient, they might act in the same way, despite having initially had quite discrepant beliefs—for example, if the evidence carried them from prior odds of 1:10 and 1:1 to posterior odds of 100:1 and 1,000:1, respectively.

When it is possible to collect more data, additional Bayesian procedures can help identify the most useful source (Raiffa, 1968). Such *value-of-information* analyses evaluate the expected impact of each possible observation on the expected utility of the actions one can take. They can also tell when the cost of further observation is greater than its expected benefit.

TABLE 6.1 Potential sources of bias in Bayesian hypothesis evaluation

Task	*Potential bias*	*Special cases*
Hypothesis formation	Untestable	Ambiguity, complexity, evidence unobservable
	Nonpartition	Nonexclusive, nonexhaustive
Assessing component probabilities	Misrepresentation	Strategic responses, nonproper scoring rules
	Incoherence	Noncomplementarity, disorganized knowledge
	Miscalibration	Overconfidence
	Nonconformity	Reliance on availability or representativeness
	Objectivism	—
Assessing prior odds	Poor survey of background	Incomplete, selective
	Failure to assess	Base-rate fallacy
Assessing likelihood ratio	Failure to assess	Noncausal, "knew-it-all-along"
	Distortion by prior beliefs	Preconceptions, lack of convergence
	Neglect of alternative hypotheses	Pseudodiagnosticity, inertia, cold readings
Aggregation	Wrong rule	Averaging, conservatism?
	Misapplying right rule	Computational error, conservatism?
Information search	Failure to search	Premature conviction
	Nondiagnostic questions	Tradition, habit
	Inefficient search	Failure to ask potentially falsifying questions
	Unrepresentative sampling	—
Action	Incomplete analysis	Neglecting consequences, unstable values
	Forgetting critical value	Confusing actual and effective certitude

Ways to stray

Bayesian inference, like other normative schemes, regards only those who adhere to its every detail as behaving optimally. Conversely, every component of the scheme offers some opportunity for suboptimality. This section catalogues possible pitfalls. For the normative minded, these possibilities might be seen as defining Bayesian inference negatively by pointing to behavior that is inconsistent with it.

For the descriptive minded, these logical possibilities suggest judgmental biases that might be observed in empirical studies. If observed in situations in which people are properly instructed and motivated to respond correctly, such deviations can be theoretically informative because they seem to reflect deep-seated judgmental processes. From a practical standpoint, such deviations suggest opportunities for constructive interventions that might lead to better inferences and, subsequently, to better decisions based on those inferences. These interventions might include training, the use of decision aids, or the replacement of intuition by formalized procedures (Edwards, 1968; Fischhoff, 1982; Kahneman and Tversky, 1979).

These potential biases are presented in Table 6.1. The left-hand column shows the task in which the problem could arise; the center column describes the possible biases; and the right-hand column points to phenomena reported in the literature that we have interpreted as special cases of these biases, which are described in detail in the following section.

Hypothesis formation

For hypothesis evaluation to begin, there must be hypotheses to evaluate. Indeed, because the diagnostic impact of data is defined only in the context of particular hypotheses, there is no systematic way that data can even be collected without such a context. In the absence of any hypotheses, the collection of data is merely idle stockpiling. Although it is a logical possibility, it does not seem to be a troublesome or common bias. In practice, even the most ambling data collection may be guided by a vague idea of the hypotheses that the collector might be asked to evaluate. Or, it may be conducted for exploratory purposes, with the goal of generating, rather than evaluating, hypotheses. If the collector's goal is to see something that fortuitously stimulates a creative insight, then the Bayesian model, or any other formal model, can offer little guidance or reprobation.

A more serious threat than the absence of hypotheses to evaluate is the possession of hypotheses that cannot be evaluated. One route to formulating hypotheses that cannot be evaluated lies through ambiguity, either intentional or inadvertent. To take a popular example, astrology columns offer hypotheses about what consequences follow what acts (e.g., "You will be better off avoiding risky enterprises"). Yet these acts and consequences are so vaguely defined that it is unclear whether what eventually happens supports the hypothesis. A more sophisticated form of ambiguity can be found in the probabilistic risk analyses used to generate detailed hypotheses about the operation of technical systems (e.g., "Toxins can be released to the atmosphere only if the following events occur"; Green and Bourne, 1972; U.S. Nuclear Regulatory Commission, Note 1). Although actual operating experience should afford an opportunity to evaluate these hypotheses, it may be unclear whether the specific events that are observed are subsumed by the generic events described in the analysis. For example, investigators disagreed over whether the fire at the Browns Ferry Nuclear Power Plant in 1975 was included among the accident sequences described in the then-definitive analysis of reactor operation (U.S. Nuclear Regulatory Commission, Note 2, Note 3).

Complexity offers a second route to formulating hypotheses that cannot be evaluated. Hypotheses that are set out clearly may have such great detail that no datum provides a clear message. For example, political advisors may escape charges of having predicted events incorrectly by noting that every last detail of their advice was not followed ("Had they only listened to me and done X and Y as well, then everything would have been all right").[3] O'Leary, Coplin, Shapiro, and Dean (1974) found that among practitioners of international relations (i.e., those working for business or government) theories are of "such complexity that no single quantitative work could even begin to test their validity" (p. 228). Indeed, some historians argue that the accounts of events that they produce are not hypotheses at all but attempts to integrate available knowledge into a coherent whole. In this light, a valid explanation accommodates all facts, leaving none to test it. This attitude toward hypotheses has its own strengths and weaknesses (Fischhoff, 1980).

Because it considers the relative support that evidence provides to competing hypotheses, the Bayesian scheme requires not only the individual hypotheses but also the set of hypotheses to be well formulated. In effect, those hypotheses must partition some space of possibilities.[4] Computationally, problems arise with nonexclusive hypotheses, which render the message of evidence ambiguous.

Whereas achieving mutually exclusive hypotheses requires precision of formulation, securing an exhaustive set of hypotheses often requires the exercise of imagination. Technically, it is easy to make a set exhaustive by defining it in terms of H and $\bar{H}$, by declaring the hypotheses that one has thought of to be the only ones of interest, or by adding a hypothesis consisting of "all other possibilities," a role that may fill. Although such specification is adequate for some purposes, wherever the alternatives to H are poorly defined, it is hard to evaluate $P(D/\bar{H})$ and, hence, the implications of D for H. Whenever the set of alternatives is incomplete, major surprises are possible. The difficulties of exhaustive enumeration are often exploited by mystery writers who unearth a neglected hypothesis that tidily accounts for available evidence. Both for generating hypotheses and for evaluating $P(D/\text{all other possibilities})$, an important skill is estimating the completeness of the set of already listed hypotheses. Evidence suggests that people tend to exaggerate the completeness of hypothesis sets (Fischhoff, Slovic, and Lichtenstein, 1978; Mehle, Gettys, Manning, Baca, and Fisher, 1981).

A final problem is that even well-formulated hypotheses may be wrong for the actions contemplated. The acid test of relevance is whether perfect information about the hypothesis (i.e., knowing whether it is true or false) would make any difference to one's actions. For example, if ethical principles proscribe incarcerating juveniles with adults, then hypotheses about the effect of prison on delinquents have no consequences for criminal policy. If their support for military spending depends upon the relative strength of the pro- and anti-arms lobbies, then legislators need not choose among competing hypotheses regarding Soviet intentions. The pejorative label for such impractical hypotheses is "academic." More charitably, they are "hypotheses relevant to possible future actions."

Assessing component probabilities

To find a place in the Bayesian model, one's beliefs must be translated into subjective probabilities of the form appearing in the model. Any difficulties in assessing such component probabilities would impair hypothesis evaluation. As mentioned, the Bayesian perspective holds probabilities to be subjective expressions, reflecting the assessor's beliefs. Accepting the subjectivist position does not, however, mean accepting any probability as an appropriate assessment of someone's state of belief. There are a number of ways in which the assessment of probabilities can go wrong.

One possible problem is lack of candor. People may misstate their beliefs, perhaps to give a response that is expected of them, perhaps to avoid admitting that an unpleasant event is likely to happen, or perhaps to achieve some strategic advantage by misrepresenting how much they know or what they believe. When the truth of statements can eventually be ascertained, the use of proper scoring rules should encourage candor (Murphy, 1972). These rules reward people as a function of both their stated beliefs and the (eventually revealed) state of the world in such a way that the probability with the highest expected value is the one expressing one's true belief. Whether these rules prove effective in practice is a moot point (Lichtenstein, Fischhoff, and Phillips, 1982; von Winterfeldt and Edwards, 1982). Where they do not prove effective or where they cannot even be applied because the truth will not be known or because no reward system is possible, other means of assuring candor are needed.

The crucial assessment problem from the subjectivist perspective is lack of coherence, failure of one's assessments to follow the probability axioms. Violations may be due to a poor job of reviewing one's knowledge. For example, $P(H)$ and $P(\bar{H})$ may not equal 1 when the two hypotheses are not evaluated simultaneously—and concentration on each evokes a different subset of one's knowledge. It is also possible that the beliefs themselves are not well thought through. In that case, working harder on probability assessment may lead to even more incoherent probabilities by revealing the underlying inconsistencies. Here, the structure of belief requires refinement (Lindley, Tversky, and Brown, 1979).

Another symptom of poor assessment is *miscalibration*, failure of one's confidence to correspond to reality. If a hypothesis with a prior probability of .2 eventually is found to be true, the initial assessment is dubious. However, it may be an accurate summary of the assessor's knowledge at that moment. Recurrent association of high prior probabilities with hypotheses that prove to be false should be cause for serious concern. Calibration formalizes this reliance on the eventually accepted truth of hypotheses for validating probability assessments. Specifically, for the well-calibrated assessor, probability judgments of .*XX* are associated with true hypotheses *XX*% of the time. Empirical studies of calibration have shown that probabilities are related, but are not identical, to proportions of correct hypotheses. The most common deviation is overconfidence—for example, being only 80% correct when 1.00 confident (Lichtenstein, Fischhoff, and Phillips, 1982).[5] As with incoherence, miscalibration may be traced to people's assessment procedures or to the knowledge base that they rely on.

A third symptom is *nonconformity*, producing a probability that differs from that of "expert" assessors for no apparent reason. The existence of such consensus is most likely when there is a statistical data base upon which to base probability assessments (e.g., public health records of mortality). In this restricted realm, the distinction between subjective and objective probabilities becomes blurred, as subjectivists would typically act as though they concur with the relative-frequency interpretation of probability, which objectivists consider to be the only meaningful one. Subjectivists would, however, never concede that frequency counts are a completely objective measure of probability, arguing that judgment is needed to establish the equivalence and independence of the set of trials from which the count was extracted and to extrapolate that frequency to future events.

Another way to look at bias is in terms of the process by which assessments are produced. There is reason for concern whenever the assessors have followed procedures that are inconsistent with the rules of statistical inference and are unaware of those inconsistencies. Two well-known deviations are reliance on the availability and representativeness heuristics when making probability assessments (Kahneman, Slovic, and Tversky, 1982; Tversky and Kahneman, 1974). Users of the former judge an event to be likely to the extent that exemplars are easy to recall or imagine; users of the latter judge an event to be likely to the extent that it represents the salient features of the process that might produce it. Although both rules can provide good service, they can also lead the user astray in predictable ways. For example, reliance on availability induces overestimation of unusually salient events (e.g., the probability of dying from flashy, hence overreported, causes such as tornadoes and homicide; Lichtenstein, Slovic, Fischhoff, Layman, and Combs, 1978).

A final process problem is the refusal to consider anything but relative frequency data when one assesses probabilities. Although subjectivists acknowledge the potential relevance of such data, they will not be bound to them. Indeed, a key selling point of Bayesian inference is its ability to accommodate diverse kinds of data. One can, for example, assess probabilities for a meaningful shrug or an off-hand comment as well as for a bead drawn from an urn or for a 40-subject experiment. The only difference is that as one moves from beads to shrugs, it becomes increasingly difficult to attest to the reasonableness of a particular assessment. An assessor who failed to seek useful and available nonfrequency evidence would, from a Bayesian perspective, be foolish. An individual who ignored nonfrequency evidence that was already on hand would be biased.[6]

Assessing the prior

Prior beliefs are captured by the ratio of the probabilities of the competing hypotheses prior to the collection of further evidence. As a result, the difficulties of assessing prior beliefs are by and large the sum of the difficulties of formulating hypotheses and assessing component probabilities, as already discussed. There seem, however, to be at least two additional (and incompatible) problems that may affect this particular stage of the assessment process.

One such problem is not treating the component probabilities equally. An extreme form of inequality is neglecting one of the hypotheses. When people act as though a hypothesis that is most probably true is absolutely true, they have effectively neglected its complement. In that case, hypothesis evaluation never begins, because one hypothesis is treated as fact. Even when both hypotheses are considered, one may be given deferential treatment. Skeptics, for example, may give undue weight to evidence that contradicts a favored hypothesis; they have a warm spot in their hearts for complements. On the other hand, Koriat, Lichtenstein, and Fischhoff (1980) found favoritism for initially favored hypotheses: Subjects who were asked to determine the relative likelihood of two possible answers to a question seemed to search primarily for supporting evidence.[7] Such directed search may serve legitimate purposes, for example, seeing if any case at all can be made for, or against, a particular hypothesis. However, it may be very difficult to estimate and to correct for the bias that it induces in the resultant sample of evidence. Indeed, it is the failure to realize the biases in the samples produced by the availability heuristic that makes it a potentially dangerous judgment rule.

These difficulties vanish in the presence of another bias that has attracted considerable attention of late: neglecting the base rate (Kahneman and Tversky, 1973). The *base-rate fallacy* refers to the tendency to allow one's posterior beliefs to be dominated by the information that one extracts from the additional datum, *D*, to the neglect of the prior beliefs, expressing what typically has been observed. The several recent reviews of this literature (Bar-Hillel, 1980; Bar-Hillel and Fischhoff, 1981; Borgida and Brekke, 1981) may be summarized as indicating that people rely most heavily on whatever information seems most relevant to the task at hand. Thus, for example, when testing a pair of hypotheses such as "John is a lawyer/John is an engineer," even weak diagnostic information relating directly to John may dominate base-rate information reporting the overall prevalence of the two professions. Base-rate information is used, however, if it can be linked more directly to the inference or if the case-specific information is palpably worthless.

The evidence for the base-rate fallacy comes primarily from studies in which base-rate information was presented, yet neglected. As such, the fallacy can be viewed as a problem of aggregation, which we treat two sections below. It is discussed here because the failure to use explicitly presented base rates strongly suggests that they will not be spontaneously sought or assessed. This suspicion is confirmed, for example, by Lyon and Slovic's (1976) finding that, when asked directly, one half of their subjects did not believe that base rates were relevant to their judgments.[8]

Assessing the likelihood ratio

In principle, people can ignore the likelihood ratio just as well as the base rate, allowing one to speak of the "likelihood-ratio fallacy" whenever people fail even to consider the likelihood ratio for a pertinent datum. This may happen, for example, when the datum provides merely circumstantial evidence (Einhorn and Hogarth, 1982), when it cannot be woven into a causal account involving the hypothesis (Tversky and Kahneman, 1980), or when it reports a nonoccurrence. A classic example of the latter is Sherlock Holmes's observation (Doyle, 1974) that his colleague, Inspector Gregory, had not considered the significance of a dog failing to bark when an intruder approached.

Failure to assess the likelihood ratio of received evidence may also be encouraged by hindsight bias, underestimating the informativeness of new data (Fischhoff, 1975, 1982). The feeling that one knew all along that D was true might make the calculation of D's likelihood ratio seem unnecessary. Denying that D has anything to add does not, however, mean that it will not have any impact—only that one will be unaware of that impact. At the same time as people deny its contribution, new information can change their thinking in ways that they cannot appreciate or undo (Fischhoff, 1977; Sue, Smith, and Caldwell, 1973). These unintended influences may or may not be those that would follow from a deliberative analysis.

When judges do assess the likelihood ratio, the sequencing of that operation may expose it to the influence of the preceding operations. In particular, the interpretation of new evidence may be affected by previous beliefs, thereby subverting the independence of the likelihood ratio and priors. Nisbett and Ross (1980) offer an impressive catalog of ways in which people can interpret and reinterpret new information so as to render it consistent with their prior beliefs. So great is people's ability to exploit the ambiguities in evidence to the advantage of their preconceptions and to discount inconsistent evidence, that erroneous beliefs can persevere long after they should have been abandoned.[9]

Such biased interpretation of evidence also thwarts the effective convergence of belief that should follow use of Bayesian inference. However discrepant the initial position of two individuals, their posterior beliefs should converge for practical purposes, providing they observe a sufficiently large set of diagnostic data about whose interpretation they agree. The ability to interpret a datum as supporting contradictory hypotheses means that convergence may never occur. Whatever data the two observers see, they become more convinced of their respective prior beliefs.

When people choose to evaluate evidence, they must compare two conditional probabilities, $P(D/H)$ and $P(D/\bar{H})$. Such comparison is essential because there is no necessary relationship between these two components of the likelihood ratio. A variety of studies suggest, however, that people consider only the numerator. That is, they are interested in how consistent the evidence is with the hypothesis they are testing, $P(D/H)$ and fail to consider its consistency with the alternative hypothesis, $P(D/\bar{H})$. As a result, the size of $P(D/H)$ determines D's support for H. Users of this strategy act as though they assume that the two conditional probabilities are inversely related, although, in principle, both may be high or low. A datum with a low $P(D/H)$ may provide strong evidence for H if $P(D/\bar{H})$ is even lower; a datum for which $P(D/H)$ is high may reveal nothing if $P(D/\bar{H})$ is equally high.

Four examples should give some flavor of the variety of guises within which incomplete appraisal of the likelihood ratio may emerge:

1. Doherty, Mynatt, Tweney, and Schiavo (1979) presented subjects with six pieces of data, D_i and allowed them to inquire about the values of 6 of the 12 conditional probabilities: $P(D_1/H)$, $P(D_1/\bar{H})$, …, $P(D_6/H)$, $P(D_6/\bar{H})$. Few subjects requested any of the pairs of conditional probabilities [e.g., $P(D_3/H)$ and $P(D_3/\bar{H})$] needed to compute likelihood ratios. The authors labeled this tendency to pick but one member of each pair, *pseudodiagnosticity*. The probabilities that subjects did solicit were primarily those, $P(D/H_i)$, involving the hypothesis, H, that a preceding manipulation had made appear more likely. The authors called this tendency *confirmatory bias* (a term to which we will return).
2. Troutman and Shanteau (1977) had subjects draw beads from a box which contained either 70 red, 30 white, and 50 blue beads or 30 red, 70 white, and 50 blue beads and infer the probability that the box was predominantly white (W). Drawing two blue beads reduced subjects' confidence in their initially favored hypothesis (W), even though that observation is equally unlikely under either hypothesis: $P(D/R) = P(D/W) = .11$. Thus, subjects considered only $P(D/H)$ for their favored hypothesis.
3. In a similar experiment, without the blue beads, Pitz, Downing, and Reinhold (1967) found that subjects who were fairly confident that the box was predominantly red would increase their confidence in that hypothesis after observing a white. Subjects apparently felt that they should see an occasional white, neglecting the fact that that event was still more likely if the box being used was predominantly white. Pitz et al. called this failure of inconsistent evidence to slow the increase of confidence in H an *inertia effect*. The inappropriate increase in faith in H here contrasts with the inappropriate decrease in it in a study by Troutman and Shanteau (1977). We would trace these two opposite effects to the same underlying cause, neglecting $P(D/\bar{H})$.
4. A favorite ploy of magicians, mentalists, and pseudopsychics who claim to read other people's minds is to provide universally valid personality descriptions (Forer, 1949; Hyman, 1977) that apply to almost everyone,

although this is not transparently so. These operators trust their listeners to assess *P* (this description/my mind is being read) and not *P* (this description/my mind is not being read).

A final threat to the validity of assessed likelihood ratios comes from the fact that the probabilities involved all concern conditional events. This added complexity seems to complicate probability assessment, with people forgetting the conditioning event, reversing the roles of the two events, or just feeling confused (Eddy, 1982; Moskowitz and Sarin, 1983).

Aggregation

Assuming that judges have attended to and assessed all components of the Bayesian model, they must still combine them to arrive at posterior odds. The two logically possible sources of error here are (a) using the wrong aggregation rule, for example, averaging, rather than multiplying, the likelihood ratio and prior odds, and (b) using the right rule, but applying it inappropriately, for example, making a computational error. Establishing whether either or both of these potential biases actually occurs was a focus of early research into intuitive Bayesian inference (reviewed by Slovic and Lichtenstein, 1971).

Most of these early studies used the un-Bayesian strategy of creating inferential tasks in which the experimenter could claim to know the correct subjective probability for all participants. This was done by using highly artificial stimuli for which all reasonable observers should have the same subjective probability. For example, subjects might be shown a series of poker chips and be asked to evaluate the hypotheses: They are being drawn from a bookbag with 70% blue chips and 30% red chips/they are being drawn from a bag with 70% red chips and 30% blue chips. The predominant result of this research was that subjects' confidence in the apparently correct hypothesis did not increase as quickly as the accumulating evidence indicated that it should.

A lively debate ensued over whether this poor performance, called *conservatism*, reflected failure to appreciate how diagnostic the evidence was, called *misperception*, or failure to combine those diagnosticity assessments according to Bayes' rule, *misaggregation*. Aside from its theoretical interest, this dispute had considerable practical importance. If people can assess component probabilities but cannot combine them, then person–machine systems may be able to relieve them of the mechanical computations. Moreover, the system could incorporate an elicitation scheme that kept users from ever forgetting any component probabilities. However, if people are the only source of probabilities and they cannot assess them very well, then the machine may be just spinning its discs (Edwards, 1962).

Although the source of this conservatism was never determined, the hypotheses that were raised reflect the problems that could interfere with aggregation. Examples, details of which may be found in Slovic and Lichtenstein (1971) are (a) anchoring—people stay stuck to their previous estimates, (b) response bias—reluctance to give extreme responses pushes answers toward the center of the response range, (c) ceiling effect—fear of "using up" the probability scale makes people hedge their responses, (d) nonlinearity—although a given *D* should produce the same ratio of posterior odds to prior odds whenever it is received, people may try instead to make the differences between prior and posterior probability of *H* constant, and (e) response mode—probability assessments may be less optimal responses than the odds (or log odds) assessments for the same task; despite their formal equivalence, one response mode may be a more natural way for people to assess and express their knowledge about a particular problem.

In the end, this line of research was quietly abandoned without establishing the relative roles of these different factors. This cessation of activity seems to be partly due to the discovery of the base-rate fallacy, which represents the antithesis of conservatism and other phenomena that led researchers to conclusions such as the following: "It may not be unreasonable to assume that ... the probability estimation task is too unfamiliar and complex to be meaningful" (Pitz, Downing, and Reinhold, 1967, p. 392). "Evidence to date seems to indicate that subjects are processing information in ways fundamentally different from Bayesian ... models" (Slovic and Lichtenstein, 1971, p. 728). "In his evaluation of evidence, man is apparently not a conservative Bayesian; he is not Bayesian at all" (Kahneman and Tversky, 1972, p. 450).

Information search

Obviously, Bayesian updating requires the collection of additional information beyond what was incorporated in the prior odds. Whether collection is contemplated at all should depend on whether one's a priori confidence in the truth of the hypothesis is adequate for deciding what to do and on the possibilities for additional data to change one's mind. When one would like to know more, the specific data collected should depend on the opportunities presented.

These opportunities can be conceptualized as questions that can elicit a set of possible answers, or D_i each of which carries a message regarding the truth of the hypotheses. All other things being equal (e.g., the cost of asking), the most valuable questions are those that are expected to produce the most diagnostic answers. Conversely, one should never ask questions all of whose possible answers have likelihood ratios of 1. Such questions should change neither one's beliefs regarding the truth of the hypotheses nor one's choice of action based on those beliefs. Value-of-information analysis includes a variety of schemes for deciding how much one should spend for information in general and how to devise the most efficient sampling strategies. It considers such factors as the cost of asking, the consequences of the possible decisions, the a priori probability of receiving different possible answers, and the likelihood ratios associated with those answers (Brown, Kahr, and Peterson, 1974; Raiffa, 1968).

Why might someone disregard these considerations and ask questions whose answers cannot be diagnostic? Tradition is one possible reason. A datum may always have been collected, without serious analysis of what has been learned from it. Official forms and graduate school applications might be two familiar loci for nondiagnostic questions. These traditions may spawn or be spawned by beliefs about the kinds of evidence that are inherently more valuable. In various circles, secret, quantitative, or introspective information might have this special status of always meriting inquiry (Fischer, 1970). Misdirected search may occur also when people's task changes and they fail to realize that the questions that helped to evaluate the old hypotheses are no longer as effective in evaluating the new ones. For example, a psychiatric social worker who moved from private practice to a large public agency might require a whole new set of question-asking skills.

The factors leading to pointless questions can, in less extreme form, lead to inefficient ones—questions that provide some information, yet less than the maximum possible. These search problems may be aggravated by difficulties with other aspects of the inferential process. Value-of-information analysis requires a *preposterior* analysis, in which one anticipates what one will believe and what one will do if various possible data are observed. If people have difficulty assessing likelihood ratios for actual data, then they are unlikely to assess them properly for hypothetical data. A further obstacle to appraising the expected value of possible questions is the very hypotheticality of the question. When seen "in the flesh," information may have more or less impact than it was expected to have, in which case it is not clear whether to trust the actual or the anticipated impact (Fischhoff, Slovic, and Lichtenstein, 1979).

A noteworthy form of efficiency is ignoring the opportunity to ask potentially falsifying questions, ones whose answers have a reasonable chance of effectively ruling out some hypothesis (e.g., a physician failing to order a test that could eliminate the possibility of a particular disease). What constitutes a "reasonable chance" depends on the usual value-of-information analysis factors (e.g., the prior probability of the disease, the importance of its detection, the cost of the test).

An obvious danger in information search is inadvertently selecting an unrepresentative set of evidence and arriving at erroneous beliefs. When a proper sampling frame is available (e.g., for the Census), one can describe a variety of specific violations of representative sampling (Kish, 1965). If those biases can be characterized, then it is possible, in principle, to correct for them (Bar-Hillel, 1979). Such situations may, however, be relatively rare with Bayesian inference, where information can come from a variety of sources and in a variety of forms. Good sense then becomes the only guide to drawing and interpreting samples. As the history of scientific progress shows, it may take a fortuitous, if unpleasant, surprise to reveal an unintended bias in sampled information.[10]

Action

Bayesian inference is embedded in statistical decision theory. Its output, posterior odds, summarizes beliefs in a way that facilitates selecting the optimal course of action. Similarly, knowledge of the possible actions and their associated consequences is essential in determining what information to gather. Two Bayesian judges who contemplated different actions, or evaluated their consequences differently, might justifiably formulate different hypotheses and collect different data even though they agreed on the interpretation of all possible data.

Many of the difficulties that frustrate attempts to take prudent action on the basis of available knowledge are not unique to Bayesian inference. These include not having well-articulated values (i.e., not knowing what one wants), failing to think through all the consequences of different actions, and allowing one's preferences to be manipulated by the way in which problems are presented (Fischhoff, Slovic, and Lichtenstein, 1980; Rokeach, 1973; Tversky and Kahneman, 1981).

With other familiar problems, the Bayesian framework may offer an illuminating nomenclature and even some assistance. For example, people may forget that rejecting one option always means accepting another (if only the inaction option) that may prove even less attractive if it is examined. Conversely, accepting any option can mean forgoing others because there are not enough resources to go around. When evaluating an option, one must consider the *opportunity costs* of doing without the net benefits that would be gained by adopting other options (Vaupel and Graham, 1981). The Bayesian framework forces one to consider at least two options.

As mentioned earlier, the critical ratio provides a threshold for translating posterior odds into action: If they are above that threshold, act as though H were true; if they are below it, act as though $\bar{H}$ were true. The ratio is set by considering the consequences of being right and of being wrong in either case. Thus, it is the Bayesian way to relate uncertain knowledge to concrete actions by showing which action seems to be in one's best interest. In evaluating these actions, it is important to remember that (a) in the presence of uncertainty, the best action may not lead to the best outcome, (b) people must act on the basis of what they themselves believe at the time of decision, not what others believe or subsequently learn (Fischhoff, 1975), and (c) when uncertainty is great, there may be little difference in the apparent attractiveness of competing actions (von Winterfeldt and Edwards, 1982).

One danger in embedding inference in a decision-making framework is that it may encourage people to confuse "acting as though H is true" with "believing that H is true." Decision makers who confuse the two may not attend to signals indicating that their best guess about H was wrong and requires revision (Gärdenfors and Sahlin, 1982). Scientists who confuse the two may forget the uncertainties that they themselves acknowledged before offering a best-guess interpretation of experimental results.

Complications

In this presentation, the interpretation of a datum is complete once one has compared the two conditional probabilities comprising the likelihood ratio. Such appraisal assumes that the datum is taken at face value. More sophisticated Bayesian models are available for situations in which that assumption seems dubious and the interpretation of data depends upon contextual factors. Two such elaborations deal with *source credibility* and *conditional independence*. These complications also point to the need for care in making claims of biased, or non-Bayesian, inferences.

Source credibility

Every datum comes from some source. Knowing that source may, in principle, have quite diverse effects on the datum's interpretation. In common parlance one speaks about sources that have unusual or limited credibility as well as those that may attempt to mislead or may have been misled themselves. On the basis of detailed modeling of the informational properties of evidentiary situations that may arise in the courtroom, Schum (1980, 1981) has shown how source credibility information may reduce, enhance, or even reverse the diagnostic impact of a particular datum. The subtlety of Schum's models suggests both the difficulty of applying Bayesian inference properly and the pitfalls awaiting those trying to rely on intuition. Without understanding the impact of source credibility information, it is hard to know how it is or how it should be interpreted.

A common task in judgment research requires participants to decide whether a target individual belongs to Category A or Category B on the basis of a brief description and some base-rate information. These descriptions vary along dimensions such as the internal consistency of the information they contain and the ratio of relevant to irrelevant information. Typically, investigators have considered only the informational content of these messages when analyzing the impact that these variations should have and do have on behavior. In principle, however, each shift in content could signal a different level of credibility. If subjects are sensitive to these signals, they may choose to respond in ways that are at odds with those dictated by the informational content—and be justified in doing so.

For example, Manis, Dovalina, Avis, and Cardoze (1980), as well as Ginosar and Trope (1980), varied the consistency of the information in a description. With consistent profiles, all information that pointed toward any category pointed toward the same category; with inconsistent profiles, such diagnostic bits pointed in both directions. Subjects relied less on inconsistent information. This might reflect sensitivity to its apparently lower diagnosticity, or it might reflect doubt about its overall credibility. If one doubts that inconsistent people exist (Cooper, 1981), one may discount sources that have produced descriptions showing inconsistency. Both responses could be justified normatively and would lead to similar judgments. However, they suggest different psychological processes. The latter interpretation would mean that this is a situation in which people do understand the need to regress predictions based on unreliable information (Kahneman and Tversky, 1973).

In situations where the content of a message provides a cue regarding its validity, failure to consider that message may lead investigators to mistake sensitivity for bias. For example, in an experimental study of manuscript reviewing, Mahoney (1977) berated his scientist subjects for being more hospitable toward a fictitious study when its reported result confirmed the dominant hypothesis in their field than when it disconfirmed it. This differential receptiveness could reflect the stodginess and prejudices of normal science (Kuhn, 1962), which refuses to relinquish its pet beliefs. However, it could also reflect a belief that investigators who report disconfirming results tend to use inferior research methods (e.g., small samples leading to more spurious results), to commit common mistakes in experimental design, or, simply, to be charlatans. Mahoney himself might set a double standard if he were told that a study did or did not confirm the existence of telekinesis.

These reinterpretations are, of course, entirely speculative. To discipline them by fact, one needs to discover (a) how subjects structure the problem (e.g., their worries about source credibility) and (b) how they appraise the different components of the inferential model that they are using. For either describing or evaluating behavior, one must establish both what people believe and what they try to do with those beliefs (see also Wetzel, 1982).

Conditional independence

The most general way of thinking about contextual effects is as an interaction between the meaning of two or more data. That is, one datum, D_i, creates a context that affects the interpretation of another datum, D_j. In Bayesian terms, such interactions are said to reflect *conditional nonindependence* because the conditional probability $P(D_j/H)$ is not necessarily equal to $P(D_j/H, D_i)$. As a result, one cannot compute the cumulative impact of a set of data simply by multiplying their respective likelihood ratios.

Source-credibility problems may be viewed as a special case of conditional nonindependence: Information about the source affects interpretation of the message. Conversely, the message may affect one's view of the credibility of the source. Conditional nonindependence is also the grounds for configural judgment, the focus of many studies of clinical diagnosis (Goldberg, 1968; Slovic and Lichtenstein, 1971). For the configural judge, the meaning of a particular cue depends upon the status of others (e.g., "That tone of voice suggests 'not suicidal' to me unless I know that it was spoken at midday"). The research record shows that, although clinicians claim to interpret cues configurally, firm evidence of configural judgment is hard to find. This discrepancy may reflect the insensitivity of the research tool, the inaccuracy of the clinicians' introspections about their own judgmental processes, or their failure to use their configural strategies consistently (Dawes, 1979; Slovic and Lichtenstein, 1971).

In studies of clinical diagnosis, both the relationships between cues and people's judgments of those relationships are modeled, typically by linear regression equations. Configural relations are represented in those equations by interaction terms. In the Bayesian model, conditional nonindependence is treated by assessing joint conditional probabilities that consider interrelated data simultaneously. Considering the subtleties of Schum's analyses of source-credibility problems, it is very difficult to model these relationships either in the world or in people's judgments. Indeed, this very complexity may mean that sets of interrelated data defy explicit normative modeling, leaving them the province of judgment (Navon, 1981).

For the expert analyst of evidence, this leads to the frustrating situation of having to take a best guess at what the data mean knowing that their mutual implications have not been understood. Because such complications are common, those frustrations will also be common. Although lay judges may face conditional nonindependence equally often, they may not always take account of it. Perhaps it is the norm to take evidence at face value unless some alarm is sounded by the evidence itself or by the source presenting it. If that is the case, then simple Bayesian models may prove as effective for hypothesis evaluation behavior as simple regression models proved for clinical judgment. As with the regression models, these Bayesian models could be interpreted literally or as "paramorphic models" capturing stimulus–response relationships without claiming to describe the underlying cognitive processes (Hoffman, 1960).

Limits

Just as there are practical limits to the informational complexities that can be treated adequately within the Bayesian framework, so there are value considerations that are best recognized and left alone. People are not always just acquiring knowledge for the sake of optimizing their actions. Some pursuits that are not sensibly accommodated in the Bayesian framework can lead to deliberately non-Bayesian behavior. For example, people may deliberately act suboptimally in the short run when they are pursuing long-run goals such as "maintaining social relations (e.g., preserving and cultivating information sources), gaining and sustaining recognition (e.g., exuding confidence where accountability is low), and being accepted (e.g., by passing up smart solutions that make one appear out of step)" (Fischhoff, 1981, p. 902). Thus, people may ask nondiagnostic questions in order to keep the conversation going, and they may pass up diagnostic ones because asking them seems untoward.

With sufficient ingenuity, such behavior could probably be translated into Bayesian terms so as to show that it is not just purposeful, but also optimal—in the sense of having the highest expected value of any course of action. Thus, for example, asking a question could be treated as an act that has consequences other than the cost of asking it and the information that it produces. These consequences might include the possible penalty of being censured for impertinence or the chance of producing a completely unexpected result leading to the creation of new hypotheses. Practically speaking, anticipating and analyzing such considerations would be very difficult. Theoretically speaking, the attempt to do so might have a very ad hoc character, as though the investigator were groping for causes that might conceivably shape people's hypothesis evaluation. To be useful, these interpretations must walk a tightrope between giving people too little credit and giving them too much. At the former extreme, any behavior for which the investigator finds no ready Bayesian expression reflects cognitive incompetence. At the other extreme, people do whatever is right for them, and the observer's task is to determine what it is that they have managed to optimize (Cohen [and commentary], 1981; Fischhoff, Goitein, and Shapira, 1982; Hogarth, 1981).

Implications and reinterpretations

Prescriptively, the Bayesian approach provides a general model of how people should make sequential inferences. Using the model descriptively requires a choice between two strategies. One strategy is to assume that people are intuitive Bayesians and explore how they use the model. Alternatively, one can assume that they do not use the model but that their judgments can be described in terms of systematic departures from it (as when we characterized a number of phenomena as special cases of failing to consider the denominator in the likelihood ratio). In the course of explicating the Bayesian model, we have used both strategies, at times applying them to the same observed behavior. In some of these cases, that behavior could be viewed either as Bayesian or non-Bayesian, depending upon what one assumes about what people believe and what problem they are solving.

We use this framework now to analyze the behavior that has been observed in a number of additional studies. Typically, it casts a somewhat different light on what the subjects in those studies were doing and should have been doing than was cast by the original authors. In some cases, this reinterpretation shows commonalities in effects that had appeared to be distinct. In others, it reveals the differences in tasks that had gone under the same label. In particular, it shows how the term *confirmatory bias* has been applied to a variety of phenomena that may be described more succinctly in terms of the Bayesian model.

Nisbett, Zukier, and Lemley (1981)

The stimuli in Nisbett et al. were thumbnail descriptions of fictional individuals. The hypotheses were of the form "the individual belongs to category A" (e.g., is a child abuser). Stimuli and hypotheses were designed to test the authors' integration of Kahneman and Tversky's (1972) work on representativeness with Tversky's (1977) work on similarity judgments. According to the former, the judged probability of an individual belonging to a category depends on the judged similarity between the salient features of the individual's description and of the category stereotype. According to the latter, judged similarity should increase with the number of salient features common to both the individual's description and the category stereotype; it should decrease with the number of features unique to each.

Nisbett et al.'s descriptions consisted of one or two diagnostic features, each pointing toward one of the two possible prediction categories, and a varying number of nondiagnostic features, pointing toward neither category. The authors found that as the number of nondiagnostic features increased, the impact of the diagnostic features decreased. Thus, the nondiagnostic data dilute the effect of the diagnostic data, leading to less confident predictions. As the authors note, this dilution could only be justified normatively if subjects perceived, what would be called here, conditional nonindependence among the descriptors; that is, if the nondiagnostic information somehow mitigated the impact of the diagnostic information, for example, by reminding subjects how complex people are and hence how unreasonable it is to make confident predictions on the basis of a single feature.

Although the possible normative interpretations of these results are quite clear, the role of similarity judgments in them is not. As described by Tversky (1977), similarity judgments concern a match between an individual and a category. If subjects rely on representativeness, then the judged probability of an individual belonging to any category depends on the apparent individual–category similarity. Nisbett et al. extend these notions to cover the case of two possible categories by assuming that people form a subjective likelihood ratio whose numerator and denominator are derived separately by representativeness from judged individual–category similarity.

Nisbett et al. designed their stimuli so that nondiagnostic data belonged to neither category stereotype. As a result, adding a non-diagnostic datum to a diagnostic datum should produce a description that is less similar to both category stereotypes than the single-datum description. The judged probability of getting such a description, given that the individual belongs to each category, should be correspondingly lower. If both conditional probabilities are reduced by equivalent amounts, then there should be no reduction in diagnosticity, hence no dilution effect. Given that the addition of nondiagnostic information did reduce the extremity of judgments, one must forfeit some component of the above account. One solution is to drop the assumption that subjects analyze evidence by forming subjective likelihood ratios. Rather, here as elsewhere, subjects ignore the denominator when considering diagnosticity. By doing so, they would not notice that the nondiagnostic information also reduced the match between the description and the competing category.

Such neglect of the denominator would have two additional implications for the interpretation of Nisbett et al.'s study. One is that pretest subjects would not be judging diagnosticity in the conventional sense when they appraised "how helpful the information would be for prediction" (p. 255). The second is that encouraging people to rely on representativeness in the manner suggested by Nisbett et al. might actually improve their inference. Asking whether *D* is more representative of A or B would mean making a judgment that has at least the elements of the likelihood ratio—a datum and two competing hypotheses—if not necessarily in the proper relationship to each other. People might also be cautioned about a problem that arises with reliance on this strategy, the tendency to neglect the priors whenever one can detect any degree of differential representativeness (Bar-Hillel and Fischhoff, 1981; Ginosar and Trope, 1980).

Snyder and Swann (1978)

In a series of studies, Snyder and Swann (1978) had subjects select questions that "would provide them with the information to best test the hypothesis" (p. 1204) that a target individual whom they were about to interview was an introvert or an extravert. In the subjects' choice of questions, the authors claimed to have demonstrated a new bias, namely "an erroneous tendency to search for evidence that would tend to confirm the hypothesis under scrutiny" (p. 1203).

From a Bayesian perspective, however, it is unclear how the choice of question could be biased toward the confirmation of a hypothesis. Mathematically, it is not possible to ask questions all of whose possible answers support a particular hypothesis. Thus, there is no confirmatory bias in the sense of selecting data that inevitably favor a particular hypothesis. Snyder and Swann apparently felt that their subjects asked questions whose answers they could anticipate and would interpret as supporting the focal hypothesis. If the answer, D_i, to a question is predictable, then $P(D_i) = 1 = P(D_i/H) = P(D_i/\bar{H})$. That is, the question is nondiagnostic and the search process is inefficient. If D_i is taken as evidence supporting *H*, then the problem lies with the interpretation, which may reflect neglect of the denominator in the likelihood ratio.

A less extreme form of this exploitation of predictability might be seen in a scientist who repeatedly replicates the same experiment, which typically produces the same observation, *D*, which is more likely given *H* than given $\bar{H}$. Each observation of *D* would not, however, provide an independent and equal contribution to affirming *H*. An observation is informative only to the extent that it is unpredictable, hence capable of affirming or disaffirming the hypothesis. With each replication of the experiment, *P*(*D*) increases. There is a corresponding decrease in the disparity between $P(D/H)$ and $P(D/\bar{H})$ that is necessary for a diagnostic likelihood ratio. After many replications, *P*(*D*) and the likelihood ratio both approach 1, making future experimentation uninformative. An analogous way to look at this problem is to view the likelihood ratio as a measure of association defined on a 2 × 2 table whose rows and columns are $(D, \bar{D})$ and $(H, \bar{H})$, respectively. As the marginals of one of the variables (here, *D*) become more extreme, the value of the measure of association tends to decrease.

Perhaps the only way to bias the search toward accepting the focal hypothesis, H, would be to ask questions for which both $P(D/H)$ and $P(D/\bar{H})$ were expected to be high, knowing that one would subsequently ignore $P(D/\bar{H})$. Such a biased search in-the-service-of biased interpretation could, of course, only be exhibited successfully when $P(D)$ is high (i.e., D is likely to be observed whether or not H is true). If subjects tried it when $P(D)$ was low, then it would represent a disconfirmation bias: Subjects would look only at $P(D/H)$, which would be low, thereby reducing their confidence in H.

In Snyder and Swann's experiment, subjects asked to test the hypothesis that the target individual was an extravert were counted as biased if they chose "confirmatory" questions such as "What would you do if you wanted to liven up a party?" rather than "disconfirmatory" questions such as "What factors make it hard for you to really open up to people?" or "neutral" questions, such as, "What are your career goals?" The nonneutral questions have two noteworthy properties. One is that they are nondiagnostic insofar as they would elicit similar responses from both introverts and extraverts (and unless there are particularly introverted ways to liven up parties). Although such questions are inefficient, their selection constitutes no bias toward confirmation. As there were only 5 neutral questions in the set of 26 possibilities, subjects still had to choose many nonneutral ones in their set of 12 test questions. A second peculiar property of the nonneutral questions is that they are phrased in a conditional way that assumes the category membership of the recipient. For example, it would seem awkward or untoward to ask an introvert "What would you do if you wanted to liven up a party?" (Trope and Basok, 1982). Perhaps the only way to get any useful information with such a question would be if respondents rejected its premises where appropriate (e.g., "What do you mean? I never want to liven up parties!").[11] Subjects' preference among nonneutral questions for those relating to the focal hypothesis ("confirmatory" questions in Snyder and Swann's terminology) might be traced to a desire for compatibility with that hypothesis, or it could be that subjects treat the focal hypothesis as true and ask questions that would be the least awkward socially.

Wason (1960, 1968)

Card task. Some analogous phenomena may be seen in studies of how people test what might be called *formal* or *logical* hypotheses (Evans, 1982; Wason, 1960, 1968). These are categorical statements, such as "All *A*s are *B*s," that may be disconfirmed by a single counterexample (e.g., an A that is not a B). In one popular experiment, subjects are told that each of four cards has either A or $\bar{A}$ on one side and either B or $\bar{B}$ on the other side. They are then shown the four cards, whose exposed sides show, respectively, A, $\bar{A}$, B and $\bar{B}$. Subjects' task is to choose the cards they would turn over in order to test the hypothesis that "All *A*s are *B*s." Even though turning over either the A or the $\bar{B}$ could falsify the hypothesis, most people "waste" one of their choices on the B card, which is necessarily inconclusive. From our perspective, this bias is an example both of asking a non-diagnostic question (B) and of failing to ask a potentially falsifying question ($\bar{B}$).

Such a task would constitute a special case of Bayesian inference in at least two senses. One is that with logical hypotheses it is possible to observe data that unambiguously falsify or verify an hypothesis (i.e., that have a likelihood ratio of 0 or infinity). With empirical hypotheses, regarding real-life events, such certitude is rare or impossible. Indeed, one might speculate that subjects fail to ask potentially falsifying questions because they are unaccustomed to testing logical hypotheses and to having falsification possible. Second, the artificiality of the problem gives subjects no basis for assigning most of the probabilities involved. Many different prior odds could be defended by imputing a likelihood that the experimenter would focus attention on a true hypothesis. Except for data that would disconfirm H or $\bar{H}$, it is hard to defend any value of the likelihood ratio, insofar as any value other than 0 and infinity requires some arbitrary assumptions about the correlation between A-ness and B-ness in the artificial universe that the experimenter created.

Triads task. In another task devised by Wason (1968), subjects are told that the numbers 2, 4, and 6 conform to a simple rule that the experimenter has in mind; their task is to guess that rule. As an aid, they may ask whether additional number triads of their own choosing conform to the rule. In such a logical task, the validity of a hypothesized rule cannot be proven. Even if a rule fits all conforming triads, there is no guarantee that some other rule might not also fit or that some future triad might not violate it. It is only subjects' suppositions about the sorts of rules that the experimenter might use and the negligible penalty for guessing wrong that would enable strictly Bayesian subjects to stop gathering information, that is, stop proposing triads and guess that a particular hypothesis is correct.

The experimenter's unannounced rule is "numbers increasing in value." The first rule that comes to most people's minds is apparently "sequential even numbers." The modal first triad is something like 8, 10, 12. Called a sign of "verification bias" by Wason, this response pattern may be explained by several different accounts.

The first account traces this apparent search bias to an interpretation bias. It argues that subjects believe that being told that 8, 10, 12 fits the rule will prove that "sequential even numbers" is the correct hypothesis, whereas being told that it does not fit will prove that "sequential even numbers" is not the correct hypothesis. Hence, they propose 8, 10, 12, expecting to get a definitive answer either way. This flaw in logical inference could be given a special name, such as *mistaking affirmation for confirmation*. However, it is most parsimoniously regarded as yet another instance of ignoring alternative hypotheses when evaluating evidence. For the datum "yes, it conforms" and the hypothesis "the rule is sequential even numbers," $P(D/H) = 1$. Hence, that answer cannot reduce one's confidence in the truth of H. The evidence becomes, of course, less conclusive when one realizes that many alternative hypotheses, including the experimenter's own, would evoke the same response and have the same associated conditional probability.

A second account holds that subjects understand the meaning of disconfirming evidence and that they view their task as comparing their favored hypothesis with its complement, all other rules. The question they ask (8, 10, 12) does, in fact, produce information that is diagnostic for this comparison, that is, its answer should change their confidence in their hypothesis. If they can be faulted, it is for inefficiency, asking a question whose expected answer will tell them very little.

In the third account, subjects compare their favored hypothesis with another specific hypothesis, acting for the moment as though those two exhausted the universe of possibilities. Although they might be faulted for making such a false assumption, subjects might still consider this simplification a useful fiction. A more "proper" strategy is to compare each hypothesis with its complement and then, should falsifying evidence be found, start afresh with a new hypothesis. However, in a situation where any one of an infinite number of hypotheses could be the true one, people may prefer to compare pairs of hypotheses sequentially; at each round, the less likely hypothesis is discarded and the more likely one is then compared with the next contender. Such a comparative strategy would be similar in spirit to *sophisticated falsificationism* (Lakatos, 1970), where one holds onto the hypothesis that seems most correct until a better candidate comes along, even to the point of retaining a hypothesis for which inconsistent data are known to exist. The "proper" strategy resembles *naive falsificationism* (Popper, 1972), where one focuses on a single hypothesis, doing everything possible to disprove it without regard for what might take its place.

Subjects using this comparative strategy might still be faulted for poor triad selection. For any pair of hypotheses, they should ask about triads for which an affirmative answer will falsify one hypothesis and a negative answer will falsify the other. Thus, 8, 10, 12 is a poor triad if the two hypotheses are "sequential even numbers" and "numbers increasing in value." The experimenter's response would be "yes" if either were the correct hypothesis. On the other hand, it would be a fine choice if the two hypotheses were "sequential even numbers" and "numbers less than 7." Without eliciting subjects' hypotheses, it is hard to tell whether to fault them for using the comparative strategy or for using it inefficiently by choosing nondiagnostic triads.

Conclusion

Bayesian inference was originally developed as a prescriptive model. Its advocates believe that when one evaluates hypotheses it is useful to identify each element in the Bayesian model with the corresponding elements in one's thinking. Such identification ensures that all necessary elements have been considered and that they have been put in the proper relationship to one another. Although the model can help people to structure their thinking, by itself it cannot show how to formulate hypotheses, assess component probabilities, or set the critical ratio. Performing these operations requires a substantive understanding of the problem at hand.

Like other prescriptive models, the Bayesian scheme assumes that people could follow its dictates if they tried and if they were given some feasible level of assistance. However, it makes no statement regarding how people actually do make decisions. The descriptive potential of the Bayesian model lies in the framework that it provides for the primitives and processes of people's intuitive inference. This article attempts to use this potential by identifying the kinds of systematic deviations from the Bayesian model that could, in principle, be observed. Illustrative examples of most of these biases are found in the research literature, some collected within a Bayesian framework, some not.

Once developed, this descriptive model was applied to interpret a variety of existing studies. Although the results of this analysis are specific to those studies, a number of general conclusions emerge:

1 In some situations, apparently diverse effects prove to be special cases of a single judgmental bias. The most powerful of these "metabiases" is the tendency to ignore $P(D/\bar{H})$ when evaluating evidence.
2 In some situations, a variety of different phenomena have been confused under a common title. *Confirmation bias*, in particular, has proven to be a catch-all phrase incorporating biases in both information search and interpretation. Because of its excess and conflicting meanings, the term might best be retired.
3 In all situations, a careful appraisal of how judges have interpreted the tasks posed to them is needed before making any assertions regarding how, if at all, their judgment is biased. When speculating about how people may have construed a task, it is important to strike a balance between exaggerating the extent to which we or they are all-knowing. Neither our understanding of people nor our ability to help them is served by uncritically assuming either that there is no way for them to justify behavior that seems suboptimal to us or that there is a hidden method to any apparent madness that they exhibit.

Acknowledgements

Support for this research came from the U.S. Office of Naval Research under Contract N00014–80–C–0150 to Perceptronics, Inc. and from the United Kingdom Medical Research Council. Our thanks go to Hal Arkes, Jack Dowie, Ward Edwards, Sarah Lichtenstein, Nils-Eric Sahlin, Paul Slovic, Amos Tversky, and an anonymous reviewer for thoughtful comments on earlier drafts of this article.

Reference notes

1. U.S. Nuclear Regulatory Commission. 1981. *Fault tree handbook* (NUREG-0492). Washington, D.C.: Author.
2. U.S. Nuclear Regulatory Commission. 1975. *Reactor safety study: An assessment of accident risks in U.S. commercial nuclear power plants* (WASH-1400, NUREG-75/014). Washington, D.C.: Author.
3. U.S. Nuclear Regulatory Commission. 1978. *Risk assessment review group report to the U.S. Nuclear Regulatory Commission* (NUREG/CR-0400). Washington, D.C.: Author.

Notes

1 The demand of coherence is what differentiates DeMorgan's "pure subjective" interpretation of probability—as whatever people actually believe—from DeFinetti's personalistic interpretation of probability as rational belief (see Kyburg and Smokler, 1980).

2 Although beyond the scope of the present article, discussion of cases in which Bayes' theorem might not be the most useful way of updating beliefs may be found in Diaconis and Zabell (1982), Good (1950), and Jeffrey (1965, 1968). One troublesome situation is receiving information that changes one's whole system of beliefs [and not just $P(H)$].

3 A topical example may be found in the runaway inflation that has followed the linking of incomes and loans to the cost-of-living index in several countries. Although his economic theories predicted the opposite result, Milton Friedman has denied that this unhappy experience constitutes evidence against his theories because the countries involved did not implement the indexation exactly as he prescribed. Even more troubling for the status of his theories about the economy is that further thought (perhaps stimulated by this irrelevant experience) has led him to conclude that his earlier derivation was wrong and that, in fact, indexation encourages inflation under some conditions ("How Indexation Builds In Inflation," 1979).

4 In the case of multiple hypotheses, the posterior probability $P(H_i/D) = P(D/H_i)P(H_i)/S_i P(D/H_i)P(H_i)$.
A diagnostic datum is then one for which $P(D/H_i) \neq S_i P(D/H_i)P(H_i)$.

5 Those who interpret subjective probabilities in terms of intuitively appropriate betting odds would never say 1.00 because that would mean willingness to bet everything on the truth of the hypothesis involved. In that light, the only reasonable interpretation of 1.00 is as "nearly 1.00" or "above .995 and rounded upward."

6 One difficulty that the Bayesian approach avoids is vagueness in expressing beliefs. People often disagree considerably about the interpretation of verbal expressions of likelihood (e.g., "probably true"). Moreover, the same individual may use a term differently in different contexts (Beyth-Marom, 1982; Reyna, 1981). The Bayesian approach requires explicitness.

7 One might argue that the collection of diverse pieces of existing evidence should not be considered the *assessment of a base rate*, a term that should be reserved for aggregating statistical data (e.g., 70% of previous cases have supported H). Following that argument would lead one to the Laplacian assumption that all hypotheses are equally likely a priori except in the presence of statistical data to the contrary. In that case, the problems discussed in the text would be relegated to the section on assessing likelihood ratios associated with the diverse data. A further argument holds that even statistical data were separate pieces until they were aggregated—meaning that their aggregation required the use of likelihood ratios. At this extreme, a priori odds would always be equal to 1.

8 Ascribing neglect to subjects requires confidence that the investigator knows what base rates really are relevant on the basis of what subjects have been told. Although this determination is relatively clear in most experiments, it can be quite difficult in real-life problems.

9 The undue influence of prior information in this context is in sharp contrast to the neglect of prior information observed with the base-rate fallacy. One difference between the two cases is that in the former case the prior beliefs are actually posterior odds that subjects had generated by actively weighing previous evidence. In many base-rate fallacy studies, the prior beliefs were only a statistical summary reporting what someone else had typically observed. A second difference is that the perseverating prior beliefs were more specific than the neglected ones, fitting Bar-Hillel's (1980) account.

10 Deliberately unrepresentative sampling seems to be another possible bias to be included in this section. However, we argue in the discussion of Snyder and Swann (1978) that

such biases are better conceptualized as problems of data interpretation than as problems of sampling.

11 Only in Experiment 2 did Snyder and Swann's subjects actually conduct the promised interview with the target person. Other subjects who listened to tapes of these interviews judged respondents to predominantly extravert questions to be more extraverted than respondents to predominantly introvert questions. One might concur with Snyder and Swann's speculation that the biased sample of questions evoked a biased sample of reported behavior, or one might speculate that the form of the interviewees' responses reflected the conditioning presumptions of the questions they were asked. That is, it may be relatively easy to tell when people are answering questions that "would typically be asked of people *already known* to be extraverts" (Snyder and Swann, 1978, p. 1204; emphasis in original).

References

Bar-Hillel, M. 1979. The role of sample size in sample evaluation. *Organizational Behavior and Human Performance, 24*, 245–257.

Bar-Hillel, M. 1980. The base-rate fallacy in probability judgments. *Acta Psychologica, 44*, 211–233.

Bar-Hillel, M., and Fischhoff, B. 1981. When do base rates affect predictions? *Journal of Personality and Social Psychology, 41*, 671–680.

Beyth-Marom, R. 1982. How probable is probable? Numerical translation of verbal probability expressions. *Journal of Forecasting, 1*, 257–269.

Borgida, E., and Brekke, N. 1981. The base-rate fallacy in attribution and prediction. In J. H. Harvey, W. J. Ickes, and R. F. Kidd (Eds.) *New directions in attribution research* (Vol. 3). Hillsdale, N.J.: Erlbaum.

Brown, R. V., Kahr, A. S., and Peterson, C. 1974. *Decision analysis for the manager.* New York: Holt, Rinehart and Winston.

Cohen, J. 1981. Can human irrationality be experimentally demonstrated? *The Behavioral and Brain Sciences, 4*, 317–370.

Cooper, W. H. 1981. Ubiquitous halo. *Psychological Bulletin, 90*, 218–244.

Dawes, R. M. 1979. The robust beauty of improper linear models in decision making. *American Psychologist, 34*, 571–582.

DeFinetti, B. 1976. Probability: Beware of falsifications! *Scientia*, 1976, *3*, 283–303.

Diaconis, P., and Zabell, S. L. 1982. Updating subjective probability. *Journal of the American Statistical Association*, 77, 822–829.

Doherty, M. E., Mynatt, C. R., Tweney, R. D., and Schiavo, M. D. 1979. Pseudodiagnosticity. *Acta Psychologica, 43*, 111–121.

Doyle, A. C. 1974. *The memoirs of Sherlock Holmes*. London: Murray and Cape. (Originally published 1893.)

Eddy, D. M. 1982. Probabilistic reasoning in clinical medicine: Problems and opportunities. In D. Kahneman, P. Slovic, and A. Tversky (Eds.) *Judgment under uncertainty: Heuristics and biases*. New York: Cambridge University Press.

Edwards, W. 1962. Dynamic decision theory and probabilistic information processing. *Human Factors, 4*, 59–73.

Edwards, W. 1968. Conservatism in human information processing. In B. Kleinmuntz (Ed.) *Formal representation of human judgment*. New York: Wiley.

Edwards, W., Lindman, H., and Savage, L. J. 1963. Bayesian statistical inference for psychological research. *Psychological Review, 70*, 193–242.

Einhorn, H. J., and Hogarth, R. M. 1982. Prediction, diagnosis and causal thinking in forecasting. *Journal of Forecasting, 1*, 23–36.

Evans, J. St. B. T. 1982. *The psychology of deductive reasoning*. London: Routledge and Kegan Paul.

Fischer, D. H. 1970. *Historians' fallacies*. New York: Harper and Row, 1970.

Fischhoff, B. 1975. Hindsight ≠ foresight: The effect of outcome knowledge on judgment under uncertainty. *Journal of Experimental Psychology: Human Perception and Performance, 1*, 288–299.

Fischhoff, B. 1977. Perceived informativeness of facts. *Journal of Experimental Psychology: Human Perception and Performance, 3*, 349–358.

Fischhoff, B. 1980. For those condemned to study the past: Reflections on historical judgment. In R. A. Shweder and D. W. Fiske (Eds.) *New directions for methodology of behavior science: Fallible judgment in behavioral research*. San Francisco: Jossey-Bass.

Fischhoff, B. 1981. Inferential interference (Review of *Human inference: Strategies and shortcomings of social judgment*, by R. Nisbett and L. Ross). *Contemporary Psychology, 26*, 901–903.

Fischhoff, B. 1982. Debiasing. In D. Kahneman, P. Slovic, and A. Tversky, (Eds.) *Judgment under uncertainty: Heuristics and biases*. New York: Cambridge University Press.

Fischhoff, B., Goitein, B., and Shapira, Z. 1982. The experienced utility of expected utility approaches. In N. Feather (Ed.) *Expectancy, incentive and action*. Hillsdale, N.J.: Erlbaum.

Fischhoff, B., Slovic, P., and Lichtenstein, S. 1978. Fault trees: Sensitivity of estimated failure probabilities to problem representation. *Journal of Experimental Psychology: Human Perception and Performance, 4*, 330–344.

Fischhoff, B., Slovic, P., and Lichtenstein, S. 1979. Subjective sensitivity analysis. *Organizational Behavior and Human Performance, 23*, 339–359.

Fischhoff, B., Slovic, P., and Lichtenstein, S. 1980. Knowing what you want: Measuring labile values. In T. Wallsten (Ed.) *Cognitive processes in choice and decision behavior*. Hillsdale, N.J.: Erlbaum.

Forer, B. 1949. The fallacy of personal validation: A classroom demonstration of gullibility. *Journal of Abnormal and Social Psychology, 44*, 118–123.

Gärdenfors, P., and Sahlin, N.-E. 1982. Unreliable probabilities, risk taking, and decision making. *Synthese, 53*, 361–386.

Ginosar, Z., and Trope, Y. 1980. The effects of base rates and individuating information on judgments about another person. *Journal of Experimental Social Psychology, 16*, 228–242.

Goldberg, L. R. 1968. Simple models or simple processes? Some research in clinical judgment. *American Psychologist, 23*, 486–496.

Good, I. J. 1950. *Probability and the weighting of evidence*. New York: Hafner.

Green, A. E., and Bourne, A. J. 1972. *Reliability technology*. New York: Wiley Interscience.

Hoffman, P. J. 1960. The paramorphic representation of clinical judgment. *Psychological Bulletin, 47*, 116– 131.

Hogarth, R. M. 1981. Beyond discrete biases: Functional and dysfunctional aspects of judgmental heuristics. *Psychological Bulletin, 90*, 191–217.

How indexation builds in inflation. 1979. *Business Week*, November 12, pp. 114–116.

Hyman, R. 1977. Cold reading. *Zetetic* (The Skeptical Inquirer), *1*(2), 18–37.

Jeffrey, R. 1965. *The logic of decision*. New York: McGraw-Hill.

Jeffrey, R. 1968. Probable knowledge. In I. Lakatos (Ed.) *The problem of inductive logic*. Amsterdam: North Holland.

Kahneman, D., and Tversky, A. 1972. Subjective probability: A judgment of representativeness. *Cognitive Psychology*, 3, 430–454.

Kahneman, D., and Tversky, A. 1973. On the psychology of prediction. *Psychological Review, 80,* 237–251.

Kahneman, D., and Tversky, A. 1979. Intuitive predictions: Biases and corrective procedures. *TIMS Studies in Management Science, 12,* 313–327.

Kahneman, D., Slovic, P., and Tversky, A. (Eds.) 1982. *Judgment under uncertainty: Heuristics and biases.* New York: Cambridge University Press.

Kish, L. 1965. *Survey sampling.* New York: Wiley.

Koriat, A., Lichtenstein, S., and Fischhoff, B. 1980. Reasons for confidence. *Journal of Experimental Psychology: Human Learning and Memory, 6,* 107–118.

Kuhn, T. 1962. *Structure of scientific revolutions.* Princeton, N.J.: Princeton University Press.

Kyburg, H. E., and Smokier, H. E. (Eds.) 1980. *Studies in subjective probability.* Huntington, New York: Krieger.

Lakatos, I. 1970. Falsification and scientific research programs. In I. Lakatos and A. Musgrave (Eds.) *Criticism and the growth of scientific knowledge.* New York: Cambridge University Press.

Lichtenstein, S., Fischhoff, B., and Phillips, L. D. 1982. Calibration of probabilities: The state of the art to 1980. In D. Kahneman, P. Slovic, and A. Tversky (Eds.) *Judgment under uncertainty: Heuristics amd biases.* New York: Cambridge University Press.

Lichtenstein, S., Slovic, P., Fischhoff, B., Layman, M., and Combs, B. 1978. Judged frequency of lethal events. *Journal of Experimental Psychology: Human Learning and Memory, 4,* 551–578.

Lindley, D. V. 1965. *Introduction to probability and statistics from a Bayesian viewpoint.* Cambridge, England: Cambridge University Press.

Lindley, D. V., Tversky, A., and Brown, R. V. 1979. On the reconciliation of probability assessments. *Journal of the Royal Statistical Society* (Series A), *142,* Pt. 2, 146–180.

Lyon, D., and Slovic, P. 1976. Dominance of accuracy information and neglect of base rates in probability estimation. *Acta Psychologica, 40,* 287–298.

Mahoney, M. J. 1977. Publication prejudices. *Cognitive Therapy and Research, 1,* 161–175.

Manis, M., Dovalina, I., Avis, N. E., and Cardoze, S. 1980. Base rates can affect individual predictions. *Journal of Personality and Social Psychology, 38,* 231–240.

Mehle, T., Gettys, C. V., Manning, C., Baca, S., and Fisher, S. 1981. The availability explanation of excessive plausibility assessments. *Acta Psychologica, 49,* 127–140.

Moskowitz, H., Sarin, R. K. 1983. Improving the consistency of conditional probability assessments for long range forecasting and decision making. *Management Science, 29,* 735–749.

Murphy, A. H. 1972. Scalar and vector partitions of the probability score: Two-stage situation. *Journal of Applied Meteorology, 11,* 273–282.

Navon, D. 1981. Statistical and metastatistical considerations in analysing the desirability of human Bayesian conservatism. *British Journal of Mathematical and Statistical Psychology, 34,* 205–212.

Nisbett, R., and Ross, L. 1980. *Human inference: Strategies and shortcomings of social judgment.* Englewood Cliffs, N.J.: Prentice-Hall.

Nisbett, R. E., Zukier, H., and Lemley, R. E. 1981. The dilution effect: Nondiagnostic information weakens the implications of diagnostic information. *Cognitive Psychology, 13,* 248–277.

Novick, M. R., and Jackson, P. E. 1974. *Statistical methods for educational and psychological research.* New York: McGraw-Hill.

O'Leary, M. K., Coplin, W. D., Shapiro, H. B., and Dean, D. 1974. The quest for relevance. *International Studies Quarterly*, *18*, 211–237.

Phillips, L. D. 1973. *Bayesian statistics for social scientists*, London: Nelson.

Pitz, G. F, Downing, L., and Reinhold, H. 1967. Sequential effects in the revision of subjective probabilities. *Canadian Journal of Psychology*, *21*, 381–393.

Popper, K. R. 1972. *Objective knowledge*. Oxford, England: Clarendon.

Raiffa, H. 1968. *Decision analysis*. Reading, Mass.: Addison-Wesley.

Reyna, V. F. 1981. The language of possibility and probability: Effects of negation on meaning. *Memory and Cognition*, *9*, 642–650.

Rokeach, M. 1973. *The nature of human values*. New York: Free Press.

Savage, L. J. 1954. *The foundations of statistics*. New York: Wiley.

Schlaifer, R. 1969. *Analysis of decisions under uncertainty*. New York: McGraw-Hill.

Schum, D. 1980. Current developments in research on cascaded inference processes. In T. Wallsten (Ed.) *Cognitive processes in choice and decision behavior*. Hillsdale, N.J.: Erlbaum.

Schum, D. 1981. Sorting out the effects of witness sensitivity and response criterion placement upon the inferential value of testimonial evidence. *Organizational Behavior and Human Performance*, *27*, 153–196.

Slovic, P., and Lichtenstein, S. 1971. Comparison of Bayesian and regression approaches to the study of information processing in judgment. *Organizational Behavior and Human Performance*, *6*, 649–744.

Snyder, M., and Swann, W. B. 1978. Hypothesis testing processes in social interaction. *Journal of Personality and Social Psychology*, *36*, 1202–1212.

Sue, S., Smith, R. E., and Caldwell, C. 1973. Effects of inadmissible evidence on the decisions of simulated jurors: A moral dilemma. *Journal of Applied Social Psychology*, *3*, 345–353.

Trope, Y., and Basok, M. 1982. Confirmatory and diagnosing strategies in social-information gathering. *Journal of Personality and Social Psychology*, *43*, 22–34.

Troutman, C. M., and Shanteau, J. 1977. Inferences based on nondiagnostic information. *Organizational Behavior and Human Performance*, *19*, 43–55.

Tversky, A. 1977. Features of similarity. *Psychological Review*, *84*, 327–352.

Tversky, A., and Kahneman, D. 1974. Judgment under uncertainty: Heuristics and biases. *Science*, *185*, 1124–1131.

Tversky, A., and Kahneman, D. 1980. Causal schemas in judgments under uncertainty. In M. Fishbein (Ed.) *Progress in social psychology*. Hillsdale, N.J.: Erlbaum, 1980.

Tversky, A., and Kahneman, D. 1981. The framing of decisions and the rationality of choice. *Science*, *211*, 453–458.

Vaupel, J., and Graham, J. 1981. Eggs in your bier. *The Public Interest*, *61*, 3–17.

Wason, P. C. 1960. On the failure to eliminate hypotheses in a conceptual task. *Quarterly Journal of Experimental Psychology*, *12*, 129–140.

Wason, P. C. 1968. Reasoning about a rule. *Quarterly Journal of Experimental Psychology*, *23*, 273–281.

Wetzel, C. G. 1982. Self-serving biases in attribution: A Bayesian analysis. *Journal of Personality and Social Psychology*, *43*, 197–209.

von Winterfeldt, D., and Edwards, W. 1982. Costs and payoffs in perceptual research. *Psychological Bulletin*, *91*, 609–622.

7

KNOWING WITH CERTAINTY

The appropriateness of extreme confidence

Baruch Fischhoff, Paul Slovic, and Sarah Lichtenstein

DECISION RESEARCH, A BRANCH OF PERCEPTRONICS EUGENE, OREGON

Abstract

How often are people wrong when they are certain that they know the answer to a question? The studies reported here suggest that the answer is "too often." For a variety of general-knowledge questions (e.g., absinthe is [a] a liqueur or [b] a precious stone), subjects first chose the most likely answer and then indicated their degree of certainty that the answer they had selected was, in fact, correct. Across several different question and response formats, subjects were consistently overconfident. They had sufficient faith in their confidence judgments to be willing to stake money on their validity. The psychological bases for unwarranted certainty are discussed in terms of the inferential processes whereby knowledge is constructed from perceptions and memories.

Introduction

Two aspects of knowledge are what one believes to be true and how confident one is in that belief. Both are represented in a statement like, "I am 70% certain that Quito is the capital of Equador." While it is often not difficult to assess the veridicality of a belief (e.g., by looking in a definitive atlas), evaluating the validity of a degree of confidence is more difficult. For example, the 70% certainty in the above statement would seem more appropriate if Quito is the capital than if Quito isn't the capital, but that is a rather crude assessment. In a sense, only statements of certainty (0% or 100%) can be evaluated individually, according to whether the beliefs to which they are attached are true or false.

One way to validate degrees of confidence is to look at the calibration of a set of such confidence statements. An individual is well calibrated if, over the long run, for all propositions assigned a given probability, the proportion that is true is equal to the probability assigned. For example, half of those statements assigned a probability of .50 of being true should be true, as should 60% of those assigned .60, and all of those about which the individual is 100% certain. A burgeoning literature on calibration has been surveyed by Lichtenstein, Fischhoff, and Phillips (1977). The primary conclusion of this review is that people tend to be overconfident, that is, they exaggerate the extent to which what they know is correct. A fairly typical set of calibration curves, drawn from several studies, appears in Figure 7.1. We see that when people should be right 70% of the time, their "hit rate" is only 60%; when they are 90% certain, they are only 75% right; and so on.

People's poor calibration may be, in part, just a question of scaling. Probabilities (or odds) are a set of numbers that people use with some internal consistency (e.g., the curves in Figure 7.1 are more or less monotonically increasing) but not in accordance with the absolute criterion of calibration. Miscalibration can have serious consequences (see Lichtenstein et al., 1977),

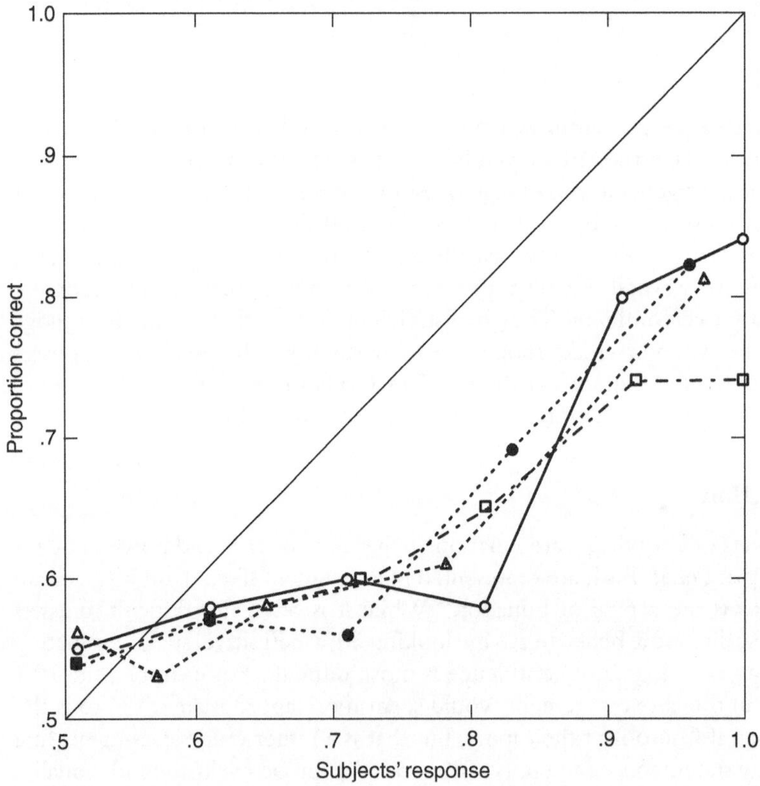

FIGURE 7.1 Some representative calibration curves. (Taken from Lichtenstein, Fischhoff, and Phillips, 1977. Copyright 1977 by D. Reidel Publishing Co. Reprinted by permission.)

yet people's inability to assess appropriately a probability of .80 may be no more surprising than the difficulty they might have in estimating brightness in candles or temperature in degrees Fahrenheit. Degrees of certainty are often used in everyday speech (as are references to temperature), but they are seldom expressed numerically nor is the opportunity to validate them often available (Tversky and Kahneman, 1974).

The extremes of the probability scale are, however, not such foreign concepts. Being 100% certain that a statement is true is readily understood by most people, and its appropriateness is readily evaluated. The following studies examine the calibration of people's expressions of extreme certainty. The studies ask, How often are people wrong when they are certain that they are right? In Experiment 1, the answer is sought in probability judgments elicited by questions posed in four different ways.

Experiment 1

Method

Stimuli. The questions covered a wide variety of topics, including history, music, geography, nature, and literature. The four formats used were the following:

1. Open-ended format. Subjects were presented with a question stem, which they were asked to complete; for example, "Absinthe is a ———." After writing down an answer, they estimated the probability that their answer was correct, using a number from .00 to 1.00.
2. One-alternative format. Subjects were asked to assess the probability (from .00 to 1.00) that simple statements were correct; for example, "What is the probability that absinthe is a precious stone?" The statement of fact being judged was sometimes true and sometimes false.
3. Two-alternative format (half range of responses). For each question, subjects were asked to choose the correct answer from two that were offered. After making each choice, they judged the probability that the choice was correct; for example, "Absinthe is (a) a precious stone or (b) a liqueur." Since they chose the more likely answer, their probabilities were limited to the range from .50 to 1.00.
4. Two-alternative format (full range of responses). Instead of having subjects pick the answer most likely to be correct as in Format 3, the experimenters randomly selected one of the two alternatives (e.g., [b] a liqueur) and had subjects judge the probability that the selected alternative was correct. Here the full range [.00, 1.00] was used. As in Format 3, one answer was correct.

Subjects and procedure. The subjects were 361 paid volunteers who responded to an ad in the University of Oregon student newspaper. They were assigned to the four groups according to preference for experiment time and date. Each group received the questions in only one of the four formats. Besides the differences in question format, the specific questions used differed somewhat from group to group. Instructions were brief and straightforward, asking subjects to choose or produce an answer and assign a probability of being correct in accordance with the format used.

Results

Lichtenstein and Fischhoff (1977) and Fischhoff and Lichtenstein (Note 1) have reported on the calibration of the entire range of probability responses of subjects in Experiment 1. Here we examine only their extreme responses. Table 7.1 shows (a) the frequency with which subjects indicated 1.00 or .00 as the probability an alternative was correct and (b) the percentage of answers associated with these extreme probabilities that were, in fact, correct. Answers assigned a probability of 1.00 of being correct were right between 20% and 30% of the time. Answers assigned a probability of .00 were right between 20% and 30% of the time. In Formats 2 and 4, where responses of 1.00 and .00 were possible, both responses occurred with about equal frequency. Furthermore, alternatives judged certain to be correct were wrong about as often as alternatives judged certain to be wrong were correct. The percentage of false certainties ranged from about 17% (Format 1) to about 30% (Format 2), but comparisons across formats should be made with caution because the items differed. Clearly, our subjects were wrong all too often when they were certain of the correctness of their choice of answer.

TABLE 7.1 Analysis of certainty responses in Experiment 1

Question format	*No. items*	*No. subjects*	*Total no. responses*	*Certainty responses (p)*	*% certainty responses*	*% correct certainty responses*
1. Open ended	43	30	1,290	1.00	19.7	83.1
2. One alternative	75	86	6,450	1.00	14.2	71.7
				.00	13.8	29.5
3. Two alternative (half range)	75	120	9,000	1.00	21.8	81.8
4. Two alternative (full range)	50	131	6,500	1.00	17.3	80.7
				.00	19.1	20.5

Experiment 2

Experiment 1 might be faulted because of the insensitivity of the response mode. With probabilities, subjects using the stereotypic responses of .50, .55, .60, and so on, have few possible responses for indicating different degrees of high certainty. At the extreme, most subjects restricted themselves to the responses .90, .95, and 1.00, corresponding to odds of 9:1, 19:1, and ∞: 1. Perhaps with a more graduated response mode, subjects would be better able to express different levels of certainty. In Experiment 2, subjects were presented with general-knowledge questions concerned with a single topic—the incidence of different causes of death in the United States—and asked to express their confidence in their answers in odds. The odds scale is open ended at the extremes, easily allowing the expression of many different levels of great certainty (e.g., 20:1, 50:1, 100:1, 500:1, etc.).

Method

Stimuli. All items involved the relative frequencies of the 41 lethal events shown in Table 7.2. They were chosen because they were easily understood and had fairly stable death rates over the last 5 years for which statistics were available. The event frequencies appearing in Table 7.2 were estimated from vital statistics reports prepared by the National Center for Health Statistics and the "Statistical Bulletin" of the Metropolitan Life Insurance Company. These frequencies provided the correct answers for the questions posed to our subjects. From among these 41 causes of death, 106 pairs were constructed according to the following criteria: (a) Each cause appeared in approximately six pairs and (b) the ratios of the statistical rates of the more-frequent event to the less-frequent event varied systematically from 1.25:1 (e.g., accidental falls vs. emphysema) to about 100,000:1 (e.g., stroke vs. botulism).

Procedure. Subjects' instructions read as follows:

> Each item consists of two possible causes of death. The question you are to answer is: Which cause of death is more frequent, in general, in the United States?

For each pair of possible causes of death, (a) and (b), we want you to mark on your answer sheet which cause you think is more frequent.

Next, we want to decide how confident you are that you have, in fact, chosen the more frequent cause of death. Indicate your confidence by the odds that your answer is correct. Odds of 2:1 mean that you are twice as likely to be right as wrong. Odds of 1,000:1 mean that you are a thousand times more likely to be right than wrong. Odds of 1:1 mean that you are equally likely to be right or wrong. That is, your answer is completely a guess.

TABLE 7.2 Lethal events whose relative frequencies were judged by subjects in Experiments 2 and 3

Lethal event	*Actual deaths per 100 million*[a]	*Lethal event*	*Actual deaths per 100 million*[a]
Smallpox	0	Firearm accident	1,100
Poisoning by vitamins	0.5	Poisoning by solid or liquid	1,250
Botulism	1	Tuberculosis	1,800
Measles	2.4	Fire and flames	3,600
Fireworks	3	Drowning	3,600
Smallpox vaccination	4	Leukemia	7,100
Whooping cough	7.2	Accidental falls	8,500
Polio	8.3	Homicide	9,200
Venomous bite or sting	23.5	Emphysema	10,600
Tornado	44	Suicide	12,000
Lightning	52	Breast cancer	15,200
Nonvenomous animal	63	Diabetes	19,000
Flood	100	Motor vehicle (car, truck, or bus) accident	27,000
Excess cold	163	Lung cancer	37,000
Syphilis	200	Cancer of the digestive system	46,400
Pregnancy, childbirth, and abortion	220	All accidents	55,000
Infectious hepatitis	330	Stroke	102,000
Appendicitis	440	All cancers	160,000
Electrocution	500	Heart disease	360,000
Motor vehicle – train collision	740	All diseases	849,000
Asthma	920		

[a] Per-year death rates are based on 100 million United States residents.

At the top of the answer sheet we have drawn a scale that looks like this:

1:1 10:1 100:1 1,000:1

10,000:1 100,000:1 1,000,000:1 etc.

This scale is used to give you an idea of the kinds of numbers you might want to use. You don't have to use exactly these numbers. You could write 75:1 if you think that it is 75 times more likely that you are right than you are wrong, or 1.2:1 if you think that it is only 20% more likely that you are right than wrong.

Do not use odds less than 1:1. That would mean that it is less likely that you are right than that you are wrong, in which case you should indicate the other cause of death as more frequent.

In case some of the causes of death are ambiguous or not well defined by the brief phrase that describes them, we have included a glossary for several of these items. Read this glossary before starting.

Subjects. The subjects were 66 paid volunteers who answered an ad in the University of Oregon student newspaper.

Results[1]

Table 7.3 shows the percentages of correct answers, grouped across subjects, for each of the most frequently used (major) odds categories. At odds of 1:1, 1.5:1, 2:1, and 3:1, subjects were reasonably well calibrated. However, as odds increased from 3:1 to 100:1, there was little or no increase in accuracy. Only 73% of the answers assigned odds of 100:1 were correct. Accuracy jumped to 81% at 1,000:1 and to 87% at 10,000:1. For the answers assigned odds of 1,000,000:1 or greater, accuracy was 90%. For the latter responses, the appropriate degree of confidence would have been odds of 9:1. The 12% of responses that are not listed in Table 7.3 because they fell between the major odds categories showed similar calibration.

As in Experiment 1, subjects in Experiment 2 exhibited great overconfidence. They were frequently wrong at even the highest odds levels. Moreover, they gave many extreme odds responses. Of 6,996 odds judgments, 3,560 (51%) were greater than 50:1. Almost one fourth of the responses were greater than 1,000:1.

Experiment 3

Although the tasks and instructions for Experiments 1 and 2 seemed reasonably straightforward, we were concerned that subjects' extreme overconfidence might be due to lack of motivation or misunderstanding of the response scale. Experiment 3 replicated Experiment 2, giving more care and attention to instructing and motivating the subjects.

TABLE 7.3 Percentage of correct answers for major odds categories

		Lethal events						General-knowledge questions		
		Experiment 2			*Experiment 3*			*Experiment 4*		
Odds	*Appropriate % correct*[a]	**N**	*%* **N**	*% correct*	**N**	*%* **N**	*% correct*	**N**	*%* **N**	*% correct*
1:1	50	644	9	53	339	8	54	861	19	53
1.5:1	60	68	1	57	108	2.5	59	210	5	56
2:1	67	575	8	64	434	10	65	455	1	63
3:1	75	189	2	71	252	6	65	157	3.5	76
5:1	83	250	4	70	322	8	71	194	4	76
10:1	91	1,167	17	66	390	9	76	376	8	74
20:1	95	126	2	72	163	4	81	66	1.5	85
50:1	98	258	4	68	227	5	74	69	1.5	83
100:1	99	1,180	17	73	319	8	87	376	8	80
1,000:1	99.9	862	13	81	219	5	84	334	7	88
10,000:1	100	459	7	87	138	3	92	263	6	89
100,000:1	100	163	2	85	23	.5	96	134	3	92
1,000,000:1	100	157	2	90	47	1	96	360	8	94
Total		6,098	88		2,981	70		3,855	75	
Overall % correct				71.0			72.5			73.1

Note. % *N* refers to the percentage of odds judgments that fell in each of the major categories. There were 66 subjects in Experiment 2, 40 in Experiment 3, and 42 in Experiment 4.

[a] For well-calibrated subjects.

Method

Experiment 3 used the 106 causes-of-death questions and odds response format of Experiment 2. The experimenter started the session with a 20-minute lecture to the subjects. In this lecture, the concepts of probability and odds were carefully explained. The subtleties of expressing one's feelings of uncertainty as numerical odds judgments were discussed, with special emphasis on how to use small odds (between 1:1 and 2:1) when one is quite uncertain about the correct answer. A chart was provided showing the relationship between various odds estimates and the corresponding probabilities. Finally, subjects were taught the concept of calibration and were urged to make odds judgments in a way that would lead them to be well calibrated. (The complete text of the instructions is available from the authors.)

The subjects for Experiment 3 were 40 persons who responded to an ad in the University of Oregon student newspaper. As in previous experiments, they were paid for participating. Group size was held to about 20 to increase the likelihood that subjects would ask questions about any facet of the task that was unclear.

Results

The proportion of correct answers for each of the most frequent odds categories is shown in the center portion of Table 7.3. The detailed instructions had several effects. First, subjects were much more prone to use atypical odds such as 1.4:1, 2.5:1, and so on. Only 70% of their judgments fell within the major odds categories of Table 7.3, as compared to 88% for Experiment 2. Second, their odds estimates tended to be smaller. About 43% of their estimates were 5:1 or less, compared to 27% for this category in Experiment 1. Third, subjects in this experiment were more often correct at odds above 10:1 and thus were better calibrated.

Nevertheless, subjects again exhibited unwarranted certainty. They assigned odds greater than or equal to 50: 1 to approximately one third of the items. Only 83% of the answers associated with these odds were correct. When subjects estimated odds of 50:1, they were correct 74% of the time and thus should have been giving odds of about 3:1. At 1,000:1, they should have been saying about 5:1.

Although only 70% of the responses fell in the major odds categories of Table 7.3, inclusion of the remaining 32% would not have changed the picture. Odds estimates falling between major categories were calibrated similarly to estimates within those categories. Elaborate instruction tempered subjects' extreme overconfidence, but only to a limited extent.

Experiment 4

Is there something peculiar to the causes-of-death items that induces such overconfidence? Experiment 4 replicated Experiment 3 using general-knowledge questions (of the type used in Experiment 1) matched in difficulty with the 106 causes-of-death items. In addition, subjects' faith in their odds judgments was tested by their willingness to participate in a gambling game based on those judgments.

Method

The questionnaire consisted of 106 two-alternative items covering a wide variety of topics; for example, "Which magazine had the largest circulation in 1970? (a) *Playboy* or (b) *Time*"; "Aden was occupied in 1839 by the (a) British or (b) French"; "Bile pigments accumulate as a result of a condition known as (a) gangrene or (b) jaundice." These items were taken from a large item pool with known characteristics. Availability of this pool allowed us to select items matched in difficulty, question by question, with the 106 items about lethal events studied in Experiments 2 and 3.

The subjects were 42 paid volunteers, recruited by an ad in the University of Oregon student newspaper. The instructions paralleled those of Experiment 3. Subjects first received the detailed lecture describing the concepts of probability, odds, and calibration. They then responded to the 106 general-knowledge items, marking the answer they thought to be correct and expressing their certainty about that answer with an odds judgment.

After responding to the 106 items, they were asked whether they would be willing to accept gambles contingent on the correctness of their answers and the appropriateness of their odds estimates. If subjects really believe in their extreme (extremely overconfident) odds responses, it should be possible to construct gambles that they are eager to accept but which, in fact, are quite disadvantageous to them. The game was described by the following instructions:

> The experiment is over. You have just earned $2.50, which you will be able to collect soon. But before you take the money and leave, I'd like you to consider whether you would be willing to play a certain game in order to possibly increase your earnings. The rules of the game are as follows.

1. Look at your answer sheet. Find the questions where you estimated the odds of your being correct as 50:1 or greater than 50:1. How many such questions were there? _______ (write number)
2. I'll give you the correct answers to these "50:1 or greater" questions. We'll count how many times your answers to these questions were wrong. Since a wrong answer in the face of such high certainty would be surprising, we'll call these wrong answers "your surprises."

3. I have a bag of poker chips in front of me. There are 100 white chips and 2 red chips in the bag. If I reach in and randomly select a chip, the odds that I will select a white chip are 100:2 or 50:1, just like the odds that your "50:1" answers are correct.
4. For every "50:1 or greater" answer you gave, I'll draw a chip out of the bag. (If you wish, you can draw the chips for me.) I'll put the chip back in the bag before I draw again, so the odds won't change. The probability of my drawing a red chip is 1/51. Since drawing a red chip is unlikely, every red chip I draw can be considered "my surprise."
5. Every time you are surprised by a wrong answer to a "50:1 or greater" question, you pay me $1. Every time I am surprised by drawing a red chip, I'll pay you $1.
6. If you are well calibrated, this game is advantageous to you. This is because I expect to lose $1 about once out of every 51 times I draw a chip, on the average. But since your odds are sometimes higher than 50:1, you expect to lose less often than that.
7. Would you play this game? Circle one. Yes No

Subjects who declined were then asked if they would play if the experimenter raised the amount he would pay them to $1.50 whenever he drew a red chip, while they still had to pay only $1 in the event of a wrong answer. Those who still refused were offered $2 and then a final offer of $2.50 for every red chip. Since the experimenters expected the game to be unfair to subjects (by capitalizing on a "known" judgmental bias), it was not actually played for money.

Results

The proportion of correct answers associated with each of the most common odds responses is shown in the right-hand column of Table 7.3. Compared with the previous studies, subjects in Experiment 4 gave a higher proportion of 1:1 odds (19% of the total responses). A few difficult items led almost all of the subjects to give answers close to 1:1, indicating that they were trying to use small odds when they felt it was appropriate to do so. However, this bit of restraint was coupled with as high a percentage of large odds estimates as was given by the untutored subjects in Experiment 2. About one quarter of all answers were assigned odds equal to or greater than 1,000:1.

Once again, answers to which extremely high odds had been assigned were frequently wrong. At odds of 10:1, subjects were correct on about three out of every four questions, appropriate to odds of 3:1. At 100:1, they should have been saying 4:1. At 1,000: 1 and at 100,000:1, estimates of about 7:1 and 9:1 would have been more in keeping with subjects' actual abilities. Over the large number of questions for which people gave odds of 1,000,000:1 or higher, they were wrong an average of about 1 time out of every 16.

The gambling game. Of the 42 subjects, 27 agreed to play the gambling game described above for $1. Six more agreed when the stakes were raised to $1.50 every time the experimenter drew a red chip. Of the holdouts, 3 subjects agreed to play at $2 for every red chip and 2 more agreed when the final offer of $2.50 was made. Only 3 subjects refused to participate at any level of payment per red chip.

After subjects had made their decisions about playing the game, they were asked whether they would change their minds if the game were to be played, on the spot, for real money. No subject indicated a desire to change his or her decision. Two subjects approached the experimenter after the experiment requesting that they be given a chance to play the game for cash. Their request was refused.

Of course, this game is strongly biased in favor of the experimenter. Since subjects were wrong about once for every eight answers assigned odds of 50:1 or greater, the game would have been approximately fair had the experimenter removed 86 of the white chips from the bag, leaving its contents at 14 white and 2 red chips.

The expected outcome of playing the game with each subject was simulated. Every wrong answer on a "50:1 or greater" question was assumed to cost the subject $1. The experimenter was assumed to have drawn 1/51 of a red chip for every answer given at odds greater than or equal to 50:1; his expected loss was then calculated in accordance with the bet the subject had accepted. For example, if a subject accepted the experimenter's first offer ($1 per red chip) and gave 17 "50:1 or greater" answers, the experimenter's simulated loss was 17/51 dollars (33¢).

The subjects who agreed to play averaged 38.3 questions with odds greater than or equal to 50:1. Thirty-six persons had expected monetary losses, and three had expected wins. Individual expected outcomes ranged between a loss of $25.63 and a gain of $1.84. The mean expected outcome was a loss of $3.64 per person and the median outcome was a loss of $2.35. Ten persons would have lost more than $5. The 39 subjects would have lost a total of $142.13 across 1,495 answers at odds greater than or equal to 50:1, an average loss of 9.5¢ for every such answer. The two persons who earnestly requested special permission to play the game had expected losses totaling $33.38 between them.

Experiment 5: playing for keeps

Subjects in Experiment 4 viewed their overconfident odds judgments as faithful enough reflections of their state of knowledge that they were willing to accept hypothetical bets more disadvantageous than many that can be found in a Las Vegas casino. Before concluding that there is money to be made in "trivia hustling," we decided to replicate Experiment 4 with real gambling at the end.

Method

Nineteen subjects participated in Experiment 5. It differed from Experiment 4 only in that the gambling game was presented as a real game. After responding to the 106 items, subjects heard the gambling game instructions and decided whether or not they would play. They were told that they could lose all the money they had earned in the experiment and possibly even more than that. After they made their decisions about playing the game, subjects were told that any earnings from the game would be added to their pay for the experiment, but that if they lost money, none of the money initially promised them for participating would be confiscated. The game was then played on those terms.

Six of the 19 subjects agreed to play the game as first specified (with a $1 payment for each "experimenter's surprise"). Three more agreed to play when the experimenter offered to increase the payment to $1.50 per red chip. Increasing the payment to $2 brought in one additional player, and three more agreed to play at $2.50. Six subjects consistently refused to participate; some because they felt they were not well calibrated, others because they did not like to gamble.

Results

When the game was actually played, the 13 participating subjects missed 46 of the 387 answers (11.9%) to which they had assigned odds greater than or equal to 50:1. All 13 subjects would have lost money, ranging from $1 to $11 (in part because, by chance, no red chips were drawn). When the experimenter's part of the game was simulated as in Experiment 4, four subjects would have lost more than $6, and the average participating subject would have lost $2.64. Thus, the hypothetical nature of the gamble in Experiment 4 apparently had minimal influence on subjects' willingness to bet.

General subject and item analyses

Is undue confidence found only in a few subjects or only for a few special items? If cases of extreme overconfidence are concentrated in only a few subjects, then the generality of our conclusions would be limited. Pathological overconfidence on the part of a small sector of the public would be worth exploring further but would not tell us much about cognitive functioning in general. The results of the gambling games reported above show that this was not the case. Most subjects were willing to play and most would have lost money because they were too often wrong when using extreme odds. The left columns of Table 7.4 show the distribution of cases of extreme overconfidence (defined as giving odds of 50:1 or greater and being wrong) over subjects for Experiments 3 and 4. The great majority of subjects had one or more cases of extreme overconfidence. The median number was 4 in Experiment

4 and between 3 and 4 in Experiment 3, well over what would be expected with well-calibrated subjects. In each experiment, one subject appeared to be an outlier (those subjects having 32 and 27 cases). Reanalyzing the data after removing those two subjects had no effect on our conclusions.

The right columns of Table 7.4 show the distribution of cases of extreme overconfidence over items. If most cases were concentrated in only a few items, the situation would be rather different than if a broad section of items fooled some of the people some of the time. It would not necessarily be less

TABLE 7.4 Frequency of extreme overconfidence (odds greater than or equal to 50:1 that were assigned to wrong answers)

No. cases of extreme overconfidence	*Number of subjects*		*No. extremely overconfident subjects*	*Number of items*	
	Experiment 3	*Experiment 4*		*Experiment 3*	*Experiment 4*
0	5[a]	3	0	33[b]	41
1	3	7	1	25	20
2	6	5	2	19	16
3	6	5	3	13	10
4	4	9	4	3	8
5	1	1	5	2	2
6	4	1	6	4	1
7	2	2	7	1	3
8	1	0	8	1	0
9	0	3	9	0	0
10	2	2	10	0	3
11	0	1	11	2	0
12	1	0	12	1	0
13	1	0	13	0	1
14	0	0	14	1	0
15	0	2	15	1	1
16	2	0			
17	1	0			
More than 17[c]	1(32)	1 (27)			

[a] There were five subjects in Experiment 3 who never showed extreme overconfidence.
[b] There were 33 items in Experiment 3 for which no subject showed extreme overconfidence.
[c] Actual number of cases is in parentheses.

interesting, for it would remain to be explained why people went astray on those few items. As the results in Table 7.4 indicate, both situations seem to have been true. There are some items on which many people gave high odds to the wrong answer, but most items did show a few such cases.

The items on which six or more subjects showed extreme overconfidence were all items that might be described as "deceptive," ones which less than 50% of the subjects answered correctly. Some correlation between deceptiveness and extreme overconfidence is inevitable; many subjects must get an answer wrong before many can get it wrong *and* be certain that they are right. There were 18 items in Experiment 3 and 17 items in Experiment 4 answered correctly by less than 50% of our subjects.

Table 7.5 shows the incidence of cases of extreme overconfidence with deceptive and nondeceptive items. Although extreme overconfidence is disproportionately prevalent with the deceptive items, it is still abundant with the nondeceptive ones. If the deceptive items are removed from the sample, then the remaining distribution of cases of extreme overconfidence over items closely resembles a Poisson distribution, which is what one would expect if such cases were distributed at random over items. One third of the easiest items, those answered correctly by 90% or more of our subjects, had at least one case of a subject answering wrongly and giving odds of being correct of 1,000:1 or greater. Deleting the one extreme subject from each of Experiments 3 and 4 had little effect on this result. Clearly, a few subjects or items are not responsible for the extreme overconfidence effect.

TABLE 7.5 Percentage wrong with deceptive and nondeceptive items

	Percentage wrong associated with odds of		
Experiment and item	*≥50:1*	*≥100:1*	*≥1000:1*
Experiment 3			
All items (106)	16.6	14.1	12.9
Deceptive items (18)	73.9	75.5	72.3
Nondeceptive items (88)	8.9	6.7	6.9
Experiment 4			
All items (106)	13.8	13.1	10.8
Deceptive items (17)	73.2	76.7	70.6
Nondeceptive items (89)	7.6	6.8	5.2
Expected with perfect calibration	≤1.96	≤.99	≤.10

General discussion

These five experiments have shown people to be wrong too often when they are certain that they are right. This result was obtained with both probability and odds responses, with minimal and extensive instructions and with two rather different types of questions. Subjects were sufficiently comfortable with their expressions of certainty that they were willing to risk money on them in both hypothetical and real gambles. Finally, cases of extreme overconfidence were widely distributed over subjects and items.

Although these studies have shown the effect to be a robust one, they have certainly not closed the topic. Further research with different subjects, different items, and different instructions would be most useful. Some moderately informed guesses at the results of such additional studies are possible. Lichtenstein and Fischhoff (1977) have found that the calibration of probability responses associated with general-knowledge questions is relatively invariant with regard to several factors not considered here, including subjects' intelligence, subjects' expertise in the subject-matter area of the questions and subjects' reliance on the stereotypic responses of .50 and 1.00. They did, however, find that calibration varies with item difficulty.

A crucial question for generality is how well the level of item difficulty found in these experiments represents the level found in the world. Although no simple answer to this question is possible, it is worth noting that the items in Experiments 2 and 3 were not constructed with the intention of eliciting extreme overconfidence. Rather, they were constructed to vary in difficulty from very hard to very easy, as defined by the ratio of the statistical frequencies of death from each of the two causes. Items in Experiments 4 and 5 were matched to these items in difficulty.

To explain these results, we must understand both how people answer questions and how they assess the validity of their answering process. Collins (Collins, Warnock, Aiello, and Miller, 1975; Collins, Note 3) has shown that people use many different strategies in answering questions. We suspect, therefore, that extreme overconfidence can come from a variety of sources. Every answering procedure may have its own ways of leading people astray and its own ways of hiding that misguidance when people try to assess answer validity. Some possible pathways to overconfidence are described below.

Many of the items we presented to our subjects are on topics for which they do not have a ready answer stored in memory. They must infer the answer from other information known to them. But people may be insufficiently critical of their inference processes. They may fail to ask "What were my assumptions in deriving that inference?" or "How good am I at making such inferences?" For example, when people draw a few instances of a category from memory to get an idea of the properties of the category, they may not realize that readily available examples need not be representative of the category (Tversky and Kahneman, 1973). Wason and Johnson-Laird (1972) have shown that people have considerable confidence in their own erroneous syllogistic reasoning. Collins et al. (1975) have described a variety of inferential strategies that people use in producing answers without realizing their limitations. Summarizing her studies on the inference process in perception, Johnson-Abercrombie (1960) concluded, "The[se erroneous] inferences were not arrived at as a series of logical steps but swiftly and almost unconsciously. The validity of the inferences was usually not inquired into; indeed, the process was usually accompanied by a feeling of certainty of being right" (p. 89). Pitz (1974), who also observed overconfidence in probability estimates, elaborated a similar hypothesis. He proposed that people tend to treat the results of inferential processes as though there was no uncertainty associated with the early stages of the inference. Such a strategy is similar to the "best-guess" heuristic that has been found to describe the behavior of subjects in cascaded inference tasks (e.g., Gettys, Kelly, and Peterson, 1973).

For other questions, people believe that they are answering directly from memory without making any inferences. People commonly view their memories as exact (although perhaps faded) copies of their original experiences. However, considerable evidence has demonstrated that memory is more than just a copying process (e.g., Neisser, 1967). According to this view, people reach conclusions about what they have seen or what they remember by reconstructing their knowledge from fragments of information, much as a paleontologist infers the appearance of a dinosaur from fragments of bone. During reconstruction, a variety of cognitive, social, and motivational factors can introduce error and distortion into the output of the process. Examples of this are the foibles of eyewitness testimony documented by Buckhout (1974), Loftus (1974), Münsterberg (1908), and others.

If people are unaware of the reconstructive nature of memory and perception and cannot distinguish between assertions and inferences (Harris and Monaco, 1978), they will not critically evaluate their inferred knowledge. In general, any process that changes the contents of memory unbeknownst to people will keep them from asking relevant validity questions and may lead to overconfidence. In his classic studies of reconstructive processes in memory, Bartlett (1932) found that subjects not only created new material but were often highly certain about that which they had invented.[2]

We present these ideas more as a framework for future research and conceptualization than as an explanation for our results. Nonetheless, if these speculations have some validity, it should be possible to find apparent examples in our data. Tables 7.6 and 7.7 present the most deceptive items from Experiments 3 and 4, respectively. Although cases of extreme overconfidence were distributed over most items, these "deceptive" items produced a disproportionate share. In the absence of detailed protocols from subjects, these cases where many people went astray may provide better clues to our intuitions than situations where just one or two subjects had trouble with an item.

TABLE 7.6 Deceptive items in Experiment 3

Causes of death compared[a]	*Percent correct*	*No. cases of extreme overconfidence*[b]
Pregnancy, abortion, childbirth versus appendicitis	15	15
All accidents versus stroke	17.5	14
Homicide versus suicide	25	12
Measles versus fireworks	25	5
Suicide versus diabetes	27.5	8
Breast cancer versus diabetes	30	1

[a] Subjects judged the first cause of death listed to be less frequent than the second.

[b] Data are the number of subjects (out of 40) who gave odds greater than or equal to 50:1 to the wrong alternative.

TABLE 7.7 Deceptive items in Experiment 4

General-knowledge question[a]	*Answers*[b]	*Percent correct*	*No. cases of extreme overconfidence*[c]
1. Three fourths of the world's cacao comes from	Africa* or South America	4.8	15
2. Which causes more deaths in the U.S.?	Appendicitis* or pregnancy, abortion, and childbirth	19.0	13
3. When was the first air raid?	1849* or 1937	26.2	10
4. Adonis was the god of	Love or vegetation*	31.0	10
5. Kahlil Gibran was most inspired by which religion?	Buddhist or Christian*	33.3	7
6. *Dido and Aeneas* is an opera written by	Berlioz or Purcell*	33.3	2
7. Potatoes are native to	Ireland or Peru*	35.7	10

[a] Some questions have been abbreviated slightly.

[b] Correct answer carries an asterisk.

[c] Data are the number of subjects (out of 42) who gave odds greater than or equal to 50:1 to the wrong alternative.

Looking at the deceptive items in Experiment 3 (see Table 7.6), we find that in many cases the cause of death incorrectly judged to be more frequent (the first one listed in each pair) is a dramatic, well-publicized event, whereas the underestimated cause is a more "quiet" killer. Considering the first three examples, (a) pregnancy, abortion, and childbirth, (b) accidents, and (c) homicide seem disproportionately more newsworthy and better reported than their comparison cause of death.[3] In these cases, people may be relying on the greater availability in memory of examples of the "flashier" causes of death without realizing that availability is an imperfect inferential rule (Tversky and Kahneman, 1973).[4] Other items suggest other answering processes. Subjects' confident—but erroneous—beliefs (not shown in Table 7.6) that there were fewer deaths from smallpox vaccine than from the disease itself may have been based on the generally valid assumption that vaccines are safer than the diseases they are meant to prevent. With smallpox, however, the vaccine has been so successful that no one has died of the disease in the U.S. since 1949, while from 6 to 10 people have died annually from complications arising from vaccination.

For the general-knowledge questions of Experiment 4 in Table 7.7, we will give a few interpretations of the varied ways that unrecognized or inadequately questioned assumptions can obscure the tenuousness of erroneous beliefs; the reader can surely provide others. Regarding Item 1, cacao is native to South America. Subjects who knew this fact (or guessed it from the Spanish-sounding name) may have been misled by assuming that the continent of origin is also the continent of greatest production. Similar reasoning may have been involved with Item 7. The potato's prominence in Irish history does not mean that it originated there. Regarding Item 3, it may not have occurred to subjects that an air raid could be conducted by balloons, which were used by Austria to bomb Venice in 1849. The fact that Adonis was a handsome youth who had an affair with Venus, the Goddess of Love, may have suggested that he, too, was a deity of love (Item 4). And so on.

Finally, let us add a warning that extreme overconfidence cuts both ways. Our sources for the answers to general-knowledge questions were a variety of encyclopedias and dictionaries. We viewed the answers they provided with great confidence. Much to our chagrin, we discovered on several occasions that these authorative sources disagreed, a possibility we had never considered. Fortunately, our own overconfidence was discovered before conducting these experiments; the offending items were deleted and the remaining ones double- and triple-checked until we were *certain* of their accuracy.

Acknowledgements

This research was supported by the Advanced Research Projects Agency (ARPA) of the Department of Defense and was monitored by the Office of Naval Research under Contract N00014-76-0074 (ARPA Order No. 3052) under a subcontract from Decisions and Designs, Inc. to Oregan Research Institute.

We would like to thank Bernard Corrigan, Robyn Dawes, Ward Edwards, and Amos Tversky for their contributions to this project.

Notes

1 A more detailed description of subjects' performances on this task and several related ones can be found in Lichtenstein, Slovic, Fischhoff, Combs, and Layman (Note 2).

2 An example of the subtle role of assumptions in the reconstruction of knowledge comes from the experience of one of the authors who became embroiled in a friendly debate with a colleague about the dates of a forthcoming conference. Both parties agreed that the conference was to last about 4 to 5 days. But the dispute centered about whether these dates were March 30 to April 3 or April 30 to May 3. The author was certain of the former dates because he specifically recalled the date March 30 in the organizer's letter. His colleague was certain of the latter period because he specifically recalled the date May 3 in the letter. Bets were placed, and the letter was consulted to resolve the dispute. To the surprise of both parties, the letter stated the dates as March 30 to May 3, an obvious mistake. Thus, both parties were correct regarding the fragment of information they recalled, but one fragment led to the wrong conclusion.

3 This speculation has been empirically affirmed by Lichtenstein, Slovic, Fischhoff, Combs, and Layman (Note 2).

4 Subjects in Experiment 3 were asked to select one answer about which they were certain and to write a short statement indicating why they were so confident. One subject explained odds of 2,000:1 that death from pregnancy was more frequent than deaths by appendicitis by writing "I've never heard of a person dying of appendicitis, but I have many times heard of persons dying during childbirth and abortion."

Reference notes

1. Fischhoff, B., and Lichtenstein, S. 1977. *The effect of response mode and question format on calibration* (Rep. No. 77–1). Eugene, Oregon: Decision Research.
2. Lichtenstein, S., Slovic, P., Fischhoff, B., Combs, B., and Layman, M. 1976. *Perceived frequency of low probability, lethal events* (Rep. No. 76–2). Eugene, Oregon: Decision Research.
3. Collins, A. 1976. *Processes in acquiring knowledge* (Rep. No. 3231). Cambridge, Mass.: Bolt Beranek and Newman, Inc.

References

Bartlett, F. C. 1932. *Remembering.* Cambridge, England: Cambridge University Press.

Buckhout, R. 1974 Eyewitness testimony. *Scientific American, 231,* 23–31.

Collins, A., Warnock, E. H., Aiello, N., and Miller, M. 1975. Reasoning from incomplete knowledge. In D. Bobrow and A. Collins (Eds.) *Representation and understanding.* New York: Academic Press.

Gettys, C. F., Kelly, C, III, and Peterson, C. R. 1973. The best guess hypothesis in multistage inference. *Organizational Behavior and Human Performance, 10,* 364–373.

Harris, R. J., and Monaco, G. E. 1978. Psychology of pragmatic implication: Information processing between the lines. *Journal of Experimental Psychology: General, 107,* 1–22.

Johnson-Abercrombie, M. L. 1960. *The anatomy of judgment.* New York: Basic Books.

Lichtenstein, S., and Fischhoff, B. 1977. Do those who know more also know more about how much they know? The calibration of probability judgments. *Organizational Behavior and Human Performance, 20*, 159–183.

Lichtenstein, S., Fischhoff, B., and Phillips, L. D. 1977. Calibration of probabilities: The state of the art. In H. Jungerman and G. de Zeeuw (Eds.), *Decision making and change in human affairs*. Amsterdam, The Netherlands: Reidel.

Loftus, E. 1974. The incredible eyewitness. *Psychology Today*, December, 116–119.

Münsterberg, H. 1908. *On the witness stand*. New York: Doubleday, Page.

Neisser, U. 1967. *Cognitive psychology*. New York: Appleton-Century-Crofts.

Pitz, G. 1974. Subjective probability distributions for imperfectly known quantities. In G. W. Gregg (Ed.) *Knowledge and cognition*. New York: Wiley.

Tversky, A., and Kahneman, D. 1973. Availability: A heuristic for judging frequency and probability. *Cognitive Psychology, 5*, 207–232.

Tversky, A., and Kahneman, D. 1974. Judgment under uncertainty: Heuristics and biases. *Science, 185*, 1124–1131.

Wason, P. C, and Johnson-Laird, P. N. 1972. *Psychology of reasoning: Structure and content*. Cambridge, Mass.: Harvard University Press.

8

FAULT TREES

Sensitivity of estimated failure probabilities to problem representation

Baruch Fischhoff, Paul Slovic, and Sarah Lichtenstein

DECISION RESEARCH, A BRANCH OF PERCEPTRONIES EUGENE, OREGON

Abstract

Fault trees represent problem situations by organizing "things that could go wrong" into functional categories. Such trees are essential devices for analyzing and evaluating the fallibility of complex systems. They follow many different formats, sometimes by design, other times inadvertently. The present study examined the effects of varying three aspects of fault tree structure on the evaluation of a fault tree for the event "a car fails to start." The fault trees studied had four to eight branches, including "battery charge insufficient," "fuel system defective," and "all other problems." Major results were as follows: (a) People were quite insensitive to what had been left out of a fault tree, (b) increasing the amount of detail for the tree as a whole or just for some of its branches produced small effects on perceptions, and (c) the perceived importance of a particular branch was increased by presenting it in pieces (i.e., as two separate component branches). Insensitivity to omissions was found with both college student subjects and experienced garage mechanics. Aside from their relevance for the study of problem solving, such results may have important implications for (a) how best to inform the public about technological risks and to involve it in policy decisions and (b) how experts should perform fault tree analyses of the risks from technological systems.

Introduction

Many problems involve some form of troubleshooting: Something goes wrong, and the problem solver attempts to figure out why by first listing and then checking possible causes. The problem solver's representation of the problem (and possible causes) can be constructed from scratch, retrieved from memory, or adopted from an external source, perhaps with supplementary information from memory. A common representation for such problems is a "fault tree," which organizes possible sources of trouble into a branching structure. Figure 8.1 presents a fault tree for the event "a car fails to start long enough to delay the driver for 1 minute." The top row presents the problem, and the next row indicates the major systems whose failure might be relevant. Below each major system are listed specific contributing failures. Sources used in constructing this particular tree were *Peterson's Basic Auto Repair Manual, The Dodge Colt Service Manual, The Chevelle Owner's Guide*, and several experienced mechanics. Using analogous sources of expertise, one could construct a tree headed by "3-month old infant cries more than 5 minutes," "bank statement does not match checkbook," "star pitcher fails to report for spring training," "SAC bomber recall system fails," or "nuclear power reactor core melts."

Once constructed, the tree serves as a guide for the problem solver who might ask: Which system is more likely to be the source of trouble? What information would allow me to check out the fault most quickly? How well do I know this system (i.e., how authoritative is the fault tree), and should I be tinkering with it myself? Fault trees are used not only in analyzing systems that have gone astray but also in attempting to design fail-safe systems. Knowing how things go wrong is a prerequisite to drafting directives on what to do right. Fault trees have been instrumental in the design of technological systems from spaceships to nuclear power plants (Atomic Energy Commission, 1975; Green and Bourne, 1972; Bryan, Note 1).

Fault trees are also used to estimate failure rates for complex systems when historical data for the system as a whole are unavailable. Probabilities are assigned to each of the pathways to failure and are then combined to provide an overall failure rate. Such analyses were the primary methodological tools in the $3 million Rasmussen study, which assessed the probability of a catastrophic loss-of-coolant accident in a nuclear power reactor (Atomic Energy Commission, 1975). Fault tree analysis has, however, come under attack from critics who question whether it is methodologically sound enough to be used as a basis for decisions of great consequence (e.g., Bryan, Note 1). One major concern of the critics is that omission of relevant pathways due to ignorance, poor memory, or lack of imagination would lead to an underestimation of failure probabilities (Kendall, 1975).

As with many other kinds of problems (Newell and Simon, 1972), any troubleshooting situation can be represented in a variety of ways. For example, in constructing the fault tree in Figure 8.1, it was necessary to decide how much detail to provide for each system; whether to present minor systems like "mischievous acts" separately or to lump them with "all other problems"; whether the four items grouped in "fuel system defective" actually belong together or whether the last two might best be listed separately with a heading like "carburetion problems"; whether to use the graphic display or an outline; and whether to use this level of specificity, or more, or less. Making these decisions will depend on considerations like the purpose of the analysis, the amount of knowledge available about the issue in question, and how much of an effect each of these aspects of presentation has on the way people evaluate the system.

The studies reported here look at the impact of three such "arbitrary" aspects of the way in which fault trees are presented and the way in which they are evaluated. These are: (a) what is listed specifically and what is left to all other problems, (b) how much detail is presented for the various "branches" (systems) of the fault tree, and (c) how various systems are grouped into branches.

Aside from their interest for students of problem solving, such questions may be important for those concerned with the management of risks in our society. The lay public is increasingly called upon to decide whether the risks from various technological systems are acceptable in the light of the accompanying benefits. The risks from many proposed projects (e.g., nuclear power, liquid natural gas, recombinant DNA research) are varied and complex. Typically, they are presented to the public by technical experts who in one way or another use fault tree representation. In preparing their presentation, they must make arbitrary decisions like those listed above.

Understanding what difference these decisions make might help us understand (a) whether the public is being properly informed (e.g., is risk information being presented in ways leading to its subjective overestimation or underestimation?), (b) how technical information can be communicated so that it is perceived most veridically, and (c) what possibilities exist for manipulating perceptions through judicious fault tree presentation. If the perceptions of experts are also affected by variations in tree representation, we might learn something about how to improve the basic methodology of a technique used for momentous decisions.

Experiment 1: pruning the tree

However exhaustive one would like to be in the design of a fault tree, at some point it is necessary to stop listing alternatives and combine the remainder into a category labeled something like *all other problems*. In theory, listing every possible cause is an endless chore, and in practice, listing many implausible causes could overwhelm the fault tree's designers and viewers with bizarre possibilities that might divert attention from the primary possibilities. In Experiment 1, we studied the sensitivity of an evaluator to fault tree components that have been omitted. In this initial investigation, we looked for gross effects, those obtained by deleting substantial portions of a tree without specifically indicating their absence. In the real world of design and disaster, elements can be left out either inadvertently (the designer lacks the appropriate knowledge or imagination) or consciously (the designer collapses certain categories into the "other" category for simplicity; the designer omits certain pathways to make the system appear safer).

Two contrary hypotheses can be advanced for the impact of deleting major components of a fault tree on a viewer's judgment of its completeness.

One hypothesis suggests that when major items are deleted, it is quite likely that the absence of at least one item will be detected. Once such an omission has been uncovered, the entire analysis is discredited and the proportion left out is exaggerated. (Imagine your reaction to discovering that a purported fault tree for "car won't start" contains no mention of battery failure.) Many public discussions regarding nuclear power and other technological risks would seem to support this hypothesis. Members of the public scrutinize a fault tree prepared by a technical expert, discover (what seems to them to be) an important omission, and doubt the quality of the entire analysis and the competence of the analyst (e.g. Barrager, Judd, and North, 1976; Birnbaum, Wong, and Wong, 1976; Settle and Golden, 1974).

In contrast with this "credibility" hypothesis, the "availability" hypothesis (Tversky and Kahneman, 1973) suggests that what is out of sight is also out of mind. Problems that are not mentioned explicitly may not be thought of. People may not realize what is missing and, therefore, may overestimate the completeness of the analysis. Suspecting this to be the case, critics of technological projects who discover omissions in fault trees have cast doubt not on the competence but on the integrity of the analysts, charging the analysts with deliberately omitting problems to induce the public to underestimate the total risks associated with the project.

A third hypothesis is that people have such well-defined frequency representations in their minds (Howell, 1973) that they will appropriately realize what is missing. It is also possible that availability and credibility effects might cancel one another, producing appropriate judgments of all other problems.

Method

Stimulus. The basic stimulus for all studies reported here is the car won't start fault tree of Figure 8.1. It was constructed by consulting a variety of shop and repair manuals, mechanics, and car buffs to make it as complete as possible.1 We distinguish between three levels of detail: *Level 1,* which presents just the system names (e.g., "battery charge insufficient"); *Level 2,* which lists three to five component problems for each system (e.g. (a) faulty ground connections; (b) terminal loose or corroded; (c) battery weak); and *Level 3,* which provides fuller detail for each of the component problems that could be elaborated, that is, everything shown in Figure 8.1. Subjects in Experiment 1 were presented with Level 2 detail.

Design. Four separate groups of subjects participated. Two groups received the full, unpruned tree of Level 2. They differed only in that the experimenter read aloud the entire tree to one group (n = 58); the other group (n = 35) read the tree by themselves, without an enforced amount of time for examining it. Each of the two other groups was given a different pruned tree; one group (n = 29) saw a tree that was missing the starting, ignition, and mischief branches, and the other group (n = 26) saw a pruned tree that lacked the battery, fuel, and other engine branches. Since the first two groups showed no difference in

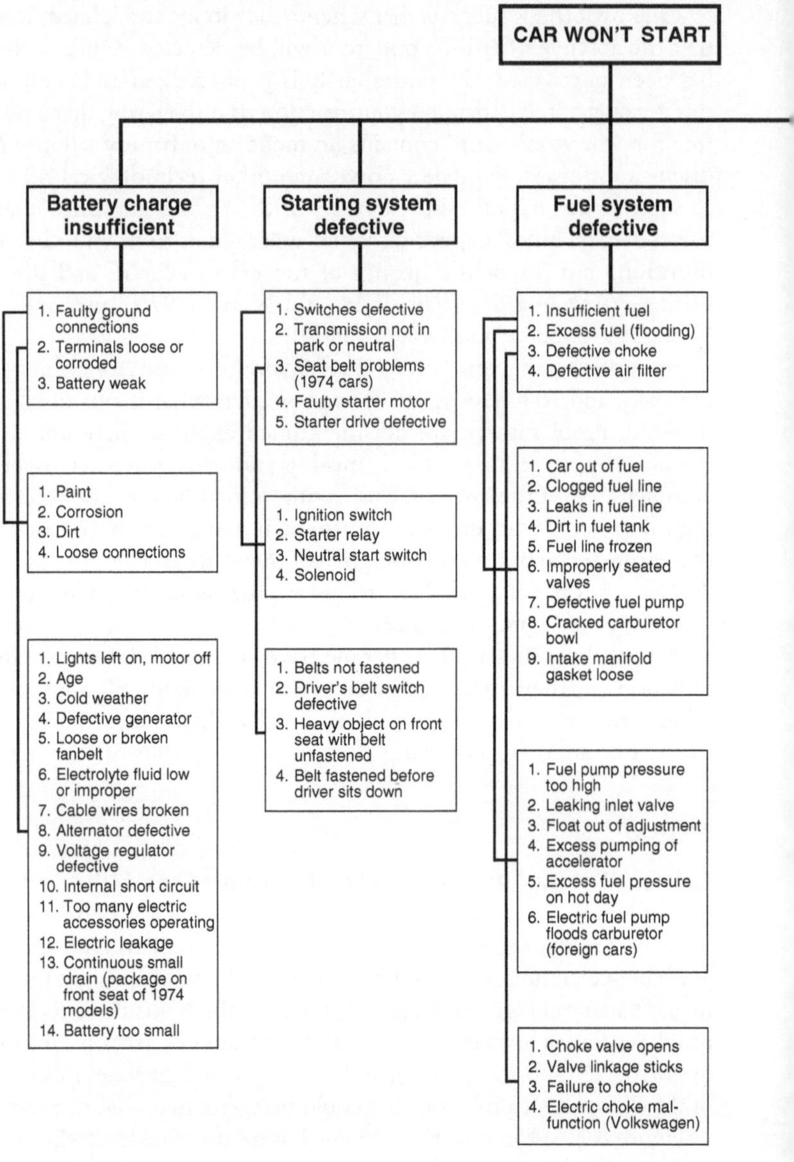

FIGURE 8.1 A possible fault tree for discovering why a car won't start

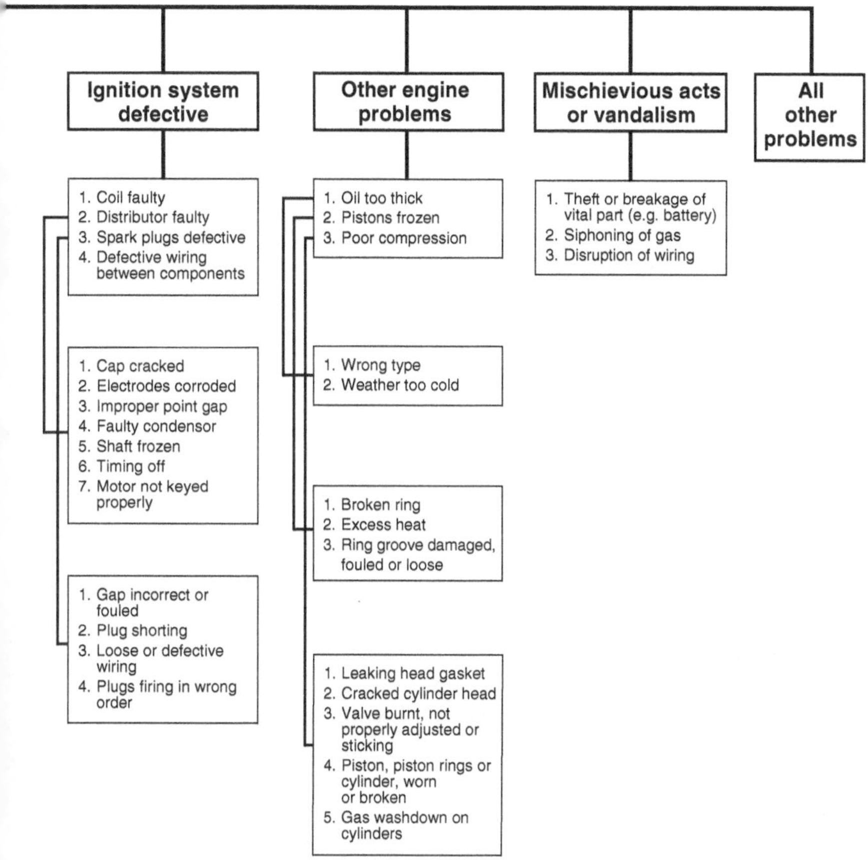

Ignition system defective
1. Coil faulty
2. Distributor faulty
3. Spark plugs defective
4. Defective wiring between components
1. Cap cracked
2. Electrodes corroded
3. Improper point gap
4. Faulty condensor
5. Shaft frozen
6. Timing off
7. Motor not keyed properly
1. Gap incorrect or fouled
2. Plug shorting
3. Loose or defective wiring
4. Plugs firing in wrong order
Other engine problems
1. Oil too thick
2. Pistons frozen
3. Poor compression
1. Wrong type
2. Weather too cold
1. Broken ring
2. Excess heat
3. Ring groove damaged, fouled or loose
1. Leaking head gasket
2. Cracked cylinder head
3. Valve burnt, not properly adjusted or sticking
4. Piston, piston rings or cylinder, worn or broken
5. Gas washdown on cylinders
Mischievious acts or vandalism
1. Theft or breakage of vital part (e.g. battery)
2. Siphoning of gas
3. Disruption of wiring
All other problems

response measures, the self-paced read-to-oneself administration was used for the two pruned tree groups.

Procedure. Before studying the fault tree, subjects were told:

> Every day, across the United States, millions of drivers perform the act of getting into an automobile, inserting a key in the ignition switch, and attempting to start the engine. Sometimes the engine fails to start, and the trip is delayed. We'd like you to think about the various problems that might be serious enough to cause a car to fail to start so that the driver's trip is delayed for at least 1 minute.
>
> The chart on page 128 is intended to help you think about this problem. It shows six [three] major deficiencies that cause a car's engine to fail to start. These major categories probably don't cover all possibilities, so we've included a seventh [fourth] category. All Other Problems.
>
> Please examine this diagram carefully and answer the following question:
>
> For every 100 times that a trip is delayed due to "starting failure," estimate, on the average, how many of those delays are caused by each of the seven [four] factors. Make your estimates on the blank lines next to the factors named below. Your estimates should sum to 100.

The numbers above in brackets were given to subjects who saw the pruned trees.

After assigning proportions, all subjects were asked:

> Please answer the following question: Over the next 1,000 times in the U.S. that drivers attempt to start their cars, how many times will the drivers experience delays in starting the engine serious enough to delay their departure by at least 1 minute?
>
> Answer: —— times out of 1,000

They were also asked a number of questions about the extent of their current and past experience with cars.

Subjects. One hundred forty-eight persons who responded to an advertisement in the University of Oregon student newspaper participated. They were assigned to an experimental group according to their preference for experiment date and hour.

Results

The difference in the mean proportion of problems assigned to each branch for the self-paced group and the read-aloud group was minimal. (Mean absolute difference across the seven branches was .018.) Results for these groups were combined.

If pruned tree subjects were sensitive to what had been omitted in the trees they studied, the proportion of problems that they attributed to "other" would equal the sum of the proportions of problems attributed to the pruned branches and to "other" by subjects who saw the full tree. If the availability hypothesis were correct, "other" would be assigned a lower proportion in the pruned tree; if the credibility hypothesis were correct, that proportion would be greater.

Table 8.1 presents the mean proportion of starting failures attributed to each branch and to "other." Pruned tree subjects clearly failed to appreciate what had been left out. For pruned tree Group 1, "other" should have increased by a factor of six (from .078 to .468) to reflect the proportion of failures due to starting, ignition, and mischief. Instead, "other" was only doubled, whereas the importance of the three systems that were mentioned was substantially increased. A similar picture emerged with the second pruned tree. Although these results clearly favor the availability over the credibility hypothesis, subjects did show some sensitivity to what had been omitted. The proportion of problems attributed to "other" was significantly greater for both pruned tree groups than for the unpruned tree group (For both t tests, $p < .001$). In addition, pruned tree Group 2, which saw a tree with more important branches deleted, assigned a higher proportion of problems to "other" than did pruned tree Group 1, $t(53) = 2.59$.

Only 1 subject of the 55 in the two pruned tree groups assigned to "other" a proportion of problems greater than or equal to the sum of the proportions of problems assigned to the missing branches (plus "other") by the unpruned tree group. There was no tendency for subjects who rated themselves higher in expertise (on a 5-point scale) to assign a higher proportion of problems to "other" in any of the groups.

TABLE 8.1 Results from Experiment 1: pruning the tree

		Mean proportion of starting failures attributed to							***Mdn** no starting failures in 1,000*
Group	*n*	*Battery*	*Starting system*	*Fuel system*	*Ignition system*	*Engine*	*Mischief*	*Other*	
Unpruned tree	93	.264	.195	.193	.144	.076	.051	.078	75
Pruned tree 1	29	.432	–	.309	–	.116	–	.140[a]	59
Pruned tree 2	26	–	.357	–	.343	–	.073	.227[b]	76

Note A dash indicates that the branch was deleted

[a] Should be .468.

[b] Should be .611.

There were no significant differences between the pruned and unpruned tree groups in the median likelihood that a randomly selected car would not start. This is a bit surprising, because the support for the availability explanation noted above would suggest that mentioning more specific sources of difficulty to unpruned tree subjects would increase the availability of starting failure. Medians were used because of the presence of some extremely high estimates (e.g., 600 starting failures in 1,000 tries), which would have unduly affected the means. The relative frequency of such unrealistic estimates (even among experienced drivers) suggests that this measure was not entirely successful, making results obtained with it somewhat dubious.

Discussion

Pruned tree subjects clearly did not appreciate how much had been left out, and, as a consequence, they overestimated the exhaustiveness of the branches they saw. One could speculate that this effect would be even more pronounced with fault trees concerning technical systems less familiar than the present.

Although the enormity of the present effect suggests that it may be quite robust, one might wonder whether it was due, at least in part, to unpruned tree subjects either (a) assuming that the experimenters were competent and thus had provided a reasonably complete tree or (b) not attending sufficiently to the "other" branch. Experiment 2 was conducted to explore these possibilities.

Experiment 2: focusing on all other problems

Method

Experiment 1 was repeated with several changes. To focus subjects' attention on what was left out, a paragraph was added to the instructions saying. "In particular, we'd like you to consider its completeness. That is, what proportion of the possible reasons for a car's not starting are left out, to be included in the category *all other problems?*" Their proportion estimation task was reduced to answering:

> For every 100 times that a trip is delayed due to "starting failure," estimate, on the average, how many of those delays are caused by factors not included in the chart; —— out of every 100 cases would fall in the category *all other problems*

Because of the difficulty that Experiment 1 subjects seemed to have had in assessing the likelihood of starting failure with a randomly selected car, the question was expanded by (a) asking them not only about a randomly selected car but also about their own car and (b) asking not only for an absolute judgment (how many times out of 1,000 attempts) but also for a relative judgment. (How many times more or less likely is a starting failure than a flat tire?).

Eighty-two subjects were recruited as in Experiment 1. The fault trees and instructions were read aloud.

Results

Focusing subjects' attention on the "other" branch of the unpruned tree had little effect on the unpruned tree group. The mean proportion of starting failures assigned to "other" was actually slightly smaller than the proportion in Experiment 1 (.067 vs. .078), $t(121) = .71$. It did, however, increase by about 50% the proportion of problems that subjects who saw the pruned trees attributed to "other." For pruned tree Group 1, the proportion attributed to "other" increased from .140 to .217, $t(49) = 1.88$; for pruned tree Group 2, from .227 to .346, $t(58) = 1.81$. In both cases, though, the proportion attributed to "other" was still much less than it should have been. These results are shown in Table 8.2.

Subjects who received the unpruned tree with its full panoply of problems thought that the absolute likelihood of a starting failure was greater than did subjects who saw the pruned trees (Mann-Whitney *U* test, *z*s = 2.42 and 1.77, for others' cars and own car, respectively). About 90% of the subjects who saw the unpruned tree thought that starting failure was more likely than a flat tire, compared with about 75% of subjects who saw the pruned tree ($p < .10$ for difference in proportions, both for others' car and for own car). Whereas the median pruned tree subject thought that a starting failure was 5 times as likely as a flat (for either kind of car), subjects who saw the unpruned tree thought that a starting failure was 20 times more likely for a randomly selected car and 60 times more likely for their own car (Mann-Whitney *U* test, *z*s = 2.26 and 4.26, respectively). Seeing the unpruned tree apparently made a starting failure seem more likely.

Discussion

Focusing subjects' attention on what is missing improved their awareness, but only partially. Because it did not lead to an exaggerated estimate of what was missing, such focusing would appear to be a valuable procedure whenever confronted with a fault tree or a similar representation. It remains to be seen how fault tree designers and evaluators can be brought to a fuller appreciation of how adequate a problem presentation is, particularly when the missing elements are ones of which they are totally or partially unaware. Perhaps a useful rule of thumb would be: The proportion of missing sources of trouble is proportional to the number of things I can think of that are missing multiplied by a measure of my general familiarity with the system (where greater familiarity is assigned a lower score). In other words, if I don't know much but still can detect something missing, then this fault tree representation is quite incomplete.

TABLE 8.2 Effects of focusing subjects' attention on "all other problems"

				Frequency of starting failures					
				Other's car			*Own car*		
		Proportion of problems attributed to "other"		*More likely than a flat*			*More likely than a flat*		
Group	***n***	*Observed*	*Should be*	*% Subjects*	***Mdn*** *response*	***Mdn*** *out of 1,000*	*% Subjects*	***Mdn*** *response*	***Mdn*** *out of 1,000*
			Experiment 1						
Unpruned tree	93	.078				75			
Pruned tree 1	29	.140	.468			59			
Pruned tree 2	26	.227	.611			76			
			Experiment 2						
Unpruned tree	30	.067		88.9	20	55	90.4	60	47
Pruned tree 1	22	.217	.468	75.0	3	25	83.3	5	35
Pruned tree 2	34	.346	.611	76.5	10	21	72.7	10	20

Note Subjects in Experiment 2 were instructed to focus on the other category. Data from Experiment 1 are included here for comparison.

Experiment 3: level of detail

Another discretionary decision faced by someone designing a fault tree for public display is how much detail to present. A designer must consider whether additional detail serves to inform the viewer or leads to confusion, feelings of incompetence, or undue apprehension over the large number of sources of trouble presented. Experiments 1 and 2 showed that omitted branches are essentially out of mind. Experiments 3 and 4 examined the effect on problem evaluation of exposing more or less detail for the displayed branches.

Method

The car-won't-start fault tree was presented to four different groups of subjects with three different levels of detail, Levels 1 and 3 described above (system names only and full detail) and another called *Level 2/plus*. This intermediate level included everything in Level 2 along with full (Level 3) detail for just one branch. Subjects receiving such a fault tree were told that a similarly detailed analysis could be performed for other branches. One group at Level 2/plus received full detail on the fuel system; another received full detail on the battery system.

One hundred ten new subjects were recruited as in Experiments 1 and 2. Questionnaires were distributed in a group setting and subjects read them at their own pace.

In this and all subsequent experiments the proportion estimation task of Experiment 1 (subjects gave proportions to all Level 1 categories) and the four final questions from Experiment 2 (number of failures out of 1,000 and comparison with a flat tire for both others' cars and own car) were used.

Results

The upper half of Table 8.3 presents results from Experiment 3, along with the results for the corresponding group given Level 2 in Experiment 1. The effects of the various manipulations of detail seem quite modest.

One might expect that when just one system receives more detail than the others, as in the two Level 2/plus groups, that system's proportion of problems would be elevated. However, neither battery nor fuel, when given a greater detail, received a proportion significantly (a = .05) larger than these systems received in the other groups.

Similarly, one might expect a lower proportion of problems attributed to mischievous acts at Level 3, when it is the only branch without additional detail (see Figure 8.1), than for the other detail levels. No such effect was found.

The battery and fuel systems, which had more detail at Level 3 than did the other systems, were judged more likely by that group than by other groups (for battery, t = 2.95 compared with Level 2 and t = .86 compared with Level 1; for fuel, t = 1.14 compared with Level 2 and t = 2.49 compared with Level 1; df = 70 in all tests).

TABLE 8.3 Effect of presenting varying degrees of detail

Level of detail	***n***	*Mean proportion of starting failures attributed to*							***Mdn** no. starting failures in 1,000 (others' cars)*
		Battery	*Starting*	*Fuel*	*Ignition*	*Engine*	*Mischief*	*Other*	
				Experiment 3—Self-paced					
System names alone (Level 1)	35	.303	.162	.149	.156	.062	.052	.116	112
Minimal detail (Level 2)[a]	35	.231	.205	.194	.151	.083	.046	.092	100
One branch with full detail									
Battery	19	.284[b]	.184	.182	.180	.066	.039	.065	100
Fuel	19	.224	.256	.174[b]	.140	.067	.051	.088	100
Full detail (Level 3)	37	.337	.126	.232	.125	.061	.057	.062	70
				Experiment 4—Read aloud					
System names alone (Level 1)	35	.330	.152	.136	.172	.067	.060	.083	27
Minimal detail (Level 2)[c]	58	.284	.189	.192	.139	.072	.054	.070	28
One branch with full detail									
Battery	22	.347[b]	.185	.175	.124	.052	.051	.066	50
Starting	26	.260	.241[b]	.120	.144	.076	.062	.095	50
Fuel	23	.304	.189	.184[b]	.127	.067	.054	.076	90
Full detail (Level 3)	67	.294	.198	.186	.151	.080	.038	.053	50

[a] Unpruned self-paced group from Experiment 1.
[b] Single branch presented in full detail.
[c] Unpruned read-aloud group from Experiment 1.

Finally, the proportion of problems attributed to all other problems decreased monotonically as level of detail increased, for linear trend. $F(1, 141) = 7.40$ $p < 01$.

Despite the expectation that the probability of a starting failure would increase as the level of detail increased, none of the four measures of this probability showed this effect. Indeed, the only such measure, shown in Table 8.3, starting failures out of 1,000 for others' cars, showed a nonsignificant trend in the opposite direction.

Experiment 4

Before concluding that detail has only a modest effect, we explored the possibility that mode of presentation mattered here (although it had not mattered in Experiment 1). We speculated that subjects that presented the rather voluminous fault trees for Levels 2/plus and 3 may have skimmed over them, not attending to the greater detail they offered. Experiment 4 replicated Experiment 3 with fault trees read aloud to 173 new subjects. A third Level 2/plus group was added, giving full detail on the starting system.

Results

The lower portion of Table 8.3 presents the results from Experiment 4, along with the appropriate Level 2 results from Experiment 1. With Levels 1 and 2, the differences between Experiments 3 and 4 were negligible. The mean absolute differences in proportions for the seven branches were .016 and .018, respectively. With Level 2/plus, there was somewhat more of a difference. As a highlighted branch, the battery system increased from .284 in Experiment 3 to .347 in Experiment 4, $t(39) = 1.33$; the fuel system, however, changed little (.174 vs .184), $t(40) = .34$.

Unlike Experiment 3, here the battery system received a somewhat (but not enormously) greater share of attributed problems when it was highlighted (Level 2/plus) than in all other conditions. So did the starting system. However, the fuel system again failed to show this effect.

Unlike Experiment 3, mischievous acts did show a significantly smaller proportion of problems at Level 3, where it is the only branch lacking full detail ($z = 2.95$ compared with Level 1; $z = 2.17$ compared with Level 2).

Again in contrast to Experiment 3, the battery and fuel systems, the ones with the greatest detail, were not more likely in Level 3, where this detail is most apparent, than in other levels.

The proportion of problems attributed to all other problems in Experiment 4 showed the same tendency to decrease as detail increased, as found in Experiment 3, except for one reversal: The weighted mean across the three Level 2/plus groups was .080, rather than a value between .053 and .070. Despite the one reversal, the linear trend in a one-way analysis of variance was significant, $F(1, 227) = 4.44$, $p < .05$.

In Experiment 4, the median starting failure rate for others' cars was similar for Levels 1 and 2 (Mann-Whitney *U* test, $z = .56$) but somewhat higher for Levels 2/plus and 3 than for Levels 1 and 2 (Mann-Whitney *U* test, $z = 1.83$). No such differences, even marginally significant, were observed for "one's own car" (not shown), suggesting that subjects' judgments regarding their own cars were less susceptible to this manipulation. Subjects in Experiment 4 also showed some effect due to detail on the likelihood of starting failure compared with that of a flat tire for both their own and others' cars. For Level 1, roughly 75% of subjects thought that starting failure was more likely; for Levels 2, 2/plus, and 3, approximately 90% thought so. The difference in proportions was significant ($z = 2.12$ for others' cars: $z = 2.47$ for own car).

Discussion

Compared with the enormous availability effect found in Experiments 1 and 2, the present results are modest. Only one of several hypothesized effects due to variation in detail was found in both Experiments 3 and 4: The proportion of problems attributed to all other problems decreased as detail increased. Further research with greater power is needed to explore the other effects that were found in one but not in both experiments. Even if real, these possible effects are probably small.

One must conclude that amounts of detail did not produce large changes in people's perceptions. The mere mention of a branch (Level 1) appeared to allow subjects to make a fairly accurate estimate of how troublesome that branch would look when fully detailed Level 2/plus subjects were somewhat (although not completely) successful in compensating for the detail missing on the more minimally presented branches. The relatively small effect of increased detail on estimated starting failure rate suggests that subjects given little detail realize their own ignorance.

Experiment 5: splitting and fusing branches

A third area in which the designer of a fault tree must make discretionary decisions is in organizing the various sources of trouble into branches. Although functional relationships introduce some constraints, the designer must make decisions like whether "disruptions of wiring" belongs under mischievous acts or "ignition system defective." A common dilemma is broad versus narrow categorization. Starting system defective and ignition system defective could be lumped together. Ignition system defective could be split into ignition system defective (Items 1 and 3 in Figure 8.1) and "distribution system defective" (Items 2 and 4). Experiment 5 examines the effect on fault tree evaluation of the way in which a fixed amount of information is organized, specifically, whether broader or narrower categorization is used.

Method

The effect of breadth of categorization on fault tree evaluation was studied by creating four new versions of the Level 2 fault tree (Figure 8.1). Two were created by splitting different existing branches into two branches containing (between them) the same information. In one, ignition system defective was split into ignition system defective with Items 1 and 3 (coil faulty and spark plugs defective) and distribution system defective with Items 2 and 4 (distributor faulty and defective wiring between components); in the second version, "fuel system defective" was split into fuel system defective (Items 1 and 2) and "[carburetion] defective" (Items 3 and 4). Thus, these two trees each had eight branches (counting other problems). Two additional trees were created by fusing two branches of the full tree. In one, starting system defective and ignition system defective were combined into one branch; in the second, fuel system defective and "other engine problems" were combined. Thus these fused versions each had six branches (counting other).

One hundred fifteen subjects were recruited as before. Each subject saw only one tree. The tasks for the subjects were the same as in Experiments 3 and 4. The trees were read aloud.

Results

Table 8.4 presents the proportion of starting failures attributed to each of the manipulated branches, along with comparable data from Experiment 1. In every case, a set of problems was perceived as more important when it was presented as two branches than when presented as one. The mean increase over the four cases was .066; in general, a set of problems was attributed about a third greater portion of the total number of starting failures when presented as two branches. The number of subjects given eight-branch trees who assigned the two new branches a combined proportion higher than the mean assigned by subjects in Experiment 1 to the comparable single branch was 20 out of 27 (sign test, $p = .010$) for the fuel-system-split group and 19 out of 26 ($p = .014$) for the ignition-system-split group. The number of subjects who assigned a proportion to the new fused branch lower than the corresponding mean sum in Experiment 1 was 24 out of 33 ($p = .007$) for the starting-ignition-combined group and 23 out of 29 ($p = .001$) for the fuel-engine-combined group.

There were no differences in either the relative (compared to a flat tire) or absolute likelihood of a starting failure for own or others' cars for subjects seeing the information presented in six- or eight-branch fault trees (not shown in table).

Discussion

Experiment 5 showed that the more pieces into which a system of failure pathways is organized, the more important that system seems. One possible

TABLE 8.4 Experiment 5: effects of splitting and fusing branches

	Separate		*Together*	
Problem	***M** proportion of starting failures*	*Sum*	***M** proportion of starting failures*	*Difference*[a]
	Splitting existing branches			
Fuel	.161	.260	.193	.067
Carburetion	.099			
$n = 27$				
Ignition	.082	.193	.144	.049
Distribution	.111			
$n = 26$				
	Fusing existing branches			
Starting	.195			
Ignition	.144	.339	.248	.091
$n = 33$				
Fuel	.193			
Other engine	.076	.269	.213	.056
$n = 29$				

Note In all places in which the number of subjects is not indicated, the results are those from the unpruned tree (Level 2) subjects in Experiment 1 (Table 8.1), for which $n = 93$.

[a] Separate sum minus together.

explanation is that people tend to assign some minimum probability to any category with which they are faced. A branch that is split in two receives two portions of this minimum probability, either because the smallest nonzero estimate allowed by the response mode was .01 or because of an assumption that "if they decided to include this branch, it must have some minimal import." An indirect way to evaluate this hypothesis is to estimate what proportion of problems is attributed to a category of "minimal import." For the 93 subjects in Experiment 1 who saw the unpruned Level 2 tree the mean proportion of problems assigned to the least important branch was .033; the mean lowest proportion excluding zero responses was .040. (Only 11% of subjects ever assigned a zero proportion of problems to any branch.) By this criterion, the increased importance (.066) garnered by splitting a branch was about twice that which might be attributed merely to increasing the number of categories.

Experiment 6: experts

Our discussion so far has been concerned with lay persons' evaluations of fault trees presented to them by technical experts. One might wonder, though, whether the effects we have found also affect the technical experts themselves. Fischhoff (1977) lists many incidents in which experts designing fault trees for important technological systems were apparently unaware of major omissions; they therefore greatly overestimated the exhaustiveness of their own analyses. Experiment 6 examined whether that which is out of sight is also out of mind for technical experts.

Experiment 6 replicated Experiment 1 using as subjects experienced mechanics in Eugene, Oregon. Thirty copies of Pruned Tree 1 (lacking the starting system, ignition system, and mischief branches) and 30 copies of the unpruned tree (Level 2) were distributed to experienced mechanics at six major garages.

In the accompanying letter, they were told:

> We are asking people like yourself who work with cars to give us their opinion about certain types of engine-starting problems. We hope you will agree to read the questionnaire and give us your opinions.
>
> Actually, we are not just interested in automobile engines. We're concerned with all kinds of complex mechanical systems ranging from automobile engines to nuclear reactors. In particular, we're interested in the ways these systems break down and the judgments about these breakdowns made by people who repair these systems. That's why we're asking for your opinions.
>
> If you are willing to participate in this study, please read the instructions on the next page; then fill out the two short questionnaires and mail them to us in the enclosed addressed and stamped envelope. We'll mail you a check for $3.
>
> There are no "right" answers to these questions. At least we don't know the answers. That's why we're interested in your opinions. Please work alone and don't refer to any books or discuss the questionnaire with anyone before you have completed it.

Otherwise, the questionnaire was like that in Experiment 1. On the final page, subjects were asked to indicate (a) the rate of starting failures per 1,000 attempts (for drivers in the United States); (b) the number of years they had made all or part of their living working with cars; and (c) how knowledgeable they were about automobile engine problems compared to other mechanics: below average, average, above average, or much above average.

Results

Twenty-nine of the 60 questionnaires were returned, 16 from the pruned tree group and 13 from the unpruned tree group. Two mechanics rated themselves as much above average in knowledge, 18 as above average, 8 as average, and 1 as below average. They had from 2 to 43 years of experience, with the mean for the unpruned tree group being 12.2 years, and for the pruned tree group, 19.8 years.

Table 8.5 presents the mean proportion of starting failures attributed to each branch by subjects in the present experiment as well as the unpruned tree results from Experiment 1 for the sake of comparison. The experts thought that battery and ignition were more serious problems than did the regular subjects, and that the starting and fuel system problems were less important. They more or less agreed about the completeness of the tree, assigning .060 to "other."

The combined proportion of problems attributed to the branches deleted from the pruned tree (plus "other") was .441. Pruned tree subjects assigned to "other" a mean proportion of only .215 (Mann-Whitney *U* test, $z = 2.87$). The respective median responses were .47 and .16.

Within the pruned tree group, the two subjects who rated themselves as much above average in knowledge assigned a mean to "other" of .075, the nine above-average mechanics assigned .278; and the five average ones, .158. There was a rank order correlation (tau) of .058 between number of years of experience and proportion assigned to "other." Thus, neither self-rated degree of knowledge nor actual experience had any systematic relation to ability to detect what was missing from the tree.

General discussion

The most dramatic result of these studies was subjects' inability to appreciate how much had been omitted from the pruned fault trees. Exaggeration of fault tree completeness was found with both "regular" subjects and experienced

TABLE 8.5 Experiment 6: judgment of experts

Group	*n*	*Mean proportion of starting failures attributed to*							***Mdn** no. starting failures in 1,000*
		Battery	*Starting system*	*Fuel system*	*Ignition system*	*Engine*	*Mischief*	*Other*	
Unpruned tree ordinary subjects	93	.264	.195	.193	.144	.076	.051	.078	75
Unpruned tree experts	13	.410	.108	.096	.248	.051	.025	.060	20
Pruned tree 1 experts	16	.483	–	.229	–	.073	–	.215[a]	100

[a] Should be .441.

mechanics. The fact that omission of major branches triggered only minimal awareness of the inadequacies of the pruned tree lent strong support to the availability hypothesis over the credibility hypothesis, as did the modest improvement when subjects' attention was focused on completeness.

How might things that are out of sight also be out of mind? One obvious reason is ignorance. There is no way to consider something that one has never heard of and that is not mentioned. In a discussion of the omissions that seem to plague technical experts performing formal risk assessments, Fischhoff (1977) suggested several other reasons: (a) failure to consider the imaginative ways in which human error can mess up a system (e.g., the Brown's Ferry fire in which the world's largest nuclear power plant almost melted down due to a technician checking for an air leak with a candle in direct violation of standard operating procedure); (b) insensitivity to the assumptions an analysis makes about constancies in the world in which the system is embedded (e.g., no major changes in government regulatory policy); (c) overconfidence in current scientific and technological knowledge (i.e., assuming that there are no new chemical, physical, biological, or psychological effects to be discovered); and (d) failure to see how the system functions as a whole (e.g., a system may fail because a backup component has been removed for routine maintenance).

Although similar problems seem likely to afflict the designers and viewers of fault trees, this list is probably incomplete. Expanding and validating such a list is important not only for our understanding of how people conceptualize complex fallible systems but also for helping them better describe and comprehend such systems. Focusing on "other" was somewhat, but only partially, successful in helping people appraise the completeness of a fault tree. Improved understanding of the reasons for omissions will help in designing better "debiasing" procedures and ways to approach an evaluation task that provide a more veridical perspective. When such "cognitive engineering" seems ineffective, people may need rules of thumb like the suggested rule for estimating how many sources of trouble one cannot think of on the basis of how many one can think of and one's familiarity with the problem area.

Because of the importance of intelligent public participation in debates about technology (Casper, 1976; Slovic, Fischhoff, and Lichtenstein, 1976), these are critical issues to which psychologists might address themselves. Such research does, however, raise serious ethical questions because of the possibility that the results will disclose ways in which public opinion can be manipulated. For example. Experiments 1 and 6 suggest that one can get the public to focus on those issues one thinks are important by never mentioning other issues. Even our "debiasing procedure" (focus on "other") in Experiment 2 was a form of manipulation, changing people's perceptions from what they would otherwise be. (Although we believe that this change is for the better, the point is moot as long as we do not know the proportion of problems in fact due to each branch). Perhaps the prime responsibility of the discoverer of such effects is to ensure that they receive the broadest possible dissemination, so that both potential manipulators and the potentially manipulated are aware of them. It may turn out that we psychologists are merely discovering "tricks" that manipulators have known about all along. Such research can also suggest aspects of presentation that may have relatively little effect on people's judgments (varying level of detail), attempted manipulations that can be readily overcome (providing more detail for one branch), and manipulations whose effects, although consistent, may be too small to have applied implications.

Aside from calling for obviously needed constructive replications (different trees, different subjects, different experts, etc.), these results suggest a variety of questions for future study: Would credibility be a stronger effect than availability if pruned fault trees were presented by distrusted technocrats rather than moderately trusted experimental psychologists? What happens when people construct their own fault trees? What happens when the detailed items from a fault tree are listed but not organized into categories? What happens with trees dealing with more technical areas or with nontechnical areas for which the assessors' feelings of competence vary? What happens when minor rather than major items are deleted? What if an omitted item is somehow brought to the subject's attention (as often happens in public debates)? What individual differences, if any, interact with these aspects of presentation in affecting evaluations? (As mentioned, the obvious covariate of years of experience was uncorrelated with the mechanics' awareness of how much had been deleted from the pruned tree.) What if detail is manipulated not by adding facts but by adding flourishes, by "fleshing out" possible sources of trouble with vividly detailed scenarios describing just how they might happen? Why were subjects' perceptions of the overall starting failure rate relatively impervious to our manipulations? Was it merely a problem of measurement?

These studies have supported the notion that availability affects frequency estimates and have helped clarify how availability mechanisms work. Failure to mention a possibility had a major effect, increasing the detail of what was mentioned had a small effect, and the packaging of what is mentioned also made a difference. Experiment 2 showed, as did Beyth-Marom and Fischhoff (1977), that forcing people to work their memories (or imaginations) harder can improve their likelihood judgments somewhat. Clearly, much more work is needed on the retrieval, perception, and representation of frequency information.

Acknowledgements

We would like to thank Barbara Combs and Gerry Hanson for their help in conducting these experiments; David Collins for technical advice on designing fault trees; and Nancy Collins, Phyllis Kemper, and Peggy Roecker for drafting the many versions of the fault tree that we used.

This research was supported by the Advanced Research Projects Agency of the Department of Defense and was monitored by the Office of Naval Research under Contract N00014 76-C-0074 to Oregon Research Institute (ARPA Order No. 2449) from Decisions and Designs, Inc., and to Perceptronics (ARPA Order No. 3052) from Decisions and Designs. In (Subcontract 76-030-0714).

Note

1 To our chagrin, the first starting failure encountered by anyone connected with this study was not included in our fault tree: an ignition key not turning because the steering wheel lock had caught the ignition switch.

Reference note

1 Bryan, W. B. 1974. *Testimony before the Subcommittee on State Energy Policy* Committee on Planning. Land Use, and Energy, California State Assembly. February.

References

Atomic Energy Commission Reactor safety study 1975. *An assessment of accident risks in U.S. commercial power plants* (*WASH-1400*) Washington, D.C.: Author.

Barrager, S. M. Judd, B. R. and North, D. W. 1976. *The economic and social costs of coal and nuclear electric generation*. Washington, D.C.: National Science Foundation.

Beyth-Marom, R, and Fischhoff, B. 1977. Direct measures of availability and judgments of category frequency. *Bulletin of the Psychonomic Society*, *9*, 236–238.

Birnbaum, M. H., Wong, R. and Wong, L. K. 1976. Combining information from sources that vary in credibility *Memory and Cognition*, *4*, 330–336.

Casper, B. M. 1976. Technology policy and democracy. *Science*, *194*, 29–35.

Fischhoff, B. 1977. Cost benefit analysis and the art of motorcycle maintenance *Policy Sciences*, *8*, 177–202.

Green, A. E. and Bourne, A. J. 1972. *Reliability technology*. New York: Wiley-Interscience.

Howell, W. C. 1973. Representation of frequency in memory. *Psychological Bulletin*, *80*, 44–53.

Kendall, H. W. 1975. *Nuclear power risks: A review of report of APS society's study group on light water reactor safety*. Cambridge, Mass.: Union of Concerned Scientists.

Newell, A., and Simon, H. A. 1972. *Human problem solving*. Englewood Cliffs, N.J.: Prentice-Hall.

Settle, R. B, and Golden, L. L. 1974. Attribution theory and advertiser credibility. *Journal of Marketing Research*, *11*, 181–185.

Slovic, P., Fischhoff, B. and Lichtenstein, S. 1976. Cognitive processes and societal risk taking. In J. S. Carroll and J. W. Payne (Eds.), *Cognition and social behavior* Potomac, Md.: Erlbaum.

Tversky, A., and Kahneman, D 1973. Availability: A heuristic for judging frequency and probability. *Cognitive Psychology*, *4*, 207–232.

9

ASSESSING UNCERTAINTY IN PHYSICAL CONSTANTS

Max Henrion

CARNEGIE MELLON UNIVERSITY, DEPARTMENT OF ENGINEERING AND PUBLIC POLICY, PITTSBURGH, PENNSYLVANIA 15213

Baruch Fischhoff

DECISION RESEARCH, A BRANCH OF PERCEPTRONICS, 1201 OAK, EUGENE, OREGON 97401

Abstract

Assessing the uncertainty due to possible systematic errors in a physical measurement unavoidably involves an element of subjective judgment. Examination of historical measurements and recommended values for the fundamental physical constants shows that the reported uncertainties have a consistent bias towards underestimating the actual errors. These findings are comparable to findings of persistent overconfidence in psychological research on the assessment of subjective probability distributions. Awareness of these biases could help in interpreting the precision of measurements, as well as provide a basis for improving the assessment of uncertainty in measurements.

I. Introduction

Accurate estimates of the fundamental constants of physics, such as the velocity of light or the rest mass of the electron, are central to the enterprise of science.[1] Like any measurements, they are subject to uncertainties from a variety of sources. Reliable assessments of this uncertainty are needed (a) to compare the precision of different measurements of the same quantity, (b) to assess the accuracy of other quantities derived from them, and, most crucially, (c) to evaluate the consistency of physical theory with the current best measurements. Thus, as Eisenhart has pointed out, "A reported value whose accuracy is entirely unknown is worthless."[2]

It is not unusual to encounter individual examples of errors in measurements of physical quantities that turn out to be disconcertingly large relative to the estimated uncertainty. One well-known case was in R. A. Millikan's oil-drop experiment in 1912 to determine *e*, whose result turned out 15 years later to be off by 0.6% or three standard deviations due to reliance on a faulty value for the viscosity of air.[3] A more recent example concerns measurements of $| \nu+- |$, the parameter that measures the degree of violation of CP (charge-conjugation–parity) invariance. The six measurements prior to 1973 agreed reasonably, but more accurate measurements since then differ consistently by about seven standard deviations from the pre-1973 mean, a discrepancy that remains unexplained in terms of experimental procedure.[4] Such extreme cases may be exceptions, but they raise the more general question of how well on the average reported uncertainties reflect actual errors, an issue on which there has been little systematic study.[5] Here we will present evidence from historical measurements of a range of physical constants to illustrate the scope of the problem of underestimation of uncertainty. A wider awareness of such results may help in interpreting reported uncertainties, and may have some important educational implications.

A *comprehensive* assessment of uncertainty cannot rest solely on statistical analysis. Unavoidably, it involves a considerable element of subjective judgment. Therefore, we shall first review some recent findings of cognitive psychology from laboratory studies of human judgment under uncertainty. After examining evidence from measurements of physical constants, we will discuss possible explanations for these problems in the light of the psychological literature, and explore the prospects for alleviating them.

II. The psychology of judgment under uncertainty

The premise of laboratory studies of human judgment is that all judgments are governed by a set of core cognitive processes.[6] If those can be understood in experimental settings, then reasonable speculations can be made about human performance in the real world.[7] This literature has revealed both strengths and weaknesses. Where people have the explicit training or where there has been the opportunity to receive clear, prompt feedback, people can assess many aspects of uncertain processes. For example, weather forecasters in the U.S. provide assessments of the probability of precipitation that are probabilistically *well calibrated*: It rains on about 70% of the occasions on which they forecast a 70% probability of rain.[8] They have developed this ability through years of hands-on experience, with guidance from computer models, with ample feedback, and within an institution that rewards them for candor (rather than, say, for exuding confidence or avoiding firm commitments). In less favorable circumstances, however, people often lack an intuitive feel for probabilistic processes, relying instead on mental "heuristics" (deterministic rules of thumb) to guide their judgments. Although often useful, these rules can lead to predictable biases.

In these studies, the intuitive asessment of uncertainty has proven to be especially problematic.[9] People seem insufficiently sensitive to how much they know, so that changes in knowledge are accompanied by inappropriate changes in confidence. The most common problem is *overconfidence*. A common way to assess the precision of an uncertain quantity is by a subjective confidence interval, indicating a range within which the assessor believes the true value has, say, a 98% chance of falling. The *probabilistic calibration* of a set of such judgments for different quantities may be measured by comparing the assessed probability for the interval with the fraction of times the true value lies within it. Cases in which the true value (once known) turns out to fall outside the assessed confidence interval, may be termed "surprises." The *surprise index* is the percent of 98% confidence intervals for which the true value is a "surprise." Having significantly more than 2% surprises indicates overconfidence, in the sense of underestimating the range of possibilities. Conversely, too few surprises means underconfidence. Another measure of calibration is the *interquartile index*, the percent of judgments for which the true value lies between the assessed 25th and 75th percentiles. An interquartile index that is much less than 50% also indicates overconfidence. The overwhelming result in laboratory studies with nonspecialist assessors has been intervals that are too tight, reflecting overconfidence: Typically the surprise index is 20%–40% instead of the ideal 2%, and the interquartile index is 30%–40% instead of 50%.[9]

The important role of probabilistic judgment is coming to be recognized in several areas involving risk analysis, including medicine, toxicology, and nuclear safety.[10] The few studies of judgments in such real-world contexts outside the psychologist's laboratory suggest that the laboratory findings of overconfidence may generalize to situations of practical importance.[11] However, such evaluations have been rare. An evaluation of the uncertainty estimates for physical constants should therefore be of interest to the study of probabilistic judgment in general, as well as to the practice of physics.

III. Uncertainty in physical measurement

It is universally accepted that every scientific measurement should be accompanied by a statement of its uncertainty, nowadays usually expressed probabilistically, as a standard error or confidence interval. However, there is less agreement about precisely what these statements are intended to represent. Typically, the terms "error" and "uncertainty" are used almost interchangeably.[12] We find it convenient to distinguish them thus: "Error" is the actual difference between a measurement and the value of the quantity it is intended to measure, and is generally unknown at the time of measurement. "Uncertainty" is a scientist's assessment of the probable magnitude of that error.

Another, conventional distinction is between *random* error due to uncontrolled variability among observations, and *systematic* error, equal to the difference between the value to which the observed mean converges and the true value. Systematic errors can arise from undetected biases in the experimental apparatus, approximations in computational procedures, errors in auxiliary variables, and deficiencies in theoretical assumptions.

Random uncertainty, being simply an estimate of the random error, may be obtained straightforwardly by statistical analysis of the unexplained variability in the measurements. Systematic uncertainty is the experimenter's estimate of the systematic error and requires the exercise of judgment. Because the random uncertainty can be reduced simply by taking more observations, the overall uncertainty often comes to be dominated by the systematic uncertainty,[13] at least in those experiments where the marginal cost of observations is relatively low.

Some have suggested that random and systematic uncertainty are qualitatively different, and so should be reported separately,[14] a suggestion followed in some recent articles. However, Mueller argues that the categorization of uncertainties is context dependent, varying from one experiment to another, and that it is appropriate to combine them forming a single expression of uncertainty, which is far more useful for evaluating each measurement.[12] Acknowledging that the terms random and systematic uncertainty are often used in ambiguous and inconsistent ways the International Committee on Weights and Measures have recommended replacing them by Category A uncertainty, evaluated by statistical means, and Category B uncertainty, evaluated by other means, including "subjective appreciation."[15]

Whatever one calls them, it is the latter that cause the most problems. Known systematic errors can usually be compensated for by experimental control or computational corrections, and much of the skill of a precision metrologist lies in such issues of experimental design. For example, Youden has urged that key features of the experimental design[16] be varied systematically, thus converting some systematic errors to random errors, which can be estimated statistically from variations over observations.

Unfortunately, such treatment is not possible for all sources of error. Innovations in experimental technique are particularly likely to produce surprising new errors. In such cases, there is no substitute for judgment in assessing the magnitude of unresolved systematic errors. Some recent discussions of error estimation recognize this, e.g., warning that estimates of systematic error "are somewhat subjective and are usually obtained from what can only be called educated guesses."[17] Where reported uncertainties are intended to be comprehensive and are expressed as standard errors, confidence intervals, or other probabilistic forms, we conclude they are equivalent to assessments of the parameters of subjective probability distributions. Hence they may appropriately be analyzed for calibration and over- or underconfidence.

IV. Measures of calibration

One can obtain some empirical insights into the calibration of a set of assessments by comparing reported uncertainties with the variability among the measurements. Suppose each experiment i, for $i = 1,2, \dots N$, reports an estimate x_i with standard error σ_i, and $\bar{x}$ is the group mean, weighted inversely by the variances. The sum of the squares of the normalized residuals, $h_i = (x_i - \bar{x})/\sigma_i$, should be distributed as the chi-squared statistic with $(N-1)$ degrees of freedom, assuming the errors are independent and normally distributed with the reported standard deviations. This statistic can then be used to test the appropriateness of these assumptions.

A related measure, the Birge ratio,[18] R_B, assesses the compatibility of a set of measurements by comparing the variability among experiments to the reported uncertainties. It may be defined as the standard deviation of the normalized residuals:

$$R_B^2 = \sum_i h_i^2 / (N-1). \tag{1}$$

Alternatively, the Birge ratio may be seen as a measure of the appropriateness of the reported uncertainties. If the uncertainty assessments are independent and perfectly calibrated, the expectation of R_B is one. If R_B is much greater than one, then one or more of the experiments has underestimated its uncertainty and may contain unrecognized systematic errors. Such insensitivity to systematic error is a sign of overconfidence. If R_B is much less than one, then either the uncertainties have, in the aggregate, been overestimated or the errors are correlated.

V. Uncertainty in *c*

The velocity of light, *c*, is perhaps the most measured fundamental physical constant,[1] starting with Galileo's unsuccessful attempts, using assistants flashing lanterns on neighboring hills. Figure 9.1 displays the results of all measurements between 1875 and 1958 with reported uncertainty from several major surveys.[13,17–20] The vertical bars represent the standard error according to the original experiment or earliest reviewer.[21] The horizontal dashed line is the 1984 value of 299 792.458 km/s.[22]

The Birge ratio for the entire set of 27 measurements with errors relative to the 1984 value is 1.42. The probability of finding such large variability by chance is less than 0.005, assuming that the errors were normally distributed with the reported standard deviations. If the standard deviations of the discrepant studies fully express the respective investigators' uncertainties regarding their estimates, then, on average, those uncertainties must be significantly underestimated.[23]

Over time, both the reported uncertainty and the actual error of these estimates have been reduced enormously. There was, however, no significant corresponding improvement in the Birge ratio, which was 1.47 for measurements up to 1941 and 1.32 since 1947 (top of Table 9.1). For both periods, the variability was significantly greater than one would expect, indicating unduly tight uncertainty estimates.

These changes in accuracy over time were accompanied by changes in the direction of error. From 1976 to 1902, the measurements overestimated the value by about 70 km/s on the average. From 1905 to 1950, they

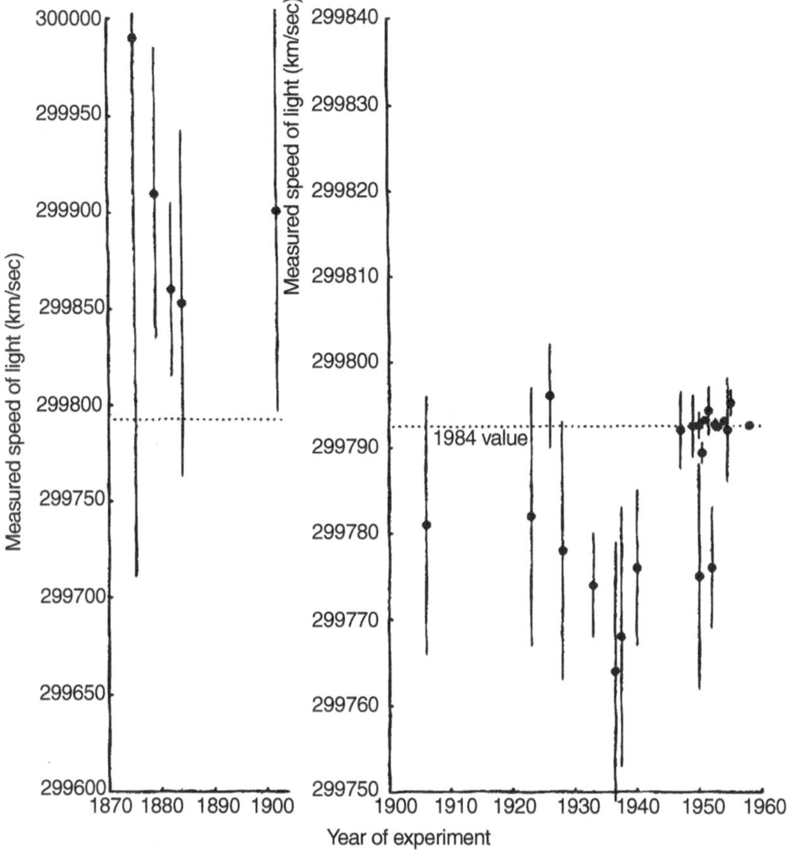

FIGURE 9.1 Measurements of the velocity of light, 1875–1958. Results are as first reported, with correction from air to vacuum where needed. The uncertainties are also as originally reported, where available, or as estimated by the earliest reviewers. Error bars show standard error (s.e. = 1.48 × probable error).

underestimated it by about 15 km/s on the average. This change in the mean caused deBray to suggest that the speed of light was not constant but decreasing by about 4 km/s/year. In 1934, Edmondson proposed that it might be varying sinusoidally with a period of 40 years. Discounting these claims, Birge[19] argued *c* has a constant value, which the overconfident uncertainty estimates too frequently failed to include. In 1941, he adjusted a set of recent estimates with untreated systematic error, producing corrected values with a Birge ratio of 0.544. Birge concluded, "Thus, after a long and, at times, hectic history, the value for *c* has at last settled down into fairly satisfactory 'steady' state."[19]

Unfortunately, as illustrated in Figure 9.2, his confidence turned out to be premature. Just nine years later, the recommended value had shifted by 2.4 of

TABLE 9.1 Calibration statistics for measurements of physical constants

		Statistic					
Quantity	*Date*	*N*[a]	*RB*[b]	*Pr*[c]	*IQ*[d]	*SI*[e]	*Refs.*[f]
c	1875–1941	13	1.47	0.01	23%	8%	13, 17–20
c	1947–1958	14	1.32	0.05	57%	14%	
c	1875–1958	27	1.42	0.005	41%	11%	
G	1798–1983	14	1.38	0.025	41%	29%	24
m'_r/mn	1949–1967	7	1.44	0.05	14%	14%	17
a^{-1} High accuracy		24	2.95	(0)	21%	38%	17
a^{-1} Low accuracy		14	1.26	0.10	64%	7%	17
$\Omega_{ABS}/\Omega_{NBS}$	1938–1968	7	0.40	0.995	100%	0%	17
Particle lives		92	1.26			9%	5, 25
Particle masses		214	1.24			6%	
Particle total		306	1.24			7%	
$S_{10}(h/1.1)$			1.24		44%	6%	26
Recommended							3, 17
values	1928–1973	40	7.42	(0)	22%	57%	19, 28, 29
Well-calibrated normal distribution			1.00		50%	2%	

[a] *N* Number of measurements analyzed.
[b] *RB* Birge ratio (standard deviation of normalized residuals).
[c] Pr Probability of getting Birge ratio that large by chance if the normalized residuals have independent unit normal distributions.
[d] IQ Interquartile index: Percent which fall between assessed 25%ile and 75%ile, or within O.675*s*.
[e] SI Surprise index: Percent that fall outside assessed 98% confidence interval, or outside 2.33s.
[f] Refs. Sources of data.

his 1941 standard deviations.[2] This 1950 value, too, was soon supplanted, by a value different by over 2 of its standard deviations. Once again, shifting estimates prompted the suggestion that *c* might be changing, this time increasing.[20] One can only speculate whether better calibrated assessments of uncertainty (showing that the varied estimates were compatible with one another) would have forestalled such paradigm-shattering proposals. Table 9.1 summarizes these data in terms of various calibration statistics, including the Birge ratio, interquartile index, and surprise index. All indicate overconfidence. For measurements of *c*, there are too few values between the 0.25 and 0.75 quartiles (41% instead of 50%), and too many surprises (11% instead of 2%).

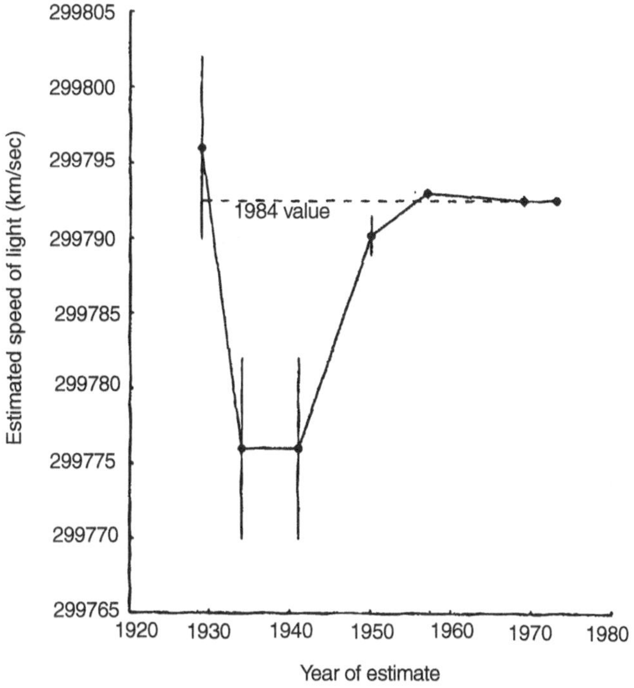

FIGURE 9.2 Recommended values for the velocity of light, 1929–1973.

VI. Uncertainty in other constants and recommended values

The remainder of Table 9.1 summarizes similar analyses for other physical constants, including the gravitational constant G, the magnetic moment of the proton μ'_p/μ'_n two sets of measures of the inverse fine structure constant, α^{-1}, and the ratio of the absolute ohm to the ohm maintained by the National Bureau of Standards, $\Omega_{ABS}/\Omega_{NBS}$. Table 9.1 reveals a similar pattern of overconfidence in all cases but the last. This shows a Birge ratio of significantly less than 1.0 and an interquartile ratio of 100%, results we shall return to later.

The "particle lives" and "particles masses" entries in Table 9.1 represent 92 measurements of Kaon and hyperon lifetimes and 214 measurements of meson resonance masses taken from the Particle Data Group's biennial *Review of Particle Properties*.[5, 25] Their thorough scrutiny of all published experimental measurements results in the rejection of about 40% on grounds of suspect assumptions, poor quality work, unreported errors, or gross inconsistency with other results. Nonetheless, the Birge ratio of 1.24 and surprise index of 6% of the 306 combined results indicate a significant degree of overconfidence in the remaining uncertainty assessments.

Periodically, reviewers propose "recommended values" for physical constants after careful consideration of all the most precise measurements. The diligence of these analyses and the incumbent obligation to confront the inconsistencies among studies should make them particularly sensitive to uncertainty. Figure 9.2 showed the residual overconfidence in such reviews for the velocity of light. Figure 9.3 shows comparable results for five other fundamental constants derived from a series of six reviews between 1952 and 1973.[3,17,28] In most cases, the best-guess value at each revision was well outside the range of possibilities defined by the uncertainty estimates for the preceding evaluation period. For example, the 1963 to 1969 shift was three to five 1963 standard deviations for all five of these constants. Table 9.1 (second to last line) gives calibration indices for these same five constants for eight reviews from 1929 to 1969,[3,17,19,28,29] with residuals calculated relative to 1973 estimates. The surprise index of 57% indicates that current estimates would have come as a surprise to earlier reviewers in more than half the cases. The 1969 review acknowledges these problems, and provides graphs similar to Figure 9.3 to caution readers to be skeptical.[17]

VII. Sources of overconfidence

In several sets of carefully analyzed measurements of physical constants, we have found consistent replication of a robust finding of laboratory studies of human judgment: Reported uncertainties are too small. How could this apparent overconfidence arise? Experimental studies of human judgment have shown that such biases can arise quite unintentionally from cognitive strategies employed in processing uncertain information. However, there are two possible ways in which such effects might be caused by deliberate decisions by the scientists:

One concerns the procedures chosen to assess the uncertainty. The recommended practice in physics is to consider all possible sources of systematic uncertainty when reporting results. However, without specific guidelines regarding what to consider and explicit recognition of the subjective elements in uncertainty assessments, one cannot be sure how comprehensively individual scientists have examined the uncertainty surrounding their own experiments. Conceivably, some of the apparent overconfidence reflects a deliberate decision to ignore the harder-to-assess sources of uncertainty.

A second possible source of bias is that, unlike laboratory experiments on judgment, which can take great care to ensure that subjects are motivated to express their uncertainty candidly, real-world settings create other pressures. Taylor *et al.* suggest that variations in attitude lead some investigators to

> ... cautiously assign unreasonably large errors so that a later measurement will not prove their work to have been "incorrect." Others tend to underestimate the sources of systematic error in their experiments, perhaps

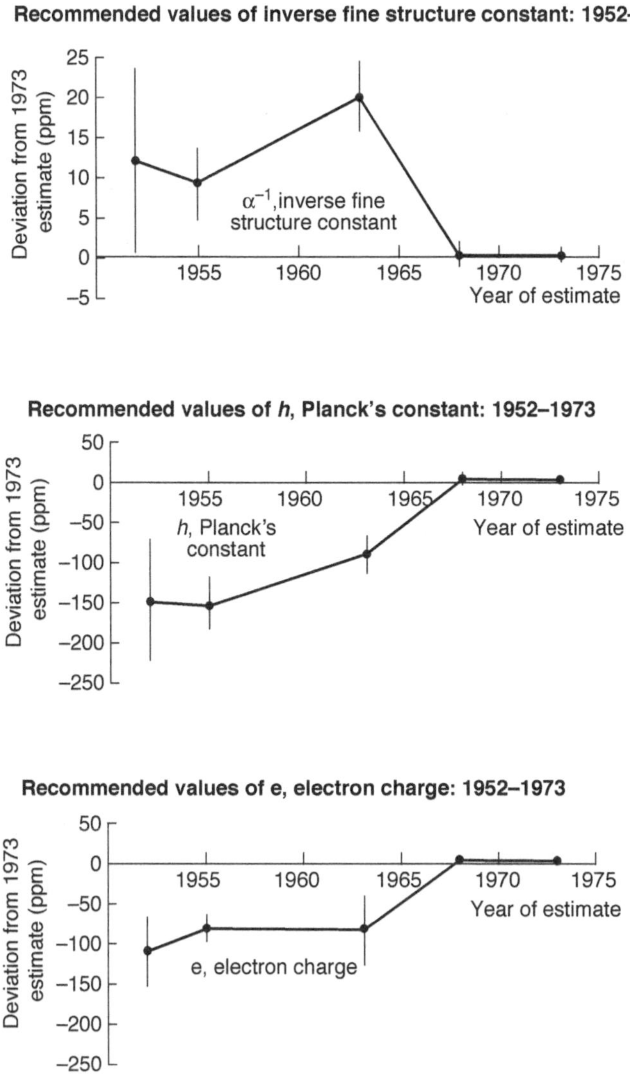

FIGURE 9.3 Recommended values for fundamental constants, 1952–1973.

Continued on next page

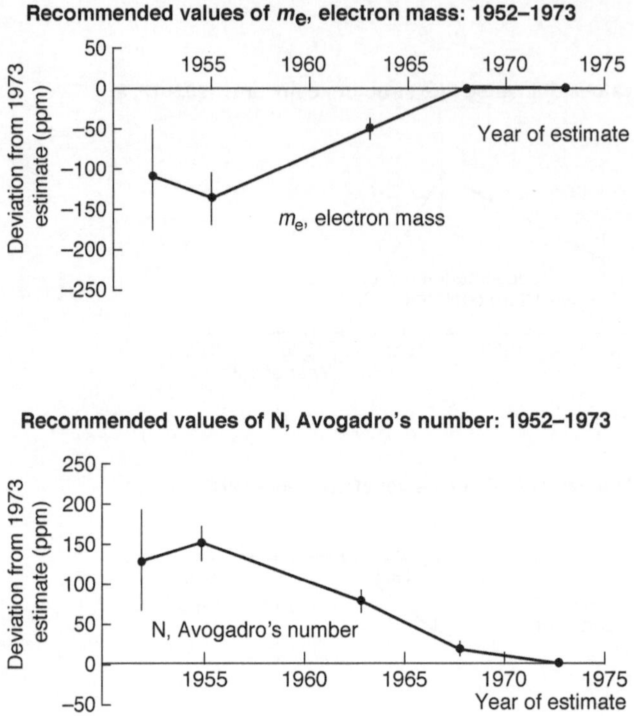

FIGURE 9.3 Recommended values for fundamental constants, 1952–1973.

> because of an unconscious (or conscious) desire "to have done the best experiments." Such variation in attitude, although out of keeping with scientific objectivity, is nevertheless unavoidable so long as scientists are also human beings.[17]

In principle, none of these problems should affect the compilations of recommended values. These analyses are intended to be comprehensive, to consider reporting practices in the field, and to capture the state of knowledge (not just the precision of particular studies).

The factors that seem to have encouraged overconfidence in laboratory studies of judgment also seem to be likely candidates for having encouraged overconfidence in the estimation of physical constants. One such factor is the difficulty of thinking of reasons why one's best guess might be wrong.[31] Supporting reasons typically come to mind more readily than contradicting ones. If ease of recollection is taken as an indicator of frequency,[32] the preponderance of reasons supporting the best guess will tend to be overestimated.

A second factor contributing to overconfidence is the unequal treatment of such confirming and disconfirming evidence as is discovered. When discrepant results are encountered, it could mean that either the new data or the old database contain undetected systematic errors. Unfortunately, people have a considerable ability to "explain away" events that are inconsistent with their prior beliefs.[33] The data elimination and adjustment procedures that precede estimation of recommended values are natural places for disproportionate skepticism regarding unexpected results to emerge. One documented example of "trimming," i.e., excessive zeal in eliminating outliers, emerged in a recent examination of Millikan's laboratory notebooks for his oil-drop experiment. Of the 107 oil drops he observed, he excluded from publication 49 that seemed discrepant, despite his claim that he reported all his observations.[34] This had little effect on the best estimate of *e*, but increased the apparent precision of the experiment.

Having a pre-existing recommended value may particularly encourage investigators to discard or adjust unexpected results, and so induce correlated errors in apparently independent experiments. The result would be initially small Birge ratios, unduly tight confidence intervals, and the exclusion of discrepant data that later prove to be more accurate than included ones. This can explain what Franklin has termed the "bandwagon effect,"[4] e.g., the tendency for the measurements of the speed of light (in Figure 9.1) to cluster around particular values at different periods. The large Birge ratio (2.95) for high accuracy measurements of α^{-1} (Table 9.1) may be due to the fact that those measurements were actually derived from experiments designed to measure other quantities, so that discrepancies in the implied value of α^{-1} would have been less obvious to the original investigators. On the other hand, measurements of the ratio of the absolute ohm to the as-maintained ohm ($\Omega_{ABS}/\Omega_{NBS}$) have a natural anchor at unity (i.e., assuming they are equal), which could explain the unusual degree of agreement, evidenced by the Birge ratio of 0.40.

VIII. Improving uncertainty assessment

In order to reduce the need for judgment in selecting data for producing recommended values, several attempts have been made to develop mathematical procedures for selecting and averaging measurements. Statisticians have long sought more robust estimators, such as trimmed means, that use all useful information without being unduly sensitive to outliers. However, in tests with real data, more sophisticated estimators have provided only marginal improvements over simple averages.[35] Roos *et al.*[5] describe a maximum likelihood technique based on their initial observation that the adjusted residual errors for particle properties had a broadtailed Student distribution. This was used in the *Review of Particle Properties* from 1976 to 1980, but was dropped after subsequent distributions of residuals differed in form.[27] There have also been attempts to develop algorithms for estimating the fundamental physical constants that deal better with apparently discrepant data. However, after tests with historical data, Taylor concluded that thoughtful and conscientious judgment about which data to exclude is still much more important than the choice of algorithm.[30]

It is no easy matter to eliminate judgmental biases.[9] A categorical recommendation to treat all results equally would, on the average, give undue credence to inaccurate observations and bad research. Nor is simple exhortation to "think harder" likely to prove any more useful here than it has in laboratory experiments on judgment. What are needed are ways to think more effectively. If the judgmental aspect of assessing uncertainty in physical measurements is explicitly acknowledged, then several techniques based on insights from psychological research[36] may be helpful.

One approach is to encourage a broader search for relevant considerations, both supportive and disconfirming, and an unbiased evaluation of those considerations. For example, referees might be asked to scrutinize the account of an experiment "blind," before knowing its outcome and whether it affirmed prior expectations.[37] To overcome the problem of anchoring on an initial best guess, investigators can be pressed to search for reasons why they might be wrong.[31] One device to encourage that search is to focus attention on extreme possibilities, for example, asking "imagine that, ten years hence, today's best estimate proves to be off by four standard deviations; how could you explain it?" Additional prompts may come from reading "horror stories," case studies describing major unsuspected errors in past experiments.

A second approach is to decompose the holistic judgment of overall uncertainty into its component sources of potential systematic errors, each to be estimated separately with the best available elicitation procedure.[36] This can be aided by checklists covering each component of the measurement apparatus, each auxiliary quantity, and each theoretical assumption and approximation employed in the calculations.

Wherever analysis of reported uncertainties reveals a systematic bias, users of the measurements may use this information to *readjust* the original reports so as to improve calibration. Along these lines, the Particle Data Group expands the standard deviation whenever the Birge ratio is greater than unity among measurements of the same quantity.[25] Their procedure usually produces stable recommended values for particle properties; however, it somewhat exaggerates the uncertainty by ignoring the possibility that some fluctuations of R_B above one are purely by chance. Roos *et al.*[26] describe another procedure for readjusting the standard errors, based on the empirical distribution of the residuals that they had observed for the particle properties.

The basis for such readjustments must be secondary analyses, such as those given here, showing probabilistic miscalibration in the estimates of uncertainty for previous similar measurements. In using such data, it is important to bear in mind that miscalibration reflects, not properties of the physical quantities themselves, but the nature of the procedures and judgmental processes used in selecting data and assessing uncertainty. Thus they are likely to vary with the training of the experimenter, the familiarity of the experimental techniques, and the maturity of the field, among other things. Compiling such data for wider classes of measurements can permit analysis of the importance of these factors. Even if uncertainty estimates are not explicitly readjusted, empirical calibration information could be helpful for anyone who uses measurements and recommended values for physical constants in their work. Even a rough estimate of overconfidence could help one interpret the significance of apparent discrepancies between measurements.

Another approach to improving uncertainty assessment is education designed to get better estimates in the first place. For example, in teaching experimental methods, greater exposure to the kind of results presented here should be helpful, together with attention to the role of judgment in the assessment process. Although one might hope that instructions in the processes and pitfalls of judgment would, by themselves, improve performance, the evidence to date suggests that people have difficulty in integrating an understanding of general principles with their own cognitive processes.[38] What *has* proved more effective in other domains is task-specific training with personal feedback.[9] For example, in laboratory classes, where students are required to measure the same quantities, it should be instructive to compile distributions of normalized residuals and measures of calibration for the class results, and to discuss together the reasons both for individual errors and for systematic miscalibration in reported uncertainties.

IX. Conclusions

The underestimation of uncertainty in measurements of physical constants and compilations of recommended values seems to be pervasive. This evidence extends previous findings of overconfidence in laboratory studies of human judgment to a task domain of great practical importance. If reported uncertainties do not reflect the magnitude of actual errors, whether due to incomplete analysis or to judgmental biases, the usefulness of those measurements is significantly diminished. Measurements are then hard to compare, and are unlikely to produce "enduring values,"[16] as we have illustrated in the repeated contradiction of accepted values by subseqent measurements. Recognizing that subjective judgment is an essential element in the assessment of systematic uncertainty enables us to use findings from cognitive research to help understand how these biases arise, and to suggest approaches for dealing with these problems more effectively.

Acknowledgments

This research was supported in part by the National Science Foundation under Grant No. SES-8213452 to Perceptronics, Inc. Any opinions or findings expressed are those of the authors and do not necessarily reflect the views of the National Science Foundation. We are grateful to George Sparling for helpful suggestions.

Notes

1 F. M. Pipkin and R. C. Ritter 1983. *Science* 219, 913.
2 C. Eisenhardt 1968. *Science* 160, 1201.
3 E. R. Cohen, K. M. Crowe, and J. W. M. Dumond 1957. *Fundamental Constants of Physics* (Interscience, New York).
4 A. D. Franklin 1984, *Am. J. Phys.* 52, 786.
5 M. Roos, M. Hietanen, and J. Luoma 1975. *Phys. Fenn.* 10, 20.
6 A. Tversky and D. Kahneman 1981. *Science* 211, 453; D. Kahneman, P. Slovic, and A. Tversky (eds) 1982. *Judgment Under Uncertainty: Heuristics and Biases* (Cambridge, New York).
7 B. Fischhoff and C. Whipple 1981. *Environmental Professional* 3, 277; M. G. Morgan, M. Henrion, and S. C. Morris 1981. *BNL No. 51358*, (Brookhaven National Laboratory).
8 A. H. Murphy and R. W. Katz (eds) 1983. *Probability, Statistics and Decision-Making in the Atmospheric Sciences* (Westview, Boulder, CO.).
9 S. Lichtenstein, B. Fischhoff, and L. D. Phillips 1982. in *Judgment Under Uncertainty: Heuristic and Biases* (Cambridge, New York); B. Fischhoff, *ibid.*
10 See, e.g., H. Raiffa 1968. *Decision Analysis* (Addison-Wesley, Reading, MA); A. E. Green and A. J. Bourne 1972. *Reliability Technology* (Wiley, New York); U.S. Nuclear Regulatory Commission 1981. *Fault Tree Handbook* (NUREG-0492, NRC, Washington, D.C.); National Academy of Sciences 1983. *Priority Mechanisms for Toxic Chemicals* (National Academy, Washington, DC).
11 See, e.g., M. Hynes and E. Vanmarcke 1976. in *Proceedings of the ASCE Engineering Mechanics Division Specialty Conference*, Waterloo, Ontario, Canada.
12 See, e.g., J. W. Mueller 1979. *Nucl. Instrum. Methods* 163, 241.
13 K. D. Froome and L. Essen 1969. *The Velocity of Light and Radio Waves* (Academic, London).
14 P. J. Campion, J. E. Burns, and A. Williams 1973. *A Code of Practice for the Detailed Statement of Accuracy* (National Physics Laboratory, H.M.S.O., London).
15 Recommendation (Cl-1981) adopted by Le Comité International des Poids et Mesures, 1982.
16 W. J. Youden 1972. *Technometrics* 1, 1.
17 B. N. Taylor, W.H. Parker, and D. N. Langenberg 1969. *The Fundamental Constants and Quantum Electrodynamics* (Academic, New York).
18 R. T. Birge 1932. *Phys. Rev.* 40, 207.
19 R. T. Birge 1942. *Rep. Prog. Phys.* 8, 90.
20 J. H. Rush 1955. *Sci. Am.* August, 62–67.
21 The mean values are those originally reported, adjusted from velocity in air to velocity in a vacuum, where necessary. Where uncertainty was reported as "probable error" it was converted to standard deviation using a factor of 1.48.
22 This value has been adopted by the General Conference on Weights and Measures as the *definition* of *c*; *c* is now used in defining the meter instead of vice versa.
23 J. W. M. Dumond and E. R.Cohen 1951. *Phys. Rev. 82*, 555.
24 From 14 measurements of *G*, from Pipkin and Ritter (Ref.1) and G. T. Gillies 1983. *The Newtonian Gravitational Constant: An Index of Measurements* (BIPM-83/1), Bureau International des Poids et Mesures), omitting three that were off by more than ten standard deviations from the current recommended value.
25 A. H. Rosenfeld 1975. *Annu. Rev. Nucl. Sci.* 25, 555; Particle Data Group: M. Roos, F. C. Porter, *et al.* 1982. *Phys. Lett.* B 111, 1.

26 Roos *et al.* (Ref. 5) fitted the normalized residuals, h_i, for the particle lives and masses to an adjusted Student distribution, $S_N(h/c)$, with $N = 10$ degrees of freedom, and an adjustment factor, $c = 1.11$. The calibration indices for this distribution are shown in Table 9.1 for comparison.

27 T. Trippe 1983. (personal communication).

28 J. W. M. Dumond and E. R. Cohen 1953. *Rev. Mod. Phys.* 25, 691; E. R. Cohen *et al.* 1955. *ibid.* 27, 363; E. R. Cohen and J. W. M. Dumond 1965. *ibid.* 37, 537; E. R. Cohen and B. N. Taylor 1973. *J. Phys. Chem* Ref. Data 2, 663.

29 R. T. Birge 1929. *Rev. Mod. Phys.* 1, 1.

30 B. N. Taylor 1982. *NBSIR 81–2426*, National Bureau of Standards.

31 A. Koriat, S. Lichtenstein, and B. Fischhoff 1980. *J. Exp. Pysch.: Hum. Learn. Mem.* 6, 107.

32 A. Tversky and D. Kahneman 1974. *Cognitive Psychology* 4, 207.

33 R. E. Nisbett and L. Ross 1980. *Human Inference: Strategies and Shortcomings of Social Judgment* (Printice-Hall, Englewood Cliffs, NJ).

34 A. D. Franklin 1981. *Hist. Studies Phys. Sci.* 11, 2.

35 S. M. Stigler 1977. *Ann. Stats.* 5,6 1055.

36 See notes (Refs. 7 and 8) and M. G. Morgan *et al.* 1984. *Risk Analysis* 4, 4 (1984); C. S. Spetzler and C.-A. S. Stael von Holstein 1975. *Man. Sci.* 22.

37 P. Slovic and B. Fischhoff 1977. *J. Exp. Psych.: Hum. Percep. Perf.* 3, 544.

38 See, e.g., G. J. Gaeth and J. Shanteau 1984. *Org. Beh. Hum. Perf.* 33, 263.

10

FIFTY–FIFTY = 50%?

Baruch Fischhoff and Wändi Bruine de Bruin

CARNEGIE MELLON UNIVERSITY, USA

Abstract

Several recent surveys have asked respondents to estimate the probabilities of relatively unlikely events, such as dying from breast cancer and smoking. Examination of their response distributions reveals a seemingly inappropriate 'blip' at 50. The two studies reported here indicate that it reflects a response artifact associated with open-ended probability scales. The blip vanishes when a response scale with explicit response options is offered. Apparently, the open-ended format leads some people to use the 50% option as 'fifty–fifty', an expression of having no idea as to the answer. As a result, the accuracy of people's reported beliefs depends on the response scale used, as well as on how it evokes and channels such feelings of epistemic uncertainty.

Introduction

People often need to know the probabilities of risk-related events. For example, deciding whether to have a mammogram requires knowing the (interrelated) probabilities of having breast cancer, detecting it, and having treatment succeed. Deciding whether to smoke requires knowing the probability of smoking-related health effects, as well as the probability of being able to quit, once having started. Deciding whether to invest in one's education requires knowing the associated change in probability of unemployment. And so on. Poorly understood probabilities can mean ineffective, even dangerous, decisions.

Unfortunately, reliable estimates of these probabilities are often hard to come by, or hard to interpret even when they are available (Freudenberg, 1988; National Research Council, 1989; Slovic, 1987). As a result, it should not be surprising that studies eliciting lay risk judgments have sometimes found troubling misunderstandings. One large national study reported that women significantly overestimate the probability of breast cancer (Black, Nease and Tosteson, 1995). Another reported that people significantly overestimate the probability of lung cancer for smokers (Viscusi, 1992). Yet another reported that respondents substantially overestimate the probability of being the victim of a home burglary at least once in the next year (Dominitz and Manski, 1997). Each of these biases could add needlessly to individuals' personal worry and to the public demand for risk-reduction policies.[1]

In the studies just cited, people were found to overestimate relatively small probabilities. Substantive interpretations for these biases are easy to find. For example, the risks in question have all received extensive, and some might argue exaggerated, media attention (Combs and Slovic, 1979; Singer and Endreny, 1994). It would be only natural for people to infer unduly high probabilities, unless they were unusually alert to the possibility of disproportionate reportage (Lichtenstein *et al.*, 1978; Tversky and Kahneman, 1973).

Nonetheless, when a pattern is observed across diverse domains, one must ask whether some common methodological feature is at least partly responsible. How people translate subjective feelings into quantitative terms is one of the oldest topics in psychology. Students of psychophysics have documented a variety of potentially relevant biases. For example, drawing people's attention to a quantitative value, however irrelevant to the task, can pull their responses in that direction (Poulton, 1989, 1994; Tversky and Kahneman, 1974). Thus, a high 'anchor' could artifactually increase probability judgments.

Because they were designed for telephone administration, most of these studies showing the '50 blip' asked respondents to fill in a blank. As a result, they offered no obvious anchor. Nor is there evidence of other common psychophysical biases, such as respondents restricting themselves to a few stereotypical values or showing reluctance to use the extremes of the response scale.

However, in several cases in which studies reported response distributions, inspection of the data revealed a seemingly strange 'blip' at 50. To some extent, this blip might reflect insensitivity to probabilities in the middle range (Kahneman and Tversky, 1979), leading respondents to use 50 for, say, everything between 40 and 60. Yet, even if the 50 responses are spread over this range, a lump remains. Possibly, it could reflect real differences in beliefs. For example, a portion of the population might see much higher risks than the rest, with their estimates coincidentally falling at 50 in each case.

Another possibility is that, for some respondents, 50 means something different than a probability. One possible alternative meaning has emerged in recent interviews regarding the risk of AIDS, which we have conducted with a diverse sample of teens (Fischhoff and Downs, 1997; Fischhoff, Downs and Bruine de Bruin, 1998; Quadrel, Fischhoff and Davis, 1993). Often, when asked about risks, these teens would answer 'fifty–fifty', in a way that suggested, not a probability, but an expression for 'I really don't know'. Table 10.1 shows several examples. Such usage might be called *Laplacian*, placing equal probability on all hypotheses – in cases where there are two possible outcomes (i.e. something might or might not happen). Or it might be called *epistemic uncertainty*, reflecting not knowing what probabilities to use to express one's state of belief (Gärdenfors and Sahlin, 1982).

Using 'fifty–fifty' to express something other than 50% would artifactually increase aggregate judgments for events typically assigned smaller values. This would be especially true when judged probabilities are otherwise very small. Then, even a modest proportion of such responses could substantially increase mean judgments. For example, eliminating all responses of 50 in the studies cited above would reduce group means by 3–7 percentage points, from the full-sample means of about 15%. Thus, 50s represent a significant proportion of the apparent overestimation of risk.

One possible contributor to such usage is the open-ended format of these studies, which requires respondents to generate their own probability responses. This procedure may increase the chances of expressing epistemic uncertainty by means of 50, relative to a procedure providing explicit response options. The present studies examine the extent to which an open-ended response mode encourages using 50 as a verbal phrase (fifty–fifty), rather than as a number (50). The studies use different response modes that vary the availability of response options, under the assumption that 'fifty–fifty' becomes less likely when explicit numerical response alternatives are made available. In Study 1,

TABLE 10.1 Some adolescents' uses of fifty–fifty

Interviewer:	OK. And how well do you think that this would really protect a person from getting HIV, if they used a new needle?
Respondent:	I wouldn't say 100% sure, I would say almost definite, but there is always that risk if you are involved, I think with that kind of life style, you know, a drug use has, ah, you are eventually going to run into that. So I say, a fifty–fifty chance.
Interviewer:	OK. Hum, OK, so you said the more people that a person has unprotected sex with [the greater the risk]. What if they have unprotected sex with the same person? Does it matter how many times?
Respondent:	It only takes once if the person is infected. It could; it could not. So I guess it's up in the air again, a fifty–fifty chance.
Interviewer:	How well do you think that would really protect a person from getting the AIDS virus, if they use a condom when they are having sex?
Respondent:	I guess it's a fifty–fifty chance, 'cause it might like break or something and then it might not.
Interviewer:	How well do you think [that using condoms] would really protect a person from getting AIDS?
Respondent:	I'd say about ... I don't know, fifty–fifty, 'cause there's always that possibility that something like the condom would break or something.

one form had respondents fill in a blank, akin to previous studies. The second form offered a 0–100 probability scale, with 101 tick marks. We expected a smaller proportion of 50 responses with the scale. Half of the respondents receiving each response mode were also given the option of answering 'absolutely no idea', instead of giving a numerical probability. If the 50 response is used to express uncertainty about which number to use, then providing an 'absolutely no idea' option might reduce the blip.

Study 1

Method

Procedure

In all versions of the survey, respondents were asked four questions concerning low-probability events, namely (a) 'What is the probability of the university being closed at least once this year because of a bomb threat?' (b) 'What is the probability of someone getting AIDS if they have sex without protection?' (c) 'What is your personal probability of developing cancer by age 80?' (d) 'What is the probability that someone will break into your room or home and steal something this year?' These events were chosen to parallel those in earlier studies, and to provide a variety of events with small statistical probabilities.

Respondents in the *blank* condition were instructed to fill in a space with 'a number between 0% (no chance) and 100% (certainty)'. In the *scale* condition, the probability scale had 101 tick marks, with endpoints labelled '0 = no chance' and '100 = certainty', and additional numbers (as well as slightly larger ticks) at 10, 20, …, 80, and 90. The response modes were presented without instructions, lest our words communicate something about epistemic uncertainty.

Approximately half of the participants in the blank and scale conditions were also offered the opportunity to state that they had 'absolutely no idea' regarding each probability. Thus, there were four conditions, differing in whether a scale was provided and whether there was an 'absolutely no idea' option.

After completing the four probability judgments, respondents were asked: (a) how confident they were in each judgment, on a scale anchored at 1 = very confident and 7 = not confident at all, and (b) how clear each probability question had been, with 1 = very clear and 7 = not at all clear. We were interested in respondents' satisfaction with the different response modes. These questions were placed after the four probability judgments, in order to avoid suggesting epistemic uncertainty.

Participants

Participants were asked to fill out a small packet of unrelated questionnaires, including the probability surveys reported here. Seventy students at the University of Pittsburgh and 48 students at Carnegie Mellon University were given the scale version of the questionnaire, half of whom received the 'absolutely no idea' option. An additional 28 aquarium owners at a meeting of Pitt Fish and 93 visitors to the spring carnival at Carnegie Mellon University received the blank condition. Again, half received the 'absolutely no idea' option.[2] Subjects recruited at the University of Pittsburgh and CMU received a candy bar for their participation, while those visiting the spring carnival were given two ride tickets. The aquarium club received $3 for each participating member. A total of 132 men and 100 women participated in this study; an additional seven respondents did not specify their gender. The average age was 23.4 (sd = 6.0). Mean years of education was 15.0 (sd = 2.2).

Results

Absolutely no idea

Thirty-three of the 120 respondents (=27.5%) in the 'absolutely no idea' condition checked that answer for at least one of the four probability questions, amounting to 53 occasions (or 11% of all responses). Most of the respondents who employed the 'absolutely no idea' option used it for one or two answers only, suggesting that they were not simply minimizing their effort. Respondents used 'absolutely no idea' almost twice as frequently when no scale was presented (13.9% versus 7.9%). However, the related chi-square tests of homogeneity did not reach significance. Cochran's test showed a question effect for the use of 'absolutely no idea' ($\chi(3) = 11.02$, $p < 0.05$), with the bomb-at-university question having as many as all the others combined. For that question, responses of 50 were less common with the 'absolutely no idea' option than without it (6.8% and 1.8%, respectively). For each question, a Kolmogorov-Smirnov test found no difference in the distributions of probability responses with and without the 'absolutely no idea' option ($\alpha = 0.05$). Neither did median tests ($\alpha = 0.05$). As a result, whether the 'absolutely no idea' option was offered is ignored in subsequent analyses.

Use of 50

Table 10.2 shows the percentage of 50s among the numerical judgments (i.e. excluding 'absolutely no idea' responses). Overall, twice as many 50 responses were provided with the blank as with the scale condition, with significant differences for the break-in and cancer-by-80 questions. A Kruskal–Wallis test, in which individuals were characterized by their total number of 50s, showed significantly more 50 responses when no scale was presented ($\chi(1) = 5.92$, $p < 0.05$). Figure 10.1 depicts the response distributions for the break-in question. (Kendall's test of concordance showed that the number of 50 responses differed by question ($\chi(3) = 16.85$, $p < 0.001$), which could reflect differences in both epistemic uncertainty and overall judged likelihood.)

TABLE 10.2 Probability responses in Study 1

	Percentage of 50 responses (excluding 'absolutely no idea')				
Question	*Blank conditions*	*Scale conditions*	*Chi-square test*	*Median response*	*Statistical estimate*
AIDS from sex	14.9	13.0	0.17	20	5[a]
Bomb at the university	5.6	2.7	1.13	5	0–9[b]
Break-in	13.0	3.5	6.94**	10	1.07[c]
Cancer by 80	15.9	5.3	6.83**	45	31–37[d]
Average	12.5	6.2	5.92*	NM	NM

*$p < 0.05$.
**$p < 0.01$.
[a] Hearst and Hulley (1988).
[b] In the past 23 years, two buildings at CMU were evacuated because of a bomb threat. The entire school was never closed for this reason (Sgt Simon, personal communication, 29 May 1998).
[c] *Commonwealth of Pennsylvania Uniform Crime Report* (1996); *Pennsylvania Abstract. A Statistical Fact Book* (1996).
[d] Seidman *et al.* (1985).
NM = Not meaningful.

Overall responses

A Friedman test showed that probability answers differed by question ($\chi(3) = 185.4$, $p < 0.001$). Probabilities were significantly higher in the blank than in the scale condition for the AIDS-from-sex question ($\chi(2) = 5.97$, $p < 0.05$).

Confidence and clarity

Perceptions of confidence and clarity showed positive correlations for each question, but reached significance only for the bomb-at-the-university question ($r = 0.18$, $p < 0.01$). A MANOVA found that confidence ratings were related to question ($F(3, 657) = 9.87$; $p < 0.001$), but not to response mode or to the presence of the 'absolutely no idea' option. Thus, respondents reported being equally comfortable with the blank and the scale conditions. Across questions, mean confidence ranged from 3.25 (for the bomb-at-the-university question) to 4.08 (for the AIDS-from-sex question) (on the 1–7 point scale). Clarity judgments similarly showed an effect for question ($F(3, 663) = 13.31$; $p < 0.001$), but not for response mode. However, an interaction between response mode and question showed that two of the questions were rated as less clear in the scale condition (AIDS from sex and break-in; $F(3, 663) = 2.90$, $p < 0.05$). For three of the four questions, clarity and confidence were unrelated to the use of 50. Kendall's tau was weakly significant for the bomb-at-the-university

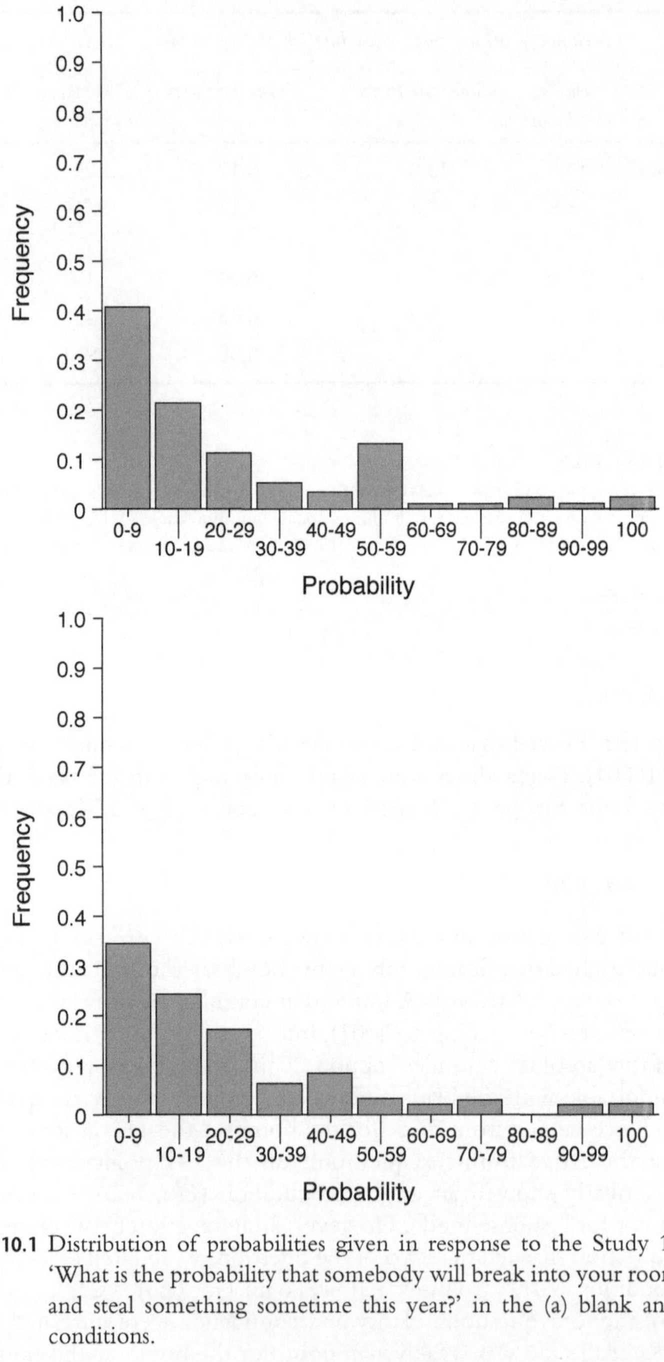

FIGURE 10.1 Distribution of probabilities given in response to the Study 1 question: 'What is the probability that somebody will break into your room or house and steal something sometime this year?' in the (a) blank and (b) scale conditions.

question, showing more 50 responses with reports of higher question clarity ($r = 0.15$, $p < 0.05$) and higher confidence ($r = 0.15$, $p < 0.05$).

Group differences

Kendall's tau showed no correlation between the use of 50 and age or gender. Higher grade levels were associated with less frequent use of 50 for one question, break-in ($r = -0.17$, $p < 0.01$). The number of 50s was unrelated to whether respondents reported having taken a class treating probabilities.

Discussion

Offering the 'absolutely no idea' option slightly decreased the use of 50 responses overall, with the difference reaching significance for one individual question. Thus, forcing uncertain respondents to give numerical probabilities, when they would rather not, may contribute to the 50 blip. The frequency with which respondents chose 'absolutely no idea' may understate the frequency with which they felt that subjective state. That response involves an admission of ignorance. Respondents who find that socially undesirable may prefer to say 'fifty–fifty', because that is at least a numerical response.

The blank condition generated twice as many 50 responses as did the linear probability scale. Thus, offering respondents an explicit set of numerical responses may have focused their attention on the numerical interpretation of probabilities, and diminished the availability of the verbal expression 'fifty–fifty'. Respondents reported similar confidence in their answers when using the probability scale and when producing their own probabilities. Two questions were judged less clear in the scale condition, perhaps because the precision implied by the 101-tick scale encouraged respondents to think harder. However, these perceptions of increased ambiguity in the questions did not contribute to any epistemic uncertainty, as expressed by using 50 as an answer.

Overall, respondents overestimated the risks addressed in Study 1, compared with our statistical estimates (Table 10.2). Estimates for the AIDS-from-sex question were higher with the blank condition, where the increased use of 50 also contributed to higher median probabilities.

Statistical power is naturally limited when observing changes in the rate of a relatively infrequent event (like the use of 50). As a result, a second study was conducted in order to examine the robustness of these results, using different subjects and items. A second probability scale was also added, in order to see the effect of offering a larger set of response options. This scale expands the 0–1% section of the linear scale form, providing additional low-probability response options (in a log format). We expected an even lower proportion of 50 responses with this log-linear scale. Because the 'absolutely no idea' manipulation proved less powerful in Study 1, we dropped it in order to concentrate on the effect of different response modes.

Study 2

Method

Procedure

In all three versions of the questionnaire, respondents assessed the probabilities of six events, in the following order: (a) [your] getting struck by lightning some time this year, (b) [your] developing cancer by age 40, (c) someone who smokes a pack or more of cigarettes a day developing lung cancer, (d) someone getting AIDS if they have sex without protection once with someone who is infected, (e) an average woman being diagnosed with breast cancer in her lifetime, and (f) [your] being alive at age 50. These events were chosen to parallel those in earlier studies, to provide a variety of events with small statistical probabilities, and to include at least one high-probability event (f). The events were specified with enough detail to allow answers to be compared to statistical estimates (Fischhoff, 1994; Fischhoff, Bostrom and Quadrel, 1996).

Respondents in the *blank* condition were instructed to fill in a space with 'a number between 0% (no chance) and 100% (certainty)'. The *linear* scale condition was the same as that in Study 1. The *log-linear* scale condition used the linear scale, but added a six-order log scale expanding upon the probability options between 0 and 1%. The response modes were, again, presented without additional instruction.[3]

After completing the six probability judgments, respondents evaluated each item on the confidence and clarity scales of Study 1.

Participants

An experimenter approached people walking on the UC Berkeley campus and asked them to complete two unrelated questionnaires, in return for $3. The present study was the first of these tasks. A total of 102 men and 58 women participated, with three more not specifying their gender. Respondents reported a mean age of 23.4 (sd = 5.8) and mean years of eduction of 14.3 (sd = 1.9), both very similar to Study 1. Respondents were randomly assigned to one of the three response modes.

Results

Use of 50

Table 10.3 shows the percentages of 50 responses. As in Study 1, they were much more common in the blank condition than with a numerical scale. These differences were statistically significant for three individual questions: AIDS from sex ($\chi(1) = 5.09, p < 0.05$), cancer by 40 ($\chi(1) = 4.20, p < 0.05$), and breast cancer ($\chi(1) = 10.94, p < 0.001$). The non-significant differences for the other three questions were in the same direction. After characterizing subjects by their number of 50s, Kruskal–Wallis tests showed that use of 50 depended on the response

TABLE 10.3 Percentages of 50 responses in Study 2

Question	*Blank*	*Linear*	*Log-linear*	*Total*
Lightning	1.9	0.0	0.0	0.7
Cancer by 40	20.4	7.4	10.3	12.9
Lung cancer	15.1	7.4	15.8	12.4
AIDS from sex	23.1	7.4	12.1	14.4
Breast cancer	24.1	7.4	2.6	12.3
Alive at 50	17.3	9.3	12.2	12.9
Average	17.0	6.5	8.8	10.9

mode ($\chi(2) = 17.14$, $p < 0.001$), with respondents giving more 50 responses in the blank condition than with the linear scale ($\chi(1) = 15.65$, $p < 0.001$) or the log-linear scale ($\chi(1) = 6.75$, $p < 0.01$). There was no appreciable difference between the two probability scales in the use of 50, even though the log-linear condition offered many more explicit response options than did the linear one. Kendall's test of concordance showed that the number of 50 responses differed by question ($\chi(5) = 19.74$, $p < 0.01$). The AIDS-from-sex question received the highest number of 50s, while the struck-by-lightning question elicited almost no 50s at all. Figure 10.2 depicts distributions for one low-probability event, cancer by 40. Visual inspection of response distributions for the blank condition shows marked blips at 50 for every event, except lightning.

(A)

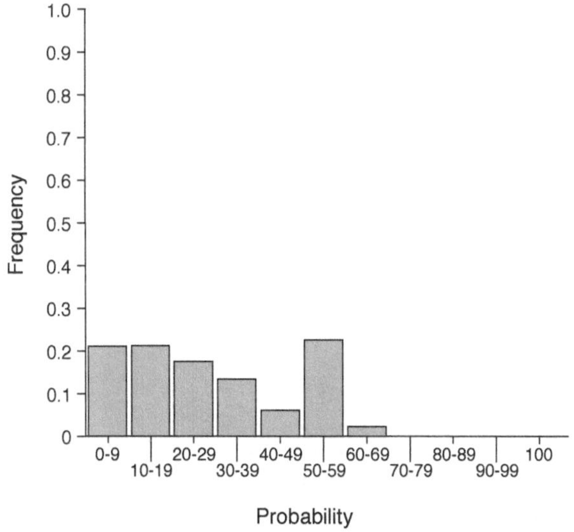

(B)

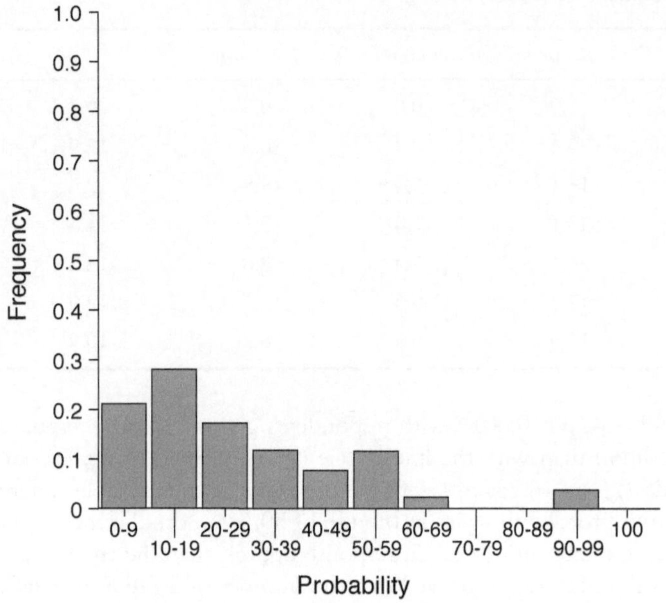

(C)

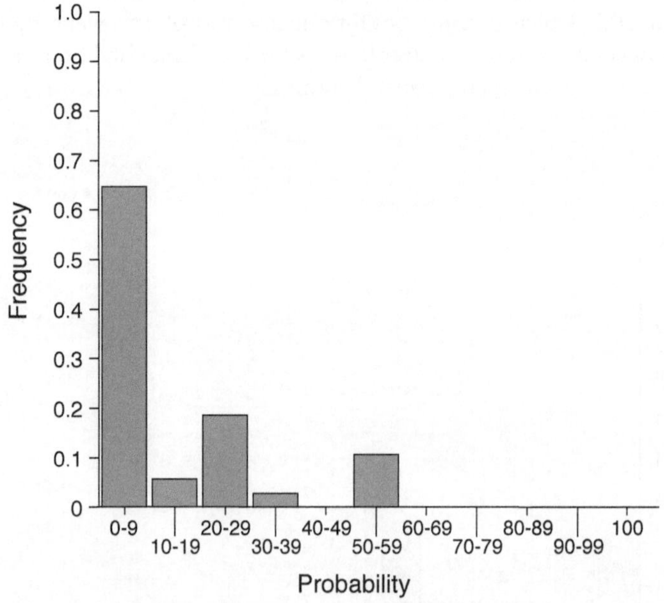

FIGURE 10.2 Distribution of probabilities given in response to the question: 'What is the probability that you will develop cancer by 40?' in the (a) blank, (b) linear, and (c) log-linear conditions.

Table 10.4 shows mean and median responses for each question and response mode. The disparity between these two measures of central tendency shows the skew in the underlying distributions. It is relatively small, except for several of the lowest probabilities. A Friedman test showed, not surprisingly, that responses differed across question ($\chi(5) = 430.67$, $p < 0.001$). Kruskal–Wallis tests comparing judgments in the three conditions reached significance for three items (cancer by 40, lung cancer from smoking, and breast cancer). A median test was also significant for these three items, as well as for a fourth ($\chi(2) = 7.9$, $p < 0.05$ for cancer by 40; $\chi(2) = 17.65$, $p < 0.001$ for lung cancer from smoking; $\chi(2) = 18.67$, $p < 0.001$ for breast cancer; and $\chi(2) = 22.99$, $p < 0.001$ for struck by lightning). The two questions without significant response-mode effects were the ones with the largest probability responses: AIDS from sex and being alive at 50. Generally speaking, the blank condition elicited the highest probabilities, followed by the linear and log-linear conditions.[4]

Confidence and clarity

All questions showed a positive correlation between perceived confidence and clarity. The correlations were significant for the lung-cancer-from-smoking ($r = 0.18$, $p < 0.05$) and AIDS-from-sex ($r = 0.27$, $p < 0.01$) questions. As in Study 1, a MANOVA found that confidence ratings were related to question ($F(5, 600) = 13.16$; $p < 0.001$), but not to response mode. Thus, respondents were

TABLE 10.4 Probability responses in Study 2

	Blank		*Linear*		*Log-linear*		*Test K–W*	
Question	*Mean*	*Median*	*Mean*	*Median*	*Mean*	*Median*	$\chi^2(2)$	*Statistical estimates*
Lightning	3.1	1.0	3.3	1.0	0.8	0.0[a]	4.43	0.000001[a]
Cancer by 40	25.2	20.0	24.5	20.0	11.4	2.0	18.52*	1–2[b]
Lung cancer	58.7	60.0	54.9	60.0	29.1	16.0	23.61*	10[c]
AIDS from sex	57.2	50.0	50.2	50.0	42.6	50.0	3.99	5[d]
Breast cancer	37.2	40.0	33.3	30.0	18.9	15.0	19.60	10[e]
Alive at 50	74.7	80.0	73.2	79.5	74.0	80.0	0.34	94.4[f]

* $p < 0.01$.

[a] 10^{-6}; *Natural Hazard Fatalities for 1994 in the United States* (1995).

[b] Seidman *et al.* (1985).

[c] Viscusi (1992).

[d] Hearst and Hulley (1988).

[e] Black, Nease and Tosteson (1995).

[f] *Vital Statistics of the United States, 1992* (1993).

equally comfortable with the (possibly more demanding) linear and log-linear scales. Across questions, mean confidence ranged from 2.60 (for lightning) to 3.92 (for breast cancer). The question-by-response-mode interaction ($F(10, 600) = 3.01$; $p < 0.01$) had no obvious interpretation. Clarity judgments similarly showed a question effect ($F(5, 575) = 5.39$; $p < 0.001$), with no effect of response mode or interaction. There was no correlation (Kendall's tau) between clarity or confidence judgments and the use of 50.

Group differences

For each question, use of 50 was unrelated to gender or to whether respondents had taken at least one class covering probabilities. Kendall's tau showed that the number of 50s was modestly correlated with overall level of education for one question (AIDS from sex; $r = -0.21$, $p < 0.05$), and with age for one question (cancer by 40; $r = -0.21$; $p < 0.05$).

Discussion

Study 2 replicated the findings of Study 1. Respondents who had to produce their own probability answers, in the blank condition, were twice as likely to use the 50 response as were the respondents offered a probability scale. Adding the log response options to the linear scale did not further reduce the use of 50. These results are consistent with the hypothesis that presenting a probability scale reduces either respondents' epistemic uncertainty or their ability to express it. Also consistent is the finding, in Study 1, that the probability scale halved respondents' use of the 'absolutely no idea' option. Within the range of variation afforded by our samples, none of these tendencies were related to reported gender, age, years of education, and instruction in probability. Epistemic uncertainty, if that is indeed the causal factor here, appears to be a general experience that occurs when individuals encounter particular items, especially when they need to produce their own numbers.

In both studies, use of 50 depended on the question's content in some, as yet to be determined, way. Examination of item differences might yield some further insight into the nature of these processes. Obviously, there were more 50s with distributions centered at 50, making detection of a blip more difficult. Less obvious is the absence of any blip with the struck-by-lightning question. Arguably, this is an event that epitomizes low probability, without evoking a feeling of 'maybe yes, maybe no'. The other events vary in so many ways that it will take further research to discipline the possible speculations regarding the feelings that they evoke. One potentially relevant feature is that all our questions involved risks to life and limb. With such fateful events, people may be particularly reluctant to commit themselves to a number when they don't really know what to say.

The frequency of 50s was unrelated to the questions' rated clarity. Thus, being less sure about what was being asked did not seem to encourage using 50 as an expression of not knowing what to say. Perhaps people are so accustomed to responding to the gist of questions that ambiguity tends not to generate epistemic uncertainty (Schwarz, 1996). If so, then epistemic uncertainty would seem to emerge when respondents are deciding what to answer, rather than when trying to figure out what they are being asked.

Wallsten, Budescu and Zwick (1993) also observed a 50 blip among respondents using an open-ended numerical response mode, in a study comparing the calibration of those judgments with ones elicited with a list of verbal probability expressions having 'toss up' at its midpoint. They suggest that uncertain respondents might answer 50 when they feel that it will be an equally defensible answer whatever actually happens. That response strategy could, it seems, be either an expression of epistemic uncertainty or a deliberate misrepresentation of another belief. If the latter, then it need not be related to response mode. Wallsten *et al.* also argued that the inherent vagueness of verbal probability terms means that they naturally incorporate an expression of epistemic uncertainty. In this light, it would be interesting to repeat their study with a list of verbal probability questions, whose midpoint is labelled 'fifty–fifty'.

It appears, thus, that the availability of the common phrase 'fifty–fifty' affects the use of the 50 response as an expression of epistemic uncertainty. The presentation of a probability scale draws respondents' attention away from such verbal answers and focuses them on probabilities as numbers. The open-ended response mode may also have increased epistemic uncertainty by requiring respondents to produce their probability answers, a task which is more difficult than simply placing a mark on a probability scale.

Whatever its source(s), reduced reliance on 50 leads to smaller estimates and, hence, to lower overestimates of the small probabilities.[5] As seen in Exhibit 10.5, responses with the log-linear scale were actually fairly accurate. However, this improvement could itself also have an element of artifact. The explicit options of the probability scales may have suggested response options that would not have occurred to respondents naturally (Fischhoff, Slovic and Lichtenstein, 1980; Poulton, 1989, 1994; Schwartz and Hippler, 1987). One reasonable expectation is that expanding the lower end of the log-linear scale encouraged responses between 0 and 1% – although it was not a powerful enough influence to affect the responses to the AIDS-from-sex and alive-by-50 questions, with their high subjective probabilities. Conversely, the linear scale may have been biased against the expression of very small values. For its part, the open-ended response mode may have biased respondents toward fifty-fifty, or the middle, relative to the linear scale. As elsewhere, how accurate people seem to be depends, in part, on how they are asked to answer (Fischhoff and MacGregor, 1983). One clear exception was with the AIDS-from-sex question. The median response of 50 suggests either a robust overestimate of this risk or resolute epistemic uncertainty, which even numerical scales cannot accommodate.[6]

Understanding individuals' risk perceptions is essential to predicting their behavior and to providing them with needed information. Doing so requires studies with well-chosen response modes and well-formulated questions. The studies reported here explore one obstacle to accurate assessment, as well as a possible remedy. By and large, judgments with a numerical scale showed a smaller 50 blip, without the reduction in confidence that more demanding scales might be suspected of inducing. One part of the acceptability of a response mode is how it evokes and deals with respondents' feelings of epistemic uncertainty. Eliminating anomalous 50 blips seems good, as long as we have confidence that those feelings are not channeled to values suggested by other features of a structured response mode. Increasing that confidence will require further study of both the conditions inducing epistemic uncertainty and the roles played by the response '50'.

Acknowledgments

Support for this research was provided by the National Institute of Alcohol Abuse and Alcoholism and the National Institute for Allergies and Infectious Diseases, and is gratefully acknowledged. The opinions expressed are those of the authors. We also appreciate the support of Susy Bielak, Stephanie Byram, Robyn Dawes, Julie Downs, Shane Frederick, Melanie Gold, Jenny Hounshell, Liz Kanabe, Chuck Manski, Claire Palmgren, Andy Parker, David Rubovits and Andrew White in various stages of this project, as well as the comments of three anonymous reviewers.

Authors' biographies

Baruch Fischhoff is University Professor, in the Department of Social and Decision Sciences and of Engineering and Public Policy, at Carnegie Mellon University. His current research interests include value elicitation, risk prioritization, integrated assessment, adolescent decision making and risk communication.

Wändi Bruine de Bruin is a PhD candidate in the Department of Social and Decision Sciences, Carnegie Mellon University. She has a Masters degree in cognitive psychology from the Free University of Amsterdam. Her research interests include knowledge elicitation, subjective probabilities, risk communication and adolescent decision making.

Notes

1 Evaluating the accuracy of judgments requires sometimes controversial calculations of 'objective' probabilities. For example, Dominitz and Manski (1997) use (with many explicit qualifications) respondents' current health insurance status as a rough approximation of the chances of not having insurance one year hence. Among their respondents, the mean estimate for being a victim of burglary in the next year was 16%, while only 5% of the sample actually became victims of this crime the next year. For such comparisons to be possible at all, assessments must be interpretable in quantitative terms. Thus, one would not avoid the problems discussed here by asking respondents to choose among verbal quantifiers, such as 'very likely' and 'rarely'. These terms have long been known to have different quantitative referents for different people and even for the same person in different contexts (e.g. 'very likely to rain' may connote a much higher number than 'very likely to kill you') (Budescu and Wallsten, 1995).

2 Initially, we ran just the scale response mode, thinking that adding the 'absolutely no idea' option would eliminate the 50 blip. When we observed few 50 responses, we realized that the earlier studies had not used a scale, but open-ended questions. As a result, we added a blank response mode. For the sake of brevity, we have combined results with the two response modes here in Study 1, rather than lead the reader through our path of discovery (and having two studies with two versions of one response mode each). This sequential process meant that questionnaires were randomly assigned with respect to the 'absolutely no idea' option, but not with respect to the response mode. We sought similar populations. Subjects receiving the two response modes were similar in age and gender. Those receiving the scale reported two additional years of education (15.9 versus 14.0; $p < 0.001$). We can see no obvious reason why these groups would differ in their use of 50; moreover, similar patterns were observed in the Study 2, with different populations (and random assignment). Nonetheless, this is an imperfect part of our procedure, which we addressed in Study 2.

3 In the data analyses, probabilities marked between the six major tick marks on the log portion of the log-linear scale were rounded up to the nearest tick mark. Thirteen of 55 respondents in the log-linear scale condition used both parts of the scale for every single question, while six additional respondents did so at least once. These respondents were excluded from the analyses. Given our previous success with this response scale, we assume that these difficulties reflect the lack of instructions, omitted in order to equate the conditions more fully and to avoid any wording that might suggest any attitude toward epistemic uncertainty. In the past, brief instructions with one or two test questions have been sufficient to allow even high-risk (low-literacy) teens to use the log-linear scale without apparent trouble (Linville, Fischer and Fischhoff, 1993; Quadrel *et al.*, 1993). Schwartz *et al.* (2000) report a 71% survey response rate and a 98% item response rate with a random national sample of women on a form making extensive use of a log-linear scale. Log-odds scales were used in Bayesian inference 'bookbag and poker chip' studies, producing some of the most accurate responses (Slovic and Lichtenstein, 1971).

4 The respective means with and without the 50s are, following the order of the events in the exhibits (3.1, 2.2; 25.2, 18.9; 58.7, 60.3; 57.2, 59.4; 37.2, 33.1; 74.7, 79.9).

5 Of course, not all feelings of epistemic uncertainty need emerge as 50. People may use 'fifty–fifty' as an anchor for their estimation process, adjusting that starting point, on the basis of whatever beliefs they have, beyond their initial feelings of not knowing what to say. If such adjustment is insufficient (Tversky and Kahneman, 1974), then such an anchor would bias judgments of small probabilities upward, without appearing as a 50 blip.

6 Linville *et al.* (1993) found similar estimates with an alternative log-linear scale, and discuss possible sources of this bias.

References

Black, W. C., Nease, R. F. and Tosteson, A. N. A. 1995. 'Perceptions of breast cancer risk and screening effectiveness in women younger than 50 years of age', *Journal of the National Cancer Institute*, 8, 720–731.

Budescu, D. F. and Wallsten, T. S. 1995. 'Processing linguistic probabilities: General principles and empirical evidence', in Busemeyer, J. R., Hastie, R. and Medin, D. L. (eds), *Decision Making from a Cognitive Perspective* (pp. 275–318), New York: Academic Press.

Combs, B. and Slovic, P. 1979. 'Newspaper coverage of causes of death', *Journalism Quarterly*, 56, 832–849.

Commonwealth of Pennsylvania Uniform Crime Report 1996. Pennsylvania: Pennsylvania State Police.

Dominitz, J. and Manski, C. F. 1997. 'Perceptions of economic insecurity. Evidence from the survey of economic expectations', *Public Opinion Quarterly*, 61, 261–287.

Fischhoff, B. 1994. 'What forecasts (seem to) mean', *International Journal of Forecasting*, 10, 387–403.

Fischhoff, B., Bostrom, A. and Quadrel, M. J. 1996. 'Risk perception and communication', in Detels, R., McEwen, J. and Omenn, G. (eds), *Oxford Textbook of Public Health* (pp. 987–1002), London: Oxford University Press.

Fischhoff, B. and Downs, J. 1997. 'Accentuate the relevant', *Psychological Science*, 8(3), 1–5.

Fischhoff, B., Downs, J. and Bruine de Bruin, W. 1998. 'Adolescent vulnerability: A framework for behavioral interventions', *Applied and Preventive Psychology*, 7, 77–94.

Fischhoff, B. and MacGregor, D. 1983. 'Judged lethality: How much people seem to know depends upon how they are asked', *Risk Analysis*, 3, 229–236.

Fischhoff, B., Slovic, P. and Lichtenstein, S. 1980. 'Knowing what you want: Measuring labile values', in Wallsten, T. (ed.), *Cognitive Processes in Choice and Decision Behavior* (pp. 117–141), Hillsdale, NJ: Lawrence Erlbaum Associates.

Freudenberg, W. 1988. 'Perceived risk, real risk', *Science*, 241, 44–49.

Gärdenfors, P. and Sahlin, N. E. 1982. 'Unreliable probabilities, risk taking and decision making', *Synthèse*, 53, 361–386.

Hearst, N. and Hulley, S. B. 1988. 'Preventing the heterosexual spread of AIDS: Are we giving our patients the best advice?' *Journal of the American Medical Association*, 259, 2428–2432.

Kahneman, D. and Tversky, A. 1979. 'Prospect theory: An analysis of decision under risk', *Econometrica*, 47, 263–281.

Lichtenstein, S., Slovic, P., Fischhoff, B., Layman, M. and Combs, B. 1978. 'Judged frequency of lethal events', *Journal of Experimental Psychology: Human Learning and Memory*, 4, 551–578.

Linville, P. W., Fischer, G. W. and Fischhoff, B. 1993. 'AIDS risk perceptions and decision biases', in Pryor, J. B. and Reeder, G. D. (eds), *The Social Psychology of HIV Infection* (pp. 5–38), Hillsdale, NJ: Lawrence Erlbaum Associates.

National Research Council 1989. *Improving Risk Communication*, Washington, DC: National Academy Press.

Natural Hazard Fatalities for 1994 in the United States 1995. Washington, DC: National Oceanic and Atmospheric Administration.

Quadrel, M. J., Fischhoff, B. and Davis, W. 1993. 'Adolescent (in)vulnerability', *American Psychologist*, 48, 102–116.

Pennsylvania Abstract. A Statistical Fact Book 1996. Harrisburg, PA: Pennsylvania State Data Center.

Poulton, E. C. 1989. *Bias in Quantifying Judgment*, Hillsdale, NJ: Lawrence Erlbaum Associates.

Poulton, E. C. 1994. *Behavioral Decision Theory: A new approach*, New York: Cambridge University Press.

Schwartz, L. M., Woloshin, S., Sox, H. C, Fischhoff, B. and Welch, H. G. 2000. 'US women's attitudes to false positive mammography results and detection of ductal carcinoma in-situ: Cross-sectional survey'. *British Medical Journal,* 320, 1635–1640.

Schwarz, N. 1996. *Cognition and Communication. Judgmental biases, research methods, and the logic of conversation*, Mahwah, NJ: Lawrence Erlbaum Associates.

Schwarz, N. and Hippler, H. J. 1987. 'What response scales may tell your respondents: Informative functions of response alternatives', in Hippler, H. J., Schwarz, N. and Sudman, S. (eds), *Social Information Processing and Survey Methodology*, New York: Springer-Verlag.

Seidman, H., Mushinski, M. H., Gelb, S. K. and Silverberg, E. 1985. 'Probabilities of eventually dying of cancer – United States, 1985', *CA, A Cancer Journal for Clinicians*, 35(1), 36–56.

Singer, E. and Endreny, P. M. 1994. *Reporting on Risk*, New York: Russell Sage Foundation.

Slovic, P. 1987. 'Perceptions of risk', *Science*, 236, 280–285.

Slovic, P. and Lichtenstein, S. 1971. 'Comparison of Bayesian and regression approaches to the study of information processing in judgments', *Organizational Behavior and Human Performance*, 6, 649–744.

Tversky, A. and Kahneman, D. 1973. 'Availability: A heuristic for judging frequency and probability', *Cognitive Psychology*, 4, 207–232.

Tversky, A. and Kahneman, D. 1974. 'Judgment under uncertainty: Heuristics and biases', *Science*, 185, 1124–1131.

Viscusi, W. K. 1992. *The Risks of Smoking*, Cambridge, MA: Harvard University Press.

Vital Statistics of the United States 1992, 1993. Centers for Disease Control and Prevention, National Center for Health Statistics, Vol 2, Section 6.

Wallsten, T. S., Budescu, D. V. and Zwick, R. 1993. 'Comparing the calibration and coherence of numerical and verbal probability judgments', *Management Science*, 39, 176–190.

PART IV

Evaluation

If people know what they what, then figuring out what they can get is the difficult part of decision making. Assuming that to be the case, behavioral decision research initially focused on judgment, trusting people to have well-articulated preferences. That strategy is consistent with psychology's long tradition of expecting people to have attitudes for whatever topics researchers raise. However, attitude research also has an undercurrent of context effects, in which seemingly minor changes in how questions are posed produces seemingly large changes in how they are answered. For example, people will sometimes not forbid an activity (e.g., profane speech) that they will also not allow. Kahneman and Tversky's prospect theory showed how context effects can arise from structural features of choices (e.g., whether outcomes are framed as gains or losses) as well semantic ones (e.g., whether actions are forbidden or allowed). The observation that preferences are sometimes stable and sometimes labile led me to study the nature of values and methods for studying them.

"Predicting frames" (Chapter 11) considers one challenge to using prospect theory for predicting choices: interpreting its abstract terms in concrete settings. Its precursor, utility theory, is often criticized for allowing advocates to describe any choice as rational, if they can posit some goals that decision makers putatively maximized. The defense against that criticism is to require stable preferences, whereby those revealed in one decision will also be revealed in all other decisions (Becker, 1976). Context effects undermine that assumption, by showing instable preferences with a single decision (when it is presented differently). Indeed, prospect theory's existential threat to utility theory comes from its powerful context (or framing) effects. The foremost of those arise from manipulating the reference point, relative to which potential outcomes are evaluated. Shifting a reference point can make an outcome feel like a gain or a loss, potentially reversing preferences. For example, a 3% raise could seem like a gain compared to one's current wages or like a loss compared to the 5% raise that one expected or that other workers received. Identifying the reference point in a choice is, therefore, a critical auxiliary assumption in using prospect theory. The experiments in this study tried, mostly in vain, to predict how people choose reference points, when more than one is possible. They suggest the value of learning more about how people intuitively frame choices.

"Value elicitation: Is there anything in there?" (Chapter 12) contrasts the philosophies of science underlying research programs that make different assumptions about the nature of preferences. At one end of a continuum, lie researchers who treat people as knowing what they want for every conceivable choice. At the other end, lie researchers who treat people as constructing their preferences, figuring out how their general values apply in specific settings. Following Kuhn (1962), Lakatos (1970), and Mills (1959), the article asks how these research philosophies can reflect different incentives, offer different insights, and produce different interpretations of behavior. Its immediate prompt was Robert Abelson's observation that behavioral decision researchers saw context effects as ubiquitous and essential, whereas survey researchers saw them as rare and inconsequential (National Research Council, 1981).

"Informed consent in eliciting environmental values" (Chapter 13) considers the ethical obligations of researchers who claim to know others' preferences. Its immediate focus is the contingent valuation method (CVM), which asks people how much they care about changes in public goods, in terms compatible with cost–benefit analyses (Mitchell and Carson, 1989). As such, it provides a way to monetize "goods" not traded in a conventional marketplace (e.g., endangered species, historic districts, scenic vistas). However, in order to satisfy cost–benefit analysts' need to evaluate specific policies, it must ask novel questions even for familiar goods (e.g., "How much will you pay for 30% less haze, 100 days per year?" – as would be achieved by installing scrubbers on two coal-fired power plants). The article describes ways to help people master such tasks and assess their success. Arrow et al. (1993) capture some of the contentious debates sparked by the application of CVM to the Exxon Valdez damage assessment (where I worked on the US Department of Justice team), focused on how seriously to take expressed preferences.

References

Arrow, K., Solow, R., Portney, P.R., Leamer, E.E., Radner, R., and Schuman, H. 1993. *Report of the NOAA panel on contingent valuation*. Washington, DC: NOAA.

Becker, G.S. 1976. *The economic approach to human behavior*. Chicago: University of Chicago Press.

Kuhn, T.S. 1962. *The structure of scientific revolution*. Chicago: University of Chicago Press.

Lakatos, I. 1970. Falsification and scientific research programmes. In I. Lakatos and A. Musgrave (eds.), *Criticism and the growth of scientific knowledge*. Cambridge: Cambridge University Press.

Mills, C.W. 1959. *The sociological imagination.* Oxford: Oxford University Press.
Mitchell, R.C., and Carson, R.T. 1989. *Using surveys to value public goods: The contingent valuation method.* Washington, DC: Resources for the Future.
National Research Council. 1981. *Surveys of subjective phenomena.* Washington, DC: Author.

11

PREDICTING FRAMES

Baruch Fischhoff

DECISION RESEARCH, A BRANCH OF PERCEPTRONICS, EUGENE, OREGON AND MEDICAL RESEARCH COUNCIL, APPLIED PSYCHOLOGY UNIT, CAMBRIDGE, ENGLAND

Abstract

Kahneman and Tversky's (1979) "prospect theory" has demonstrated that the way in which a decision problem is formulated, or "framed," can have strong and predictable effects on the perceived attractiveness of the options it offers. At times, the relative attractiveness of two options may be reversed as the result of a reframing that should make no difference at all, according to traditional economic theories of choice. To predict people's behavior with the theory, one must be able to predict what frame they will impose on a particular problem. The seven studies reported here explored different ways of predicting frames, with results that were generally discouraging for the prediction of individuals' choices, generally encouraging for the prediction of group choices.

Introduction

With their "prospect theory," Kahneman and Tversky (1979; Tversky and Kahneman, 1981) have provided a significant new tool for the analysis of choices made under conditions of uncertainty. The primitives of this descriptive theory of decision making are a value function, $\upsilon(x)$, which attaches a subjective worth to each possible outcome of a gamble or prospect, and a weighting function, $\pi(\rho)$, which expresses the subjective importance attached to the probability of obtaining a particular outcome. The attractiveness, V, of a gamble that offers a chance of r to gain (or to lose) x and a chance of q to gain (or to lose) y would equal $\pi(\rho)\upsilon(x) + \pi(q)\upsilon(y)$.

One important feature of the value function is that it assesses outcomes in terms of the change they represent from some reference point, which could represent one's current status, one's anticipated status, or some other psychologically significant point. A second feature is that the function is steeper for losses than for gains, meaning that a given change in one's status hurts more as a loss than it pleases as a gain. A third feature is that it is concave above that reference point and convex below it. Such negatively accelerated curves mean, for example, that the subjective difference between gaining (or losing) $10 and $20 is greater than the difference between gaining (or losing) $110 and $120.

Perhaps the most notable feature of the probability weighting function is the great importance attached to outcomes that will be received with certainty. Thus, for example, the prospect of losing $50 with probability 1.0 is more than twice as aversive as the prospect of losing the same amount with probability .5. For intermediate (i.e., noncertainty) values, the weighting function is somewhat insensitive to changes in probability. For example, a .5 chance of winning $50 would not be 25% more attractive than would be a .4 chance of winning $50.

The nonlinearities of the value and probability weighting functions mean that people's choices should be sensitive to the way in which decisions are formulated or "framed." In their expositions, Kahneman and Tversky have offered powerful demonstrations of how the same decision problem may be framed in ways that are formally equivalent in terms of classical (utility theory) models of choice behavior but that produce reversals of preference that can be predicted on the basis of prospect theory.

In these demonstrations, each decision problem is presented in a format that is consistent with one particular frame. The prospect theory analysis proceeds under the assumption "either that the original formulation of prospects leaves no room for further editing, or that the edited prospects can be specified without ambiguity" (Kahneman and Tversky, 1979, p. 275). In such studies, the theorist's ability to control the way in which stimuli are presented and interpreted is crucial to being able to predict how the various portions of a stimulus will be integrated into the production of a response (here, a choice). In order to predict behavior in less controlled situations, one must be able to anticipate how problems will be represented and what frames people will use to interpret them.

The present studies examine people's choices between prospects on which it is possible to impose several different frames whose usage should lead to different choices of options, according to prospect theory. These studies ask, in effect, how do we predict which frame people will adopt, so that we can then predict their choices (in accordance with the theory). Because it is so difficult to study simultaneously the interpretations that people give to stimuli and the (decision) rules that they use to evaluate and choose between those stimuli, these studies assume the truth of prospect theory and use it to discern what interpretations have been given. This research strategy cannot lead to rejection of the descriptive validity of the theory, for any failure is seen as lying with the interpretation. Nonetheless, repeated difficulty in finding the interpretation needed to make the theory work can suggest limits to its practicality (as well as to the imagination of the interpreter).

Basic study

Although prospect theory allows considerable latitude in framing, it is by no means mute with regard to the formulations that people will adopt:

> Prospect theory describes two phases in the choice process: an early phase of editing and a subsequent phase of evaluation. The editing phase consists of preliminary analysis of the offered prospects which *often* yields a simpler representation of these prospects. (Kahneman and Tversky, 1979, p. 274; emphasis added)

The three central editing operations are (a) *coding*, describing each outcome in terms of changes from a neutral reference point (usually one's current asset position); (b) *segregation*, isolating riskless components of risky decisions (e.g., a gamble that either gave $300 with .8 probability or $200 with .2 would be "naturally decomposed into a sure gain of $200 and the risky prospect" [Kahneman and Tversky, 1979, p. 274] of winning $100 with .8 or $0 with .2); and (c) *cancellation*, "discarding of components that are shared by the offered prospects"1 (Kahneman and Tversky, 1979, p. 274).

However, as indicated by the word *often* in the quote above, the theory, although not mute, is deliberately not altogether explicit about how these editing operations are performed. Regarding coding, the reference point "can be affected by the formulation of the offered prospects and by the expectations of the decision maker" (Kahneman and Tversky, 1979, p. 274) and, hence, need not be one's current asset position. The segregation operation will only be carried out where the riskless component is "readily seen" by decision makers. It would not occur, for example, where people were overwhelmed by the surface structure of a problem. Regarding cancellation, "a pair of prospects can be decomposed into common and distinctive components in more than one way, and different decompositions sometimes lead to different preferences" (Kahneman and Tversky, 1979, p. 271).

The present studies consider three frames that reasonable individuals might impose on or derive from the decision problem described in the introductory section of Table 11.1. The three "ways one might attempt to think about the problem" attempt to capture these three perspectives:

Frame 1. Here, one accepts the problem as it is presented in the description, performing no editing beyond the necessary act of choosing a reference point (coding). The point chosen is the present situation (no lives lost yet).

In terms of the theory, the value of the sure-loss option is simply $V(\text{B}) = \pi(1)\upsilon(-50) = \upsilon(-50)$, given the convention that $\pi(1) = 1$ and $\pi(0) = 0$. The value of the gamble option would be: $V(\text{A}) = \pi(.5)\upsilon(-40) + \pi(.5)\upsilon(-60)$, or $V(\text{A}) = \pi(.5)\ [\upsilon(-40) + \upsilon(-60)]$. Option A will then be preferred to Option B if $\pi(.5)\ [\upsilon(-40) + \upsilon(-60)] > \upsilon(-50)$ or, equivalently, if $\pi(.5) < \upsilon(-50)/[\upsilon(-40) + \upsilon(-60)]$ (remembering that the sum in brackets is negative, which results in reversal of the inequality when it is used as a divisor). Given the convexity of the value function for losses, the ratio of the values will be greater than .5. One way of seeing this is to note that convexity means that $\upsilon(-50)$ is closer to $\upsilon(-60)$ than to $\upsilon(-40)$. Hence, twice $\upsilon(-50)$ will be more than the sum of $\upsilon(-60)$ and $\upsilon(-40)$. Given the assumed shape of the probability weighting function, $\pi(.5)$ will be less than .5, meaning that the inequality holds. Hence, prospect theory predicts that anyone adopting this perspective will be risk seeking, in the sense of preferring a gamble to a sure thing with the same expectation.

Frame 2. Here, one first isolates and then cancels the certain loss of 40 lives that is common to both options. Having done this, one proceeds to evaluate a sure loss of 10 (more) lives and a gamble involving equal chances of losing 0 and 20 (more) lives.

Despite this cancellation procedure, the evaluation and comparison of the two prospects would be similar to that for the prospects as seen from Frame 1. The value of the sure-loss option is $V(B) = \pi(1)\upsilon(-10) = \upsilon(-10)$. The value of the gamble option is $V(A) = \pi(.5)\upsilon(0) + \pi(.5)\upsilon(-20) = \pi(.5)\upsilon(-20)$. Option A will be preferred to Option B if $\pi(.5)\upsilon(-20) > \upsilon(-10)$ or, equivalently, if $\pi(.5) < \upsilon(-10)/\upsilon(-20)$. Given the convexity of the value function for losses, the right-hand side of this inequality will be more than .5 (because losing 20 lives

TABLE 11.1 Basic stimulus

A civil defense committee in a large metropolitan area met recently to discuss contingency plans in the event of various emergencies. One emergency under discussion was the following: "A train carrying a very toxic chemical derails and the storage tanks begin to leak. The threat of explosion and lethal discharge of poisonous gas is imminent."

Two possible actions were considered by the committee. These are described below. Read them and indicate your opinion about the relative merits of each.

Option A: Carries with it a .5 probability of containing the threat with a loss of 40 lives and a .5 probability of losing 60 lives. It is like taking the gamble:

.5 lose 40 lives

.5 lose 60 lives.

Option B: Would result in the loss of 50 lives:

lose 50 lives.

Here are three ways one might think about this problem:

1. This is a choice between a 50–50 gamble (lose 40 or lose 60 lives) and a sure thing (the loss of 50 lives).
2. Whatever is done, at least 40 lives will be lost. This is a choice between a gamble with a 50–50 chance of either losing no additional lives or losing 20 additional lives (A) and the sure-loss of 10 lives (B).
3. Option B produces a loss of 50 lives. Taking Option A would mean gambling over a .5 chance to *save* 10 lives and a .5 chance to *lose* 10 additional lives.

is not twice as bad as losing 10). Given the shape of the probability weighting function, the left-hand side will be less than .5, leading to a preference of the gamble (Option A) for those adopting this frame.

Frame 3. Here, one realizes that the best (in the sense of least bad) outcome that one can guarantee is losing 50 lives. Taking this guarantee as something in hand, the sure-loss option now involves staying put; the gamble is over saving 10 lives or losing 10 more. Such a focus on the minimax outcome could be modeled by prospect theory in one of two ways, the first of which seems much more plausible: (a) shifting one's reference point to the 50 lives that one has written off or (b) canceling 50 lives lost from both prospects. In either case, the sure-loss option becomes $V(B) = \upsilon(0) = 0$, whereas the gamble is now $V(A) = \pi(.5)\upsilon(10) + \pi(.5)\upsilon(-10)$. Here Option A will be preferred to Option B if $\pi(.5)[\upsilon(10) + \upsilon(-10)] > 0$ or, equivalently, if $\upsilon(10) > \upsilon(-10)$. However, because the value function for losses is steeper than is the value function for gains, the loss of 10 lives should loom larger than should the gain of 10. Hence those who adopt this perspective should prefer the sure loss to the gamble (i.e., Option B to Option A).

Although all three of these frames can be modeled by prospect theory, it is not clear that they are equally attractive. For example, if one believes that people tend to be captive of whatever superficial problem representation they receive (e.g., Slovic, Fischhoff, and Lichtenstein, 1982; Tversky and Kahneman, 1981), then the most natural of these frames should be Frame 1, which seems most compatible with the problem description and which requires the least editing before the prospects can be evaluated. Similar considerations would suggest that the relative popularity of Frames 2 and 3 might depend upon whether cancellation (2) seems more or less natural than reference points shifts (3).

Experiment 1 explores these questions by asking subjects to rank the frames in terms of naturalness. Whatever the overall results of this popularity poll, one would expect the gamble option to be preferred by subjects who judge Frames 1 and 2 as most natural, whereas the sure loss will be preferred by those (few or many) subjects who judge Frame 3 to be most natural.

Experiment 1

The first study presented the problem appearing in Table 11.1. Participants were asked to indicate (a) the phrasing that seemed to them to be most natural, (b) the phrasing that seemed least natural, (c) the option that they would select, and (d) the strength of that preference (on a 4-point scale). According to the present interpretation of prospect theory, those who adopt Frames 1 and 2 should also prefer the gamble option (see Option A in Table 11.1), whereas those who adopt Frame 3 should prefer the sure-loss option (see Option B in Table 11.1).

Method

Subjects received a one-page form labeled "Civil Defense." After presenting the problem appearing in Table 11.1 verbatim, the form asked them:

> Which of these three phrasings seems to be the *most* natural way of thinking about this problem?
>
> 1. 2. 3. (circle one)
>
> Which of these three phasings seems to be the *least* natural way of thinking about this problem?
>
> 1. 2. 3. (circle one)
>
> Which option would you select?
>
> —— Option A.
> —— Option B.

How strong is your preference for the alternative you just checked? Check one:

Slight preference ☐
Moderate preference ☐
Strong preference ☐
Very strong preference ☐

For this study and the following ones, the relevant form was one of several given in a self-paced session involving unrelated judgmental tasks. Subjects were recruited through advertisements in the University of Oregon student paper and a state employment service office. Typically, subjects recruited this way are divided evenly between men and women, with men averaging about 24 years in age and women about 21.

In the present study, 42 individuals received the form described above, while another 42 received a simpler form omitting the discussion of the three ways of looking at the problem.

Results

The bottom row of Table 11.2 presents the option choices of subjects who received the simple form, which had no discussion of or questions about frames. Of these 42 individuals, 36 preferred the gamble to the sure loss. Such risk seeking in the realm of losses is quite contrary to standard utility theory, which assumes that large losses receive disproportionate weight, leading to risk aversion. This pattern of preferences is, however, quite compatible with prospect theory, assuming that a frame such as 1 or 2 is adopted. From these option preferences, one would expect Frame 3 to be much less popular than Frame 1 or 2, with Frame 3 being adopted by about 15% of all individuals.

TABLE 11.2 Frame and option preferences in Experiment 1

	Option preference		
Frame preference	*Gamble*	*Sure loss*	*Total*
Subjects with long form			
Choice as most natural			
1	12	4	16
2	12	0	12
3	12	2	14
Choice as least natural			
1	11	0	11
2	9	3	12
3	16	3	19
All subjects	36	6	42
Subjects with simple form	36	6	42

The right-hand column of the top section of Table 11.2 shows the popularity of the frames, as judged by subjects receiving the long form. The frames were judged equally attractive, from the perspectives of choices as most natural frame and choices as least natural frame.

According to these popularity ratings, one third of all subjects in the long form group should prefer the sure-loss option, being those subjects who preferred Frame 3. However, as the remainder of the table shows, the sure-loss option was just as unpopular with the long form as with the simple form; only 6 of 42 subjects preferred it. Moreover, the sure loss was no more popular among subjects who found Frame 3 to be most natural, nor was it any less popular among subjects who found Frame 3 to be least popular.

The strength of preference ratings provided no additional information. If the verbal ratings are converted into a scale from 1 (slight preference) to 4 (very strong preference), mean ratings were approximately 2.2 for all frame and option preferences.

Experiment 2

In Experiment 1, prospect theory provided no guide to predicting subjects' choice preferences on the basis of their frame preferences. Although general principles of experimental design might prompt one to prefer a pair of options that are more balanced in terms of their overall attractiveness, the unpopularity of the sure loss still does not explain the relative naturalness of Frame 3 or the lack of a relation between frame and option choices.

One aspect of the experimental manipulation to which these difficulties might be attributed is the fact that the civil defense problem is initially presented in a manner that seems most compatible with Frame 1. Perhaps that perspective leads readers to an initial preference for the gamble. When asked about the most natural frame, one third of the subjects may still like Frame 3; however, as far as choices go, their minds have already been made up in favor of the gamble indicated by Frame 1. Experiment 2 tests this hypothesis by offering revisions of the basic form that used initial presentations more compatible with Frames 2 and 3.

Method

Three new forms were created. One initially described the options in a manner that seemed to reflect Frame 2:

> *Option A*: Would result in the loss of at least 40 lives. However, it does offer a .5 probability of containing the threat with the loss of 40 lives. On the other hand, there is also a .5 probability of losing 20 additional lives. It is like taking the gamble:
>
> .5 lose no more lives (in addition to losing 40 lives)
> .5 lose 20 more lives.
>
> *Option B*: Would result in the loss of 50 lives.
> lose 50 lives

A second new form tried to capture Frame 3 in its description of options:

> *Option A*: Would result in the loss of 50 lives.
> lose 50 lives
>
> *Option B*: Carries with it a .5 probability of saving 10 lives compared with Option A and a .5 probability of losing 10 lives. It is like taking the gamble:
>
> .5 saving 10 lives
> .5 losing 10 additional lives.

Because the Frame 3 form reversed the order of the options for the sake of a less awkward presentation, a new version of the original form was also devised. This form was identical to the original except that the order of the options was reversed to facilitate comparisons with Frame 3. Contrasting this order-reversed form with the original form should reveal whether the order in which options are presented affects their attractiveness.

A total of 155 individuals judged either one of these new forms or additional copies of the original long form (from Experiment 1).

Results

Order reversal. Reversing the order of the options in the Frame 1 form made no difference in either frame preference or option preference. With each order, subjects were divided approximately equally in terms of which frames they thought were most natural and least natural. With each order, most subjects found the sure-loss option to be less attractive; 16.7% picked the sure loss when it was presented first, compared with 20.8% who picked it when it was listed second (Z = .54). The two groups were pooled.

Alternative problem presentations. The left-hand side of Table 11.3 shows the percentage of subjects judging each frame to be most natural, within the three experimental groups, each of which received an initial problem presentation designed to be particularly compatible with one of the frames. These naturalness ratings suggest that the presentation manipulation was only mildly successful. All three frames received a higher proportion of "most natural" ratings in the formulation designed to highlight them than with the two other formulations. However, this difference approached statistical significance only with Frame 3 (Z = 1.47). Regarding choices of least natural frame (not shown), Frames 1 and 2 were as likely to be rated least natural in the forms designed to highlight them as in the other forms. Frame 3, however, was much less likely to be rated least natural in the form highlighting it (13.5% vs. 33.8%, $Z = 2.42$).

The weakness of this manipulation reduces the likelihood that the pattern of results observed in Experiment 1 was due to the particular problem wording used there. People's frame preferences seem to be relatively independent of the frame they see first. The robustness of those frame preferences also reduces the chances that subjects' option choices would be altered by the present manipulation.

TABLE 11.3 Frame and option preferences in Experiment 2 (civil defense problem)

		% of subjects choosing frame as most natural			*% of subjects choosing sure-loss option*			
		Frame preference			*Frame preference*			
Frame presented	*n*	*1*	*2*	*3*	*1*	*2*	*3*	*Ave %*
1	114	32	35	33	25	15	18	19
2	46	22	37	41	10	12	26	17
3	37	30	22	49	27	25	17	22
Ave %		29	33	38	23	15	20	19

Note. The number of subjects involved in each cell equals the entry in the left-hand part of the table multiplied by *n* for that row and divided by 100. For example, 38 people (33 × 114/100) were presented Frame 1 and preferred Frame 3; 7 of these 38 chose the sure-loss option, producing the 18% in the corresponding cell on the right-hand side.

As the right-hand side of Table 11.3 shows, the sure-loss option was as unattractive here as it was in Experiment 1, garnering only 19% of all choices, half of what it should have received had it been preferred by all subjects who preferred Frame 3. The popularity of the sure loss was unrelated to either the frame that subjects received (right-hand column) or the frame that they preferred (bottom row). Indeed, it was chosen by only 17% of the subjects who both received and preferred Frame 3.

Experiment 3

Although participants in Experiment 2 saw the problem in only one formulation, they read descriptions of all three frames prior to choosing an option. Conceivably, this reading exercise weakened the experimental manipulation by diluting the impact of the frame used in the problem description. Experiment 3 examines this possibility by reversing the order of the option-preference and frame-preference tasks, so that subjects will have seen only one formulation at the time they make their option choices. One possible drawback of this ordering is that subjects will not have made a deliberative choice between frames when they choose options. However, before considering further which order is superior, it is worth seeing whether order makes any difference.

Method

Forms were identical to those used in Experiment 2, except that the option-selection task preceded the naturalness-judgment task. That task was moved to a separate page in order to reduce the chances that subjects might still read about the three frames prior to selecting an option. All three forms presented the sure loss first. As an indirect manipulation check, subjects were asked later in the experimental session to remember which of the three frames they had received and which option they had selected. Eighty-two individuals judged this set of forms.

Results

Responses in Experiment 3 were, in general, very similar to those observed in Experiment 2. The sure loss remained unpopular; it was chosen by 26.8% of subjects (compared with 19.3% in Experiment 2). As before, it was no more popular among subjects who received the Frame 3 version; it was selected by 27.6% of those subjects compared with 26.4% of subjects receiving Frame 1 or 2. Nor was the sure-loss option any more popular among subjects who judged Frame 3 to be most natural after making their choice; it was chosen by 24.0% of those who preferred Frame 3 and by 28.7% of those who preferred the other frames.

In addition to showing the robustness of Experiment 2's results, Experiment 3 provided some subsidiary information about frame judgment. Postponing the naturalness judgments had no effect on subjects' (minimal) propensity to judge the presented frame as being most natural; 36.6% did so here, versus 36.0% in Experiment 2. Nonetheless, when asked unexpectedly after 40 min. of unrelated tasks to remember which of the three frames had been used in their problem description, 75.0% did so correctly (compared with the 34.0% that would be expected by chance). When asked after the memory task which frame they now preferred, 75.3% expressed the same preference as they had earlier. This consistency of preference was quite similar for those who had initially preferred Frames 1, 2, and 3 (71.4%, 76.7%, and 77.3%, respectively). Although these recall tasks focused additional attention on the frame that was presented, they did not increase its popularity: Still only 36.0% of subjects judged the presented frame to be most natural. Apparently, people do attend to frames; however, their frame preferences are not readily manipulated.

Ninety-one percent of subjects remembered which option they had chosen. This percentage was somewhat higher among those who chose the gamble (94.6% vs. 81.0% for those who chose the sure loss, $Z = 1.86$), perhaps reflecting a somewhat weaker strength of preference among subjects choosing that option.

Experiment 4

A natural question to ask at this juncture is how general these results are. Experiments 4, 5, and 6 test their generality by varying different parameters of the research design. Experiment 4 replaces the civil defense context with analogous options using dollars. It asks, in effect, whether there is something special about losses of life that makes people's choices insensitive to frames. Although the lead examples in Tversky and Kahneman's (1981) presentation of framing effects involved losses of life and subsequent examples used dollars, they did not undertake any systematic variation of the kinds of stakes.

Method

Forms comparable to those used in Experiment 2 were created by replacing all losses of life by losses of an equivalent number of dollars. Instead of the civil defense introduction, subjects were told simply to "imagine that you are faced with two unattractive options." One hundred and seventeen individuals judged these forms.

Results

Two versions of the Frame 1 form were again used. Again, reversing the order of options had no discernible effect. In both cases, approximately one third of subjects chose the sure loss. Their choices of most and least natural frame were essentially identical. Hence, responses to those two forms were combined.

The left-hand side of Table 11.4 shows frame preferences as a function of frame presented. As before, frame preferences were quite robust, showing little response to frame presentation. Frame 2 was somewhat more popular when its formulation was used; Frame 1 was somewhat less popular. There seems to be some shift in the overall popularity of the different frames with the two kinds of stakes. Frame 1 was judged most popular 29% of the time in the lives context, compared with 40% with the dollars context (Z = 2.05). Frame 3's share of choices as most popular dropped correspondingly from 38% to 29% (Z = 1.62). If robust and common, such context dependency could further complicate the task of predicting what frames people will adopt (and what choices they will make).

Given the reduced popularity of Frame 3, one would expect the sure loss to be even less popular here than it was in the lives context. However, it was actually somewhat more popular, being chosen by 26% of subjects here, compared with 19% in Experiment 2 (Z = 1.49). Perhaps limiting one's losses is a more attractive (or palatable) option when those losses are dollars rather than lives.

As shown by the right-hand side of Table 11.4, there was again no relationship within experiments between option choice and either frame preference or frame presentation.

Experiment 5

For subjects who are unversed in prospect theory, it is necessary to devise a verbal equivalent of each frame, both for presenting the problem in a particular formulation and for inquiring about the naturalness of each frame. One possible concern

TABLE 11.4 Frame and option preferences in Experiment 4 (dollar problem)

		% of subjects choosing frame as most natural			*% of subjects choosing sure-loss option*			
		Frame preference			*Frame preference*			
Frame presented	*n*	*1*	*2*	*3*	*1*	*2*	*3*	*Ave %*
1	38	29	32	39	36	17	40	32
2	40	43	43	15	41	24	0	28
3	39	49	18	33	26	0	23	21
Ave%		40	31	29	34	17	26	23

Note. The number of subjects involved in each cell equals the entry in the left-hand part of the table multiplied by *n* for that row, divided by 100. For example, 6 people (15 × 40/100) preferred Frame 3 when they were presented Frame 2; none of these 6 preferred the sure loss, producing the 0% on the corresponding cell on the right-hand side.

about Experiments 1–4 might be that the manner in which each frame was described verbally somehow failed to capture its formal properties. Experiment 5 used different descriptions designed to highlight the reference point in each frame and to increase the similarity between the way that each frame was described and the way that the corresponding problem description was formulated.

In Experiments 1–4, neither frame preference nor frame presentation predicted option choice. If one could choose which of those two variables would be related to option choice, frame presentation would be the more useful predictor. As an experimenter, one can control which frame is used. As an observer, one might hope to discern the frame latent in the presentation of a real-world problem. Even if option choices were related to frame preferences, we know so little about how those preferences are determined that it would be difficult to exploit that relationship. The modest shifts in the overall popularity of Frames 1 and 3 between the dollars and lives contexts suggest that these preferences may not even be stable across content areas. Experiment 5 explored the stability of frame preference by presenting each subject with both a dollars and a lives problem (both using the same new frame formulations).

Method

Three new civil defense forms were created, each cast in the perspective of one frame. Because the order in which options were presented had no effect in Experiments 2 and 4, the same ordering was used here for each form (sure loss first).

In the description of the "three ways one might think about the problem," Frame 1 was left untouched. Frame 2 was changed nominally to mention the sure loss first. Frame 3 was reworded to make the reference point more independent conceptually of the sure-loss option. It read:

Frame 3. One can guarantee that no more and no less than 50 lives will be lost. Option A assures that that will happen. Option B means gambling over a .5 chance to *save* 10 lives and a .5 chance to *lose* 10 additional lives.

The Frame 1 form was unchanged except for this rewording of the frame descriptions.

The introduction of the Frame 2 form was changed in order to accentuate the 40 lost lives that were common to both options. Specifically, a sentence was added stating that "Forty people's lives are going to be lost whatever you do." The option descriptions were then changed to:

> *Option A:* Would result in the loss of 10 additional lives.
> lose 10 additional lives.
> *Option B:* Offers a .5 probability of containing the threat with the loss of 40 lives. On the other hand, there is also a .5 probability of losing 20 additional lives. It is like taking the gamble:
> .5 lose no additonal lives
> .5 lose 20 additional lives.

Analogous changes attempted to highlight the reference point in the Frame 3 form. To the introduction was added "Fifty lives are in imminent danger." The options were described as:

> *Option A*: Would constrain the loss of life to those 50 people. That is, it would result in
>
> saving 0 lives
> losing 0 additional lives.
>
> *Option B*: Carries with it a .5 probability of saving 10 of those people whose lives are in danger and a .5 probability of losing 10 additional lives. It is like taking the gamble:
>
> .5 saving 10 lives
> .5 losing 10 additional lives.

A set of three forms for the dollars context was constructed making comparable changes. Each subject received a dollars form using the same frame some 15 min. after completing the civil defense form, following several unrelated intervening tasks. This within-subjects design was intended to give some preliminary idea of the consistency of option and frame preferences across these two domains. However, given the transparent similarity of the two tasks and the proximity of their presentation, the comparisons must be treated cautiously.

A total of 127 individuals completed these forms, with roughly equal numbers receiving each frame formulation.

Results

Civil defense. The left-hand side of Table 11.5 reports the frame preferences of subjects shown for each form. Over all forms, the re-wordings reduced Frame 3's share of most natural judgments from 38% to 25% ($Z = 2.41$) while increasing selections of Frame 1 (from 29% to 34%, $Z = 0.94$) and of Frame 2 (from 33% to 41%, $Z = 1.45$). Whether these changes suggest small or large sensitivity of frame preference to verbal representation depends on how large one judges the present wording changes to be. The attempt to increase the similarity between the frame descriptions shown to all subjects and the problem descriptions designed to capture each frame was moderately successful. Each frame was judged to be more natural by subjects whose problem description was intended to match that frame than by subjects receiving one of the other two problem descriptions. Overall, 45.7% of subjects preferred the frame they saw.

The right-hand side of Table 11.5 shows option choice as a function of frame given and frame preferred. Even though the popularity of Frame 3 dropped by one third (compared with Experiment 2), the popularity of the sure-loss option increased from 19% to 30% of all choices ($Z = 2.20$). Within Experiment 5, the frequency of choosing the sure loss was unrelated to the frame used or to the frame chosen as most natural (or to choices of least natural—not shown).

TABLE 11.5 Frame and option preferences in Experiment 5

Frame presented	*n*	*% of subjects choosing frame as most natural*			*% of subjects choosing sure-loss option*			
		Frame preference			*Frame preference*			
		1	*2*	*3*	*1*	*2*	*3*	*Ave %*
				Civil defense problem				
1	43	51	28	21	36	25	33	33
2	41	15	59	27	33	25	45	32
3	43	35	37	28	40	19	17	26
Ave%		34	41	25	37	23	31	30
				Dollars problem				
1	40	63	18	20	44	14	38	38
2	41	39	39	22	38	31	33	34
3	42	50	24	26	57	50	27	48
Ave%		50	27	23	47	33	32	40

Dollars. Although they should be used cautiously, responses to the dollar forms were similar to those seen with the civil defense forms in the present study and the dollar forms of Experiment 4: (a) In Experiments 2 and 4, the sure loss was more popular with the dollar problem than with the civil defense problem (31% vs. 19%). Here, the respective percentages were 40% and 30%. (b) The increased popularity of the sure loss was achieved without a corresponding increase in the popularity of Frame 3. (c) There was no relation between frame preference and option choice. (d) As with the civil defense forms, Experiment 5's rewording increased Frame l's popularity (from 40% to 50% of "most natural" choices). (e) As with the civil defense forms, all three frames were most often judged most natural by subjects whose problem had been presented in terms of that frame.

The only deviation from this pattern of similarities was that subjects who received the Frame 3 form chose the sure-loss option more often than did subjects who received the other forms (48% vs. 36%, $Z= 1.27$). Although this change is in the anticipated direction, given the welter of comparisons showing no difference or modest changes in the opposite direction, it is hard to attribute very much significance to it.

Consistency. Comparing responses to the dollars and the lives versions revealed that the great majority of subjects also preferred the same frame for both problems (69.0% vs. the 36.8% that would be expected by chance). Their option choices were somewhat less consistent, with 63.1% choosing the same option both times, compared with 54.0% expected by chance. If one assumes that subjects addressing the dollar problem had no memory of the civil defense problem, these results might be taken to indicate that they had more stable (or better articulated) frame preferences than option preferences. If one assumes that subjects remembered everything, the result at least indicates that they did not feel compelled to be consistent.

Experiment 6

An implicit assumption of the preceding experiments is that people interpret the naturalness question as an enquiry regarding the frame that they themselves have used. Conceivably, however, they might have interpreted it as a judgment of consensus (how most other people would judge the problem) or of normative status (how the problem should be considered). Because prospect theory is not a normative theory, it makes no statement as to which frame subjects should use. That does not mean that subjects do not decide spontaneously that the experimenter has some "right answer" in mind when asking about naturalness. If subjects did interpret "most natural frame" as "proper frame," then the absence of any relationship between frame and option preference would mean rejection of prospect theory as a normative guide (a finding that would not trouble advocates of the theory unless one believed that a descriptively valid theory must be normatively appealing). Or, one could argue that subjects did give the theory normative status, but were unable to follow its dictates in linking a frame preference with an option choice. To clarify the meaning of the naturalness judgment, subjects in Experiment 6 were asked explicitly about what frame had guided them.

Method

Forms were identical to those of Experiment 3, except that subjects were asked "which of these three phrasings most closely captures the way in which you thought about the problem when making your choice between the two options." Following description of the three frames, this question preceded the two questions regarding the most and least natural ways of thinking about the frame. In all, 180 subjects participated, with approximately equal numbers receiving forms that emphasized each of the three frames.

Results and discussion

In the aggregate, the frame-you-used wording produced patterns of results like those produced by the most-natural-frame wording. Again, only a minority of subjects (41.7%) chose the frame highlighted in their questionnaire. Pooling across the three forms, each frame again had a substantial body of adherents, with Frames 1, 2, and 3 chosen by 18.3%, 45.6%, and 36.1% of subjects in Experiment 6, respectively, compared with 28.0%, 41.5%, and 30.5% of subjects in Experiment 3, $\chi^2(2) = 3.22$, $p > .10$.

The sure-loss option was for some reason a bit more popular here than in Experiment 3; it was chosen by 35.0% of subjects compared with 26.8% ($Z = 1.33$). Preference for the sure loss was, however, still unrelated to choice of frame. It was chosen by 35.4% of those who reported using Frame 3 and by 34.8% of those who reported relying on Frame 1 or 2. Thus, subjects either did not know what frames they used or did not use prospect theory to translate that frame into a choice. The sure loss was chosen slightly, but not significantly, more often by those who received Frame 3 than by those who received Frame 1 or 2 (42.6% vs. 31.1%, $Z = 1.54$).

As mentioned, subjects chose the most (and least) natural frame after pointing to the one they used. The juxtaposition of these tasks would seem to force subjects to find some way to distinguish between the "most natural" frame and the one they had just reported choosing (otherwise, why would they be asked two separate questions). Because their sequential position renders the naturalness judgments suspect, they are reported only cursorily. In general, they looked much like those observed elsewhere: The presented frame was chosen by a minority of subjects (40.0%) as most natural, and there was no relation between frame preference and option preference, $\chi^2(1) = .34$. Some 51.6% of subjects chose as most natural the frame that they reported using (compared with the 31.1% that would be expected by chance). Even among these "consistent subjects," who might be seen as having the most robust frame preferences, there was no relation between frame and option choice.

Experiment 7

All of the forms used in Experiments 1–6 have involved one particular gamble, albeit presented in a variety of ways and contexts. The "hidden story" of these studies has been the persistent popularity of the gamble relative to the sure loss, independent of the frame used or preferred. One step toward establishing the generality of the present results is clarifying the generality of that aversion to sure losses (or preference for risk seeking). Experiment 7 varies the parameters of the civil defense problem, always keeping the sure loss equal to the expected value of the gamble.

If we had more detailed knowledge of the shapes of the value function and the probability weighting function, it would be possible to derive precise predictions regarding the way in which these changes should affect the option preferences of individuals adopting various frames. Conversely, one could use the changes in option preferences observed with such changes in the options as a means to assess the shapes of the functions. As such detailed analyses are beyond the scope of the present article, the new problems will be exploited primarily as indicators of the robustness of the preference for risk seeking observed above.

Method

Eight variants of the basic problem were created and presented to groups of approximately 40 subjects drawn from the same pool as the previous studies. The problems are described on the left-hand side of Table 11.6. The "gamble parameters" are the probabilities and magnitudes of the two outcomes of the gamble; "sure loss" is the magnitude of the sure loss, which was always set equal to the expected value of the gamble. The variations may be briefly described as:

- Problem 1 changes the gamble's outcomes from 40 and 60 (in the basic problem) to 0 and 100. This change raises the maximum possible loss considerably but also provides a .5 chance of escaping with no loss at all.
- Problem 2, like Problem 1, offers the chance of no loss at all with the gamble. However, the maximum possible loss and sure loss were both raised considerably.
- Problems 3 and 4 use the gamble losses of Problems 1 and 2, respectively. However, they shift the gamble's probabilities for (no loss, loss) from (.5, .5) to (.8, .2). The sure loss is correspondingly reduced by a factor of 2½.
- Problems 5 and 6 use the gamble losses of Problems 3 and 4 but reduce the probability of the loss outcome even more. The sure loss is also reduced.
- Problems 7 and 8 are like the basic problem in offering no hope of escaping without loss. They differ from the basic problem in the variance of the gamble's outcomes.

These problems were all presented in the same format as was the basic problem. Alternative frames were neither used nor asked about because the present data were collected some years ago, prior to the development of prospect theory. Subjects were, however, asked how the number of lives lost with the sure-loss option would have to change for them to change their preferences. Specifically, they were asked (where A represents the gamble):

1. If you chose A over B, how few lives would have to be lost in the certain-loss option (B) before you would take it?
   I would choose B if "only" ______ lives were lost for certain.
2. If you chose B over A, how high could the life toll in the certain-loss option go before you would shift to A?
   I would shift to A if the certain loss with B rose to ______ lives.

Such judgments could, in principle, be used to parameterize the various curves in prospect theory. However, they are used here only to get some idea of how strong subjects' preferences were for the options they chose.

Table 11.6 Option preferences for variations of civil defense problem

	Problem description					*Preferences*			
	Gamble parameters							*Mdn switching value*	
Problem number	P_1	L_1	P_2	L_2	*Sure loss*	*n*	*% of subjects choosing sure loss*	*Chose gamble*	*Chose sure loss*
1	.5	0	.5	100	50	39	3	2	51
2	.5	0	.5	100,000	50,000	34	9	50	75,000
3	.8	0	.2	100	20	39	13	0	25
4	.8	0	.2	100,000	20,000	36	11	100	60,000
5	.99	0	.01	100	1	41	29	0	5
6	.99	0	.01	100,000	1,000	46	4	50	37,500
7	.5	1	.5	99	50	39	13	1	75
8	.5	25	.5	75	50	38	18	25	60
Basic[a]	.5	40	.5	60	50	42	14	–	–
Total						312	13		

Note. P = probability; L = loss of lives.
[a] Results from simple form of Experiment 1—not included in totals.

Results

Choices. The clearest pattern in Table 11.6 is the preponderance of subjects who preferred the gamble in every variation of this civil defense problem. Sure-loss choices ranged from 3% (Problem 1) to 29% (Problem 5), averaging 12.5% over all conditions. This suggests that the results obtained with the particular values used in the basic problem were not unrepresentative of what would have been obtained with a wide variety of other values. Although such consistency of option choices does not guarantee that the relations between choices and frames would be the same with all these problem variants, it does suggest that replication of Experiments 1–6 with different parameter values is not the most promising way to explore the effects of framing.

Both because the variations in results across problems are fairly small and because the problems were not designed to discriminate between frames, a detailed exposition is inappropriate here. Attention to a few examples may help show the kind of predictions that prospect theory makes in such cases (whatever frame is adopted) and the modest support for those predictions that is found here. For example, contrast Problem 1 with the basic problem; The convexity of the loss function means that reducing L_1 from 40 to 0 should have more impact than increasing L_2 from 60 to 100. As the sure loss is the same in both cases, the gamble should be more attractive in Problem 1, which it is ($Z = 1.88$). Similar logic leads one to expect the gamble to be more attractive in Problem 7 than in the basic problem, which it is not. On the other hand, compare Problems 1 and 7: Adding a first life lost to L_1 should do more to reduce the gamble's attractiveness in Problem 7 than subtracting 1 life from the 100 associated with L_2 does to increase it. Problem 7's gamble is, indeed, less attractive ($Z = 1.70$). Within the limits of the present method and sample sizes, the combination of the basic problem with 1 and 7 suggests that qualitative effects, such as crossing the threshold from 0 to 1 life lost, may be more important than quantitatively large changes.

As a final example, consider the contrast between Problems 5 and 6. The relatively great popularity of the sure loss in Problem 5 is surprising because of the relatively great weight given both to the first unit of loss and to outcomes known with certainty. Prospect theory would presumably account for this popularity by noting that small probabilities are overweighted, meaning that a substantial portion of subjects gave enough weight to P_2 (equal to .01) to compensate for the aversiveness of the sure loss. In Problem 6, the probabilities remain the same, but the stakes have escalated considerably. The great drop-off in support for the sure loss ($Z = 3.16$) suggests that the value function for losses is much less steep in the 1,000 to 100,000 range than it is in the 1 to 100 range.[2]

Strength of preference. The switching values, indicating the value that the sure loss would have to assume for subjects to change their preferences, offer some hints as to the strength and nature of subjects' preferences. In general, those who chose the gamble initially were very reluctant to switch. With Problem 1, for example, the median switching value of 2 means that they wanted to have the sure loss cut by 96%, from 50 to 2. With Problems 3, 5, 7 and 8, the median subject wanted the sure loss reduced to L_1, the lesser of the gamble's possible losses, before switching. If taken literally, these responses would reflect risk seeking in the extreme. It remains to be seen, however, whether numerical judgments, such as switching values, will prove as robust as the categorical judgments of option preference. Two specific qualifications of the present method are that (a) these were values for switching, not values indicating when the two options were equally attractive; (b) these judgments were given by individuals who had already made an explicit commitment to one option. Nonetheless, it seems as though these subjects were willing to incur quite large risks in order to avoid a sure loss of life.

So few subjects chose the sure loss that it is difficult to reach any firm conclusions on the basis of their switching values. It appears, though, that these individuals were much less dogged in their choices. For most problems, only a modest increase in the sure loss was needed for these subjects to switch to the gamble. In no case was a subject willing to let the sure loss rise to the value of L_2 before switching.

Like subjects in Experiments 1–6, these subjects also indicated the strength of their preference for the chosen alternative on a 4-point scale. These ratings provided the same picture as did the switching values. For seven of the eight problems, subjects who preferred the gamble were more confident in their choices than were subjects who chose the sure loss. Over all problems, the mean preferences were 2.5 versus 1.9, respectively (using the same scoring system as in Experiment 1).

Discussion

If one assumes that people adopt Frames 1 or 2, then the present results provide strong support for one of prospect theory's strongest predictions. According to the theory, people should and, according to these data, people do prefer the gamble to the sure loss. Moreover, they do so despite a variety of changes in parameters, context, wording, instructions, and order of presentation. Although these variations do not exhaust all possibilities, they do cover several local ranges fairly well.

This prediction (and set of behaviors) runs contrary to the predictions of the standard expected utility model. That theory applies no weighting function to probabilities and assumes that people are risk averse for losses. From that perspective, the expected utility of the gamble will be equal to $.5v(-40) + .5v(-60)$, whereas the expected utility of the sure loss will be simply $v(-50)$. Risk aversion assures that the average of $v(-40)$ and $v(-60)$ will be less than $v(-50)$. Hence, people will prefer the sure loss, which has less negative utility.

Why, however, should one assume that people adopt Frame 1 or 2 rather than Frame 3? Frame 3 is not only a fairly popular perspective among subjects but also quite a legitimate one from prospect theory's point of view. Indeed, one could not even treat such a perspective within expected utility theory, which assumes that prospects are evaluated by looking at their effect on one's total asset position, not by comparison with a reference point reflecting where one could be by accepting a sure loss.

According to Tversky and Kahneman (1981):

> A diversity of factors determine the reference outcome in everyday life. The reference outcome is usually a state to which one has adapted; it is sometimes set by social norms and expectations; it sometimes corresponds to a level of aspiration, which may or may not be realistic. (p. 456)

In this light, the present studies can be viewed as having explored—and established—what reference point people tend to adopt in this sort of problem. The robustness of this dominant perspective is perhaps demonstrated most clearly by the negligible impact of trying to present the problem so as to accentuate Frame 3's perspective. Unlike some of the problems used as demonstrations by Kahneman and Tversky (1979; see also Fischhoff, Slovic, and Lichtenstein, 1980; Tversky and Kahneman, 1981), this kind of problem does not evoke labile preferences. If one wished to extract a general principle, it might be that even though shifts to a minimax loss reference point are theoretically possible, they are not commonly undertaken. That is, people do not readily adapt to absorbing losses. Such a principle would be one step toward developing the substantive theory of how people respond to particular decision problems that is needed to give prospect theory greater predictive validity.

The difficulty with this interpretation is that it does not account for the fact that Frame 3 was judged to be the most natural perspective by roughly one third of all subjects in each condition of Experiments 1–6. Nor does it account for the absence of any relation within those studies between frame preference and option preference. Indeed, to the extent that overall preference levels for Frame 3 did vary across experiments (from 23% to 38%), its popularity was negatively correlated ($\gamma = -.80$) with variations in the popularity of the sure loss (from 14% to 40%). Clearly, one would like to be able to predict people's choices from their appraisal of what frame seems to fit a problem best, particularly when those naturalness judgments seem quite robust (as suggested by Experiment 5). If one can only infer frames from preferences after assuming the truth of the theory, one runs the risk of making the theory itself untestable. The dissatisfaction that many people have with utility theory often arises from observing the convoluted lengths to which one must go in reinterpreting decision problems so as to find something whose utility people were maximizing (Fischhoff, Goitein, and Shapira, 1981).

One quick way to dispel these doubts is to dismiss naturalness judgments as a research tool, arguing that people cannot introspect accurately about the factors influencing their decisions and choices. An analysis of the situations in which introspections are more and less likely to be accurate is beyond the present article (see, e.g., Ericsson and Simon, 1980; Nisbett and Wilson, 1977; Smith and Miller, 1978; White, 1980). Tversky and Kahneman's (1981) feeling that "individuals who face a decision problem … are normally unaware of alternative frames and of the potential effects on the relative attractiveness of options" (p. 457) might be seen as supporting the mistrust of introspections about decision processes such as those studied here. If one accepts prospect theory, then the present results might encourage one to agree with them.

Acknowledgements

This research was supported by the Office of Naval Research under Contract N00014-82-C-0643 to Perceptronics, Inc.

I wish to thank to Ruth Beyth-Marom, Daniel Kahneman, Don MacGregor, Paul Slovic, Amos Tversky, Thomas Wallsten, and three anonymous reviewers for comments on earlier drafts of this article.

Notes

1 In addition to these three substantive operations, there are also three neatening or stylistic operations: (d) *combination,* adding the probabilities associated with identical outcomes. Thus, a gamble that gave one $200 with .25 probability, $200 with .25, or $0 with .5, would be translated to receiving $200 with .5 and $0 with .5; (e) *simplification*, rounding of probabilities or outcomes; (f) *detection of dominance*, eliminating from further consideration any choice option that is inferior to some other option in all respects. These operations are not considered here.

2 Alternatively, one could argue that something about the size of the stakes makes the .01 probability much more likely to be considered an impossibility in Problem 6 than in Problem 5. Such an influence of values on probability weighting is not only intellectually unsatisfying but also is at odds with the great importance that people seem to attach to unlikely, but potentially catastrophic, events (Slovic, Fischhoff, and Lichtenstein, 1979).

References

Ericsson, K. A., and Simon, H. A. 1980. Verbal reports as data. *Psychological Review, 87*, 215–251.

Fischhoff, B., Goitein, B., and Shapira, Z. 1981. The experienced utility of expected utility approaches. In N. Feather (Ed.), *Expectancy, incentive and action*. Hillsdale, N.J.: Erlbaum.

Fischhoff, B., Slovic, P., and Lichtenstein, S. 1980. Knowing what you want: Measuring labile values. In T. Wallsten (Ed.), *Cognitive processes in choice and decision behavior*. Hillsdale, N.J.: Erlbaum.

Kahneman, D., and Tversky, A. 1979. Prospect theory. *Econometrica, 47*, 263–292.

Nisbett, R. E., and Wilson, T. D. 1977. Telling more than we can know: Verbal reports on mental processes. *Psychological Review, 84*, 231–259.

Slovic, P., Fischhoff, B., and Lichtenstein, S. 1979. Rating the risks. *Environment, 21*(3), 14–20, 36–39.

Slovic, P., Fischhoff, B., and Lichtenstein, S. 1982. Response mode, framing, and information-processing effects in risk assessment. In R. Hogarth (Ed.), *New directions for methodology of social and behavioural science: Question framing and response consistency*, No. 11. San Francisco: Jossey-Bass.

Smith, E. R., and Miller, F. S. 1978. Limits on perception of cognitive processes: A reply to Nisbett and Wilson. *Psychological Review, 85*, 355–362.

Tversky, A., and Kahneman, D. 1981. The framing of decisions and the psychology of choice. *Science, 211*, 453–458.

White, P. 1980. Limitations on verbal reports of internal events: A refutation of Nisbett and Wilson and of Bem. *Psychological Review, 87*, 105–112.

12

VALUE ELICITATION

Is there anything in there?

Baruch Fischhoff

DEPARTMENT OF SOCIAL AND DECISION SCIENCES AND DEPARTMENT OF ENGINEERING AND PUBLIC POLICY, CARNEGIE MELLON UNIVERSITY

Abstract

Eliciting people's values is a central pursuit in many areas of the social sciences, including survey research, attitude research, economics, and behavior decision theory. These disciplines differ considerably in the core assumptions they make about the nature of the values that are available for elicitation. These assumptions lead to very different methodological concerns and interpretations, as well as to different risks of reading too much or too little into people's responses. The analysis here characterizes these assumptions and the research paradigms based on them. It also offers an account of how they arise, rooted in the psychological and sociological contexts within which different researchers function.

> Taken all together, how would you say things are these days—would you say that you are very happy, pretty happy, or not too happy?
>
> National Opinion Research Center (NORC), 1978

> Think about the last time during the past month that you were tired easily. Suppose that it had been possible to pay a sum of money to have eliminated being tired easily immediately that *one* time. What sum of money would you have been willing to pay?
>
> Dickie, Gerking, McClelland, and Schulze, 1987, p. 19 (Appendix 1)

> In this task, you will be asked to choose between a certain loss and a gamble that exposes you to some chance of loss. Specifically, you must choose either: Situation A. One chance in 4 to lose $200 (and 3 chances in 4 to lose nothing). OR Situation B. A certain loss of $50. Of course, you'd probably prefer not to be in either of these situations, but, if forced to either play the gamble (A) or accept the certain loss (B), which would you prefer to do?
>
> Fischhoff, Slovic, and Lichtenstein, 1980, p. 127

> 600 people are ill from a serious disease. Physicians face the following choice among treatments: Treatment A will save 200 lives. Treatment B has 1 chance in 3 to save all 600 lives and 2 chances in 3 to save 0 lives. Which treatment would you choose, A or B?
>
> Tversky and Kahneman, 1981, p. 454

Problematic preferences

A continuum of philosophies

A critical tenet for many students of other people's values is that "If we've got questions, then they've got answers." Perhaps the most ardent subscribers to this belief are experimental psychologists, survey researchers, and economists. Psychologists expect their "subjects" to behave reasonably with any clearly described task, even if it has been tortuously contrived in order to probe esoteric theoretical points. Survey researchers expect their "participants" to provide meaningful answers to items on any topic intriguing them (or their clients), assuming that the questions have been put into good English. Economists expect "actors" to pursue their own best interests, thereby making choices that reveal their values, in whatever decisions the marketplace poses (and economists choose to study).

This article examines this *philosophy of articulated values* both in its own right and by positioning it on a continuum of philosophies toward value formation and measurement. At the other end of this continuum lies what might be called the *philosophy of basic values*. It holds that people lack well-differentiated values for all but the most familiar of evaluation questions, about which they have had the chance, by trial, error, and rumination, to settle on stable values. In other cases, they must derive specific valuations from some basic values through an inferential process.

Perhaps the clearest example of this latter perspective might be found in the work of decision analysts (Raiffa, 1968; von Winterfeldt and Edwards, 1986; Watson and Buede, 1988). These consultants lead their clients to decompose complex evaluation problems into basic dimensions of concern, called *attributes*. Each attribute represents a reason why one might like or dislike the possible outcomes of a decision. For example, the options facing someone in the market for a car are different vehicles (including, perhaps, none at all), whose attributes might include cost, style, and reliability.

The relative attractiveness (or unattractiveness) of different amounts of each attribute is then captured in a *utility function*, defined over the range of possible consequences (e.g., Just how much worse is breaking down once a month than breaking down twice a year?). After evaluating the attributes in isolation, the decision maker must consider their relative importance (e.g., Just how much money is it worth to reduce the frequency of repairs from annual to biennial?). These tradeoffs are expressed in a multiattribute utility function. Having done all of this, the consequences associated with specific actions are then evaluated by mapping them into the space spanned by that function.

Between the philosophies of articulated values and basic values, lie intermediate positions. These hold that although people need not have answers to all questions, neither need they start from scratch each time an evaluative question arises. Rather, people have stable values of moderate complexity, which provide an advanced starting point for responding to questions of real-world complexity. Where a particular version of this perspective falls on the continuum defined by the two extreme philosophies depends on how well developed these partial perspectives are held to be.

Each of these philosophies directs the student of values to different sets of focal methodological concerns. For example, if people can answer any question, then an obvious concern is that they answer the right one. As a result, investigators adhering to the articulated values philosophy will worry about posing the question most germane to their theoretical interests and ensuring that it is understood as intended. On the other hand, if complex evaluations are to be derived from simple evaluative principles, then it is essential that the relevant principles be assembled and that the inferential process be conducted successfully. That process could fail if it required too much of an intellectual effort and, also, if the question were poorly formulated or inadequately understood. If people have thought some about the topic of an evaluation question, then they have less far to go in order to produce a full answer. Yet, even if people hold such partial perspectives, there is still the risk that they will miss some nuances of the question and, as a result, overestimate how completely they have understood it and their values regarding the issues that it raises.

A choice of paradigms

The effort to deal with these different worries in a systematic fashion has led to distinct research paradigms (Kuhn, 1962). Each such paradigm offers a set of methods for dealing with its focal worries, along with empirical tests of success in doing so. Each has evolved some theory to substantiate its approach. As paradigms, each is better suited to answering problems within its frame of reference than to challenging that frame. Thus, for example, the articulated values paradigm is better at devising additional ways to improve the understanding of questions than at determining whether understanding is possible.

This is, of course, something of a caricature. Many investigators are capable of wearing more than one hat. For example, survey researchers have extensively studied the properties of the *don't know* response (T. Smith, 1984). Still, when one is trying to get a survey (or experiment or economic analysis) out the door, it is hard to address these issues at length for every question. It may be easier to take *no answer* for an answer in principle than in practice. At the other extreme, it may be unprofitable for a consulting decision analyst to deal with situations in which the answer to a complex evaluation question is there for the asking, without the rigmarole of multiattribute utility elicitation.

To the extent that studies are conducted primarily within a single paradigm, it becomes critical to choose the right one. Table 12.1 summarizes the costs of various mismatches between the assumed and actual states of people's values. Above the diagonal are cases in which more is expected of people than they are prepared to give. The risk here is misplaced precision, reading too much into poorly articulated responses and missing the opportunity to help people clarify their thinking. Below the diagonal are cases in which too little is expected of people. The risk here is misplaced imprecision, needlessly complicating the task and casting doubt on already clear thinking.

The choice of a paradigm ought to be driven by the perceived costs and likelihoods of these different mismatches. Thus, one might not hire a survey researcher to study how acutely ill individuals evaluate alternative medical procedures, nor might one hire a philosopher to lead consumers through the intricacies of evaluating alternative dentifrices. Evaluation professionals should, in turn, devote themselves to the problems most suited to their methods.

Yet, it is in the nature of paradigms that they provide clearer indications of relative than of absolute success. That is, they show which applications of the set of accepted methods work better, rather than whether the set as a whole is up to the job. After describing these paradigms in somewhat greater detail, I will consider some of the specific processes by which work within them can create an exaggerated feeling for the breadth of their applicability.

As a device for doing so, I will highlight how each paradigm might interpret several sets of potentially puzzling results, namely those produced by the studies posing the four evaluation questions opening this article. In each case,

TABLE 12.1 Risk of misdiagnosis

	Proper assumption		
Assumption made	*Articulated values*	*Partial perspectives*	*Basic values*
Articulated values	—	Get incomplete values Inadvertently impose perspective	Get meaningless values Impose single perspective
Partial perspectives	Promote new perspectives Distract from sharpening	—	Impose multiple perspectives Exaggerate resolvability
Basic values	Shake confidence Distract from sharpening	Discourage Distract from reconciliation	—

Note. Above diagonal: misplaced precision, undue confidence in results, missed opportunity to help. Below diagonal: needless complication, neglect of basic methodology, induced confusion.

two apparently equivalent ways of formulating the question produced rather different evaluations. Assuming that the studies were competently conducted, an articulated values perspective would hold that if the answers are different, then so must the questions have been. Any inconsistency is in the eye of the beholder, rather than in the answers of the respondents.

A basic values philosophy leads to quite a different interpretation: If their responses are buffetted by superficial changes in question formulation, then people must not know what they want. As a result, none of the evaluations should be taken seriously. At best, they reflect a gut level response to some very general issue. According to the intermediate, partial perspectives philosophy, each answer says something about respondents. However, neither should be taken as fully representing their values.

A sample of problems

Happiness. Surveys sometimes include questions asking respondents to evaluate the overall state of their affairs. Answers to these questions might be used, for example, as barometers of public morale or as predictors of responses on other items (i.e., for statistical analyses removing individual mood as a covariate). In reviewing archival data, Turner and Krauss (1978) discovered the apparent inconsistency revealed in Figure 12.1. Two respected survey organizations, asking virtually identical happiness questions, produced substantially different proportions of respondents evaluating their situation as making them *very happy*. If the temptation of naive extrapolation is indulged, then quite different societies seem to be emerging from the two surveys (happinesswise, at least).[1]

After a series of analyses carefully examining alternative hypotheses, Turner and Krauss (1978) concluded that the most likely source of the response pattern in Figure 12.1 was differences in the items preceding the happiness question. In the NORC survey, these items concerned family life; in the Survey Research Center (SRC) survey (Campbell, Converse, and Rodgers, 1976), they were items unrelated to that aspect of personal status.[2]

If respondents have fully articulated values, then different answers imply different questions. Inadvertently, the two surveys have created somewhat different happiness questions. Perhaps Happiness_1 (from the NORC survey) emphasizes the role of family life, whereas Happiness_2 (from the SRC survey) gives respondents more freedom in weighting the different facets of their lives.

From the opposing perspective, the same data tell quite a different story. If a few marginally related questions can have so great an impact, then how meaningful can the happiness question (and the responses to it) be? Conceivably, it is possible to take all things together and assess the happiness associated with them. However, as long as assessments depend on the mood induced by immediately preceding questions, that goal has yet to be achieved.

According to the partial perspectives philosophy, the two responses might be stable. However, neither should be interpreted as a thoughtful expression

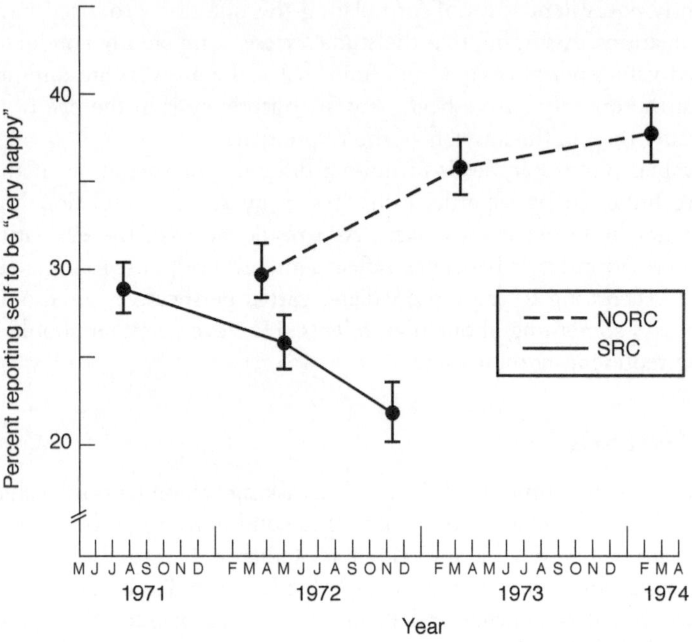

FIGURE 12.1 Trends in self-reported happiness, 1971–1973

Note. Estimates are derived from sample surveys of noninstitutionalized population of the continental United States, aged 18 and over. Error bars demark ±1 standard error around sample estimate. NORC = National Opinion Research Center; SRC = Survey Research Center. Questions were "Taken all together, how would you say things are these days—would you say that you are very happy, pretty happy, or not too happy?" (NORC); and "Taking all things together, how would you say things are these days—would you say you're very happy, pretty happy, or not too happy these days?" (SRC). From "Why Do Surveys Disagree? Some Preliminary Hypotheses and Some Disagreeable Examples" (p. 166) by C. F. Turner, 1984, in C. F. Turner and E. Martin, *Surveying Subjective Phenomena*, New York: Russell Sage Foundation. Copyright 1984 by the Russell Sage Foundation. Reprinted by permission.

of respondents' happiness. Achieving that would require helping respondents generate and evaluate alternative perspectives on the problem, not just the one perspective that happens to have been presented to them.

Headache. According to Executive Order 12291 (Bentkover, Covello, and Mumpower, 1985), cost–benefit analyses must be conducted for all significant federal actions. Where those actions affect the environment, that often requires putting price tags on goods not customarily traded in any marketplace. For regulations governing ozone levels, one such good is a change in the rate of subclinical health effects, such as headaches and shortness of breath. In order to monetize these consequences, resource economists have conducted surveys asking questions like the second example in the set of quotations at the beginning of this article (Cummings, Brookshire, and Schulze, 1986; V. K. Smith and Desvousges, 1988).

In Dickie et al.'s (1987) survey, people who reported having experienced being tired easily estimated that they would be willing to pay $17, on average, to eliminate their last day of feeling tired easily. Later in the same survey, the interviewer computed the overall monthly cost of eliminating each respondent's three most serious ozone-related health effects. This was done by multiplying how much people reported being willing to pay to eliminate the last occurrence of each effect by the number of reported episodes per month, then summing those products across symptoms. Respondents were then asked, "On a monthly basis is [—] what you would be willing to pay to eliminate these three symptoms?" (p. 20, Appendix 1). If respondents recanted, they were then asked what monthly dollar amount they would pay for the package. The markedly reduced dollar amount that most subjects provided was then prorated over the individual health effects. By this computation, respondents were now willing to pay about $2 to eliminate a day of being tired easily.

From a regulatory perspective, these strikingly different estimates indicate markedly different economic benefits from reducing ozone levels. (Indeed, the Office of Management and Budget [A. Carlin, personal communication, 1987] has seriously criticized the Dickie et al., 1987, study as a basis for revising regulations under the Clean Air Act.) From an articulated values perspective, they imply that the two questions must actually be different in some fundamental ways. For example, people might be willing to pay much more for a one-time special treatment of their last headache than for each routine treatment. From a basic values perspective, these results indicate that people know that symptomatic relief is worth something, but have little idea how much (even after an hour of talking about health effects). As a result, respondents are knocked about by ephemeral aspects of the survey, such as the highly unusual challenge to their values embodied by the request to reconsider. The investigators in this study seem to have adopted a partial perspectives philosophy. They treat respondents' values seriously, but not seriously enough to believe that respondents have gotten it right the first time. Rather, respondents need the help provided by showing them the overall implications of their initial estimates (Furby and Fischhoff, 1989).

Gamble. In samples of people shown the third example (Fischhoff et al., 1980), most people have preferred the gamble to the sure loss. However, they reverse this preference when the sure loss is described as an insurance premium, protecting them against the potentially greater loss associated with the gamble (Fischhoff et al., 1980; Hershey and Schoemaker, 1980). This difference is sufficiently powerful that it can often be evoked within subject, in successively presented problems.

From an articulated values perspective, the appearance of equivalence in these two versions of the problem must be illusory. Observers who see inconsistency in these responses must simply have failed to realize the differences.

Perhaps, as a matter of principle, people refuse both to accept sure losses and to decline insurance against downside risks. In that case, these seemingly superficial differences in description evoke meaningful differences in how people judge themselves and one another. People want both to preserve a fighting chance and to show due caution. How they would respond to a real-world analog of this problem would depend on how it was presented.[3]

From a basic values perspective, these results show that people know that they dislike losing money, but that is about it. They cannot make the sort of precise tradeoffs depicted in such analytical problems. As a result, they cling to superficial cues as ways to get through the task.

In this case, some subsidiary evidence seemingly supports the intermediate perspective. When both versions are presented to the same person, there is an asymmetrical transfer effect (Poulton, 1968, 1989). Specifically, there are fewer reversals of preference when the insurance version comes first than when it comes second. This suggests that viewing the sure loss as an insurance premium is a relevant perspective, but not one that is immediately available. By contrast, respondents do realize, at some level, that premiums are sure losses. Studies of insurance behavior show, in fact, some reluctance to accept that perspective. For example, people prefer policies with low deductibles, even though they are financially unattractive. Apparently, people like the higher probability of getting some reimbursement, so that their premium does not have to be viewed as a sure loss (Kunreuther et al., 1978).

Disease. About two thirds of the subjects responding to the fourth problem (Tversky and Kahneman, 1981) have been found to prefer Treatment A, with its sure saving of 200 lives. On the other hand, about the same portion prefer the second treatment when the two alternatives are described in terms of the number of lives that will be lost. In this version, Treatment A now provides a sure loss of 400 lives, whereas Treatment B gives a chance of no lives lost at all.

Applying the alternative philosophies to interpreting these results is straightforward. One difference in this case is that there is not only some independent evidence but also some theory to direct such interpretations. The discrepancies associated with the three previous problems were discovered, more or less fortuitously, by comparing responses to questions that happened to have been posed in slightly different ways. In this case, the discrepancies were generated deliberately. Kahneman and Tversky (1979) produced the alternative wordings as demonstrations of their *prospect theory*, which predicts systematic differences in choices as a function of how options are described, or *framed*. The shift from gains (i.e., lives saved) to losses is one such framing difference.

Prospect theory embodies a partial perspectives philosophy. It views these differing preferences as representing stable derivations of intermediate complexity from a set of basic human values identified by the theory. The sources of these differences seem ephemeral, however, in the sense that people would be uncomfortable living with them. Adopting an articulated values philosophy here would require arguing that people regard the different frames as meaningfully different questions—and would continue to do so even after thoughtful reflection.

In the absence of a theoretical account (such as prospect theory) or converging evidence (such as the asymmetrical transfer effect with the sure-loss-premium questions), one's accounting of seemingly inconsistent preferences becomes a matter of opinion. Those opinions might reflect both the particulars of individual problems and the general orientation of a paradigm. The next section describes these paradigms. The following section considers how they could sustain such different views on the general state of human values.

The paradigms

However the notion of paradigm is conceptualized (Lakatos and Musgrave, 1970), it is likely to involve (a) a focal set of methodological worries, (b) a corresponding set of accepted treatments, (c) a theoretical basis for justifying these treatments and directing their application, and (d) criteria for determining whether problems have been satisfactorily addressed. Table 12.2 characterizes the three paradigms in these terms. This section elaborates on some representative entries in that table.

Philosophy of articulated values

Investigators working within this paradigm have enormous respect for people's ability to articulate and express values on the most diverse topics. Indeed, so great is this respect that investigators' worrying often focuses on ensuring that evaluative questions are formulated and understood exactly as intended. Any slip could evoke a precise, thoughtful answer to the wrong question (Fischhoff and Furby, 1988; Mitchell and Carson, 1989; Sudman and Bradburn, 1982).

A hard-won lesson in this struggle involves recognizing the powerful influence that social pressures can exert on respondents (DeMaio, 1984). As a result, investigators take great pains to insulate the question–answerer relationship from any extraneous influences, lest those become part of the question. To prevent such complications, interviewers and experimenters stick to tight scripts, which they try to administer impassively in settings protected from prying eyes and ears. Lacking the opportunity to impose such control, economists must argue that marketplace transactions fortuitously have these desirable properties, in order to justify interpreting purchase decisions as reflecting just the value of the good and not the influences, say, of advertising or peer pressure.

At first blush, this protectiveness might seem somewhat paradoxical. After all, if people have such well-articulated preferences, why do they need to be shielded so completely from stray influences? The answer is that the investigator cannot tell just which stray influence will trigger one of those preferences. Indeed, the more deeply rooted are individuals' values, the more sensitive they should be to the nuances of how an evaluation problem is posed.

TABLE 12.2 Three paradigms for eliciting values

Worry	*Treatment*	*Theoretical base*	*Test of success*
Assumption: People know what they want about all possible questions (to some degree of precision)			
Inappropriate default assumptions (for unstated part of question)	Examine interpretation, specify more, manipulate expectations	Nonverbal communication, experimenter–interviewer effects, psycholinguistics	Full specification, empathy with subjects
Inappropriate interpretation of stated question	Use good English, consensual terms	Survey technique, linguistics	Sensible answers, consensual interpretation of terms
Difficulty in expressing values	Choose correct response mode	Psychometrics, measurement theory	Consistency (reliability of representation)
Strategic response	Proper incentives, neutral context	Microeconomics, demand characteristics	Sensible answers, nonresponse to "irrelevant" changes
Assumption: People have stable but incoherent perspectives, causing divergent responses to formally equivalent forms			
Deep consistency in methods across studies (failing to reveal problem)	"Looking for trouble": multiple methods in different studies; "asking for trouble": open-ended questions	Framing theory, new psychophysics, multiple disciplines, anthropology	Nonresponse to irrelevant changes
Eliciting values incompletely (within study)	Multiple methods within study, open ended	Same as above, counseling skills	Inability to elicit more
Inability to reconcile perspectives	Talking through implications	Normative analysis, counseling skills	Unpressured consistent response to new perspectives
Assumption: People lack articulated values on specific topic (but have pertinent basic values)			
Pressure to respond	Measure intensity, allow no response, alternative modes of expression	Survey research, social psychology	Satisfaction, stability among remainder
Instability over time	Accelerate experience	Attitude formation, behavioral decision theory	Stable convergence
Inability to relate	Client-centered process	Normative (re)analysis	Full characterization
Undetected insensitivity	Ask formally different questions	Normative analysis	Proper sensitivity

For example, it is considered bad form if the demeanor of an interviewer (or the wording of a question) suggests what the investigator expects (or wants) to hear. Respondents might move in that direction (or the opposite) because they aim to please (or to frustrate). Or, they might be unmoved by such a hint because they are indifferent to the information or social pressure that it conveys. Because a hint becomes part of the evaluation question, its influence is confounded with that of the issues that interested the investigator in the first place.

Unfortunately, the logical consistency of this position can border on tautology, inferring that a change is significant from respondents' sensitivity to it and inferring that respondents have articulated values from their responses to changes in questions now known to be significant. Conversely, responding the same way to two versions of a task means that the differences between them are not irrelevant and that people know their own minds well enough not to be swayed by meaningless variations.

The potential circularity of such claims can be disrupted either by data or by argument. At the one extreme, investigators can demonstrate empirically that people have well-founded beliefs on the specific questions that they receive. At the other extreme, they can offer theoretical reasons why such beliefs ought to be in place (bolstered, perhaps, by empirical demonstrations in other investigations). Developing these data and arguments in their general form has helped to stimulate basic research into nonverbal communication, interviewer effects, and even the psycholinguistics of question interpretations (e.g., Jabine, Straf, Tanur, and Tourangeau, 1984; Rosenthal and Rosnow, 1969; Turner and Martin, 1984).

Within this paradigm, the test of success is getting the question specified exactly the way that one wants and verifying that it has been so understood. A vital service that professional survey houses offer is being able to render the questions of diverse clients into good English using consensual terms. This very diversity, however, ensures that there cannot be specific theory and data for every question that they ask. As a result, the test of success is often an intuitive appeal to how sensible answers seem to be. The risks of circularity here, too, are obvious.[4]

Assuming that respondents have understood the question, they still need to be able to express their (ready) answer in terms acceptable to the investigator. The great edifice of psychometric theory has evolved to manage potential problems here by providing elicitation methods compatible with respondents' thought processes and investigators' needs (Coombs, 1964; Nunnally, 1968). The associated tests of success are, in part, external—the ability to predict responses to other tasks—and, in part, internal—the consistency of responses to related stimuli. The risk in the former case is that the theoretical tie between measures is flawed. The risk in the latter case is that respondents have found some internally consistent way to respond to questions asked within a common format and varying in obvious ways (Poulton, 1989).

Perhaps surprisingly, the main concern of early contingent valuation investigators was not that respondents would have difficulty expressing their values in dollar terms. On the contrary, they feared that subjects would be able to use the response mode all too well. Knowing just what they want (and how to get it), subjects might engage in strategic behavior, misrepresenting their values in order to shift to others the burden of paying for goods that they value (Samuelson, 1954). In response, investigators developed sophisticated tasks and statistical analyses. Applications of these methods seem to have allayed the fears of many practitioners (Brookshire, Ives, and Schulze, 1976).[5]

Philosophy of basic values

From the perspective of the philosophy of basic values, people's time is very limited, whereas the set of possible evaluative questions is very, very large. As a result, people cannot be expected to have articulated opinions on more than a small set of issues of immediate concern. Indeed, some theorists have argued that one way to control people is by forcing them to consider an impossibly diverse range of issues (e.g., through the nightly news). People who think that they can have some opinion on every issue find that they do not have thoughtful opinions on any issues (Ellul, 1963). The only way to have informed opinions on complex issues is by deriving them carefully from deeply held values on more general and fundamental issues (Rokeach, 1973).

Taking the headache question as an example, a meaningful answer is much more plausible from someone who has invested time and money in seeking symptomatic relief, which can serve as a firm point of reference for evaluating that special treatment. (Economists sometimes call these *averting behaviors* [Dickie et al., 1987].) Otherwise, the question seems patently unanswerable—and the wild discrepancies found in the research provide clear evidence of respondents' grasping at straws.

From the perspective of this paradigm, the existence of such documented discrepancies means that not all responses can be taken seriously. As a result, investigators adhering to it worry about any aspects of their methodology that might pressure respondents to produce unthoughtful evaluations. In this regard, an inherent difficulty with most surveys and experiments is that there is little cost for misrepresenting one's values, including pretending that one has them. By contrast, offering no response may seem like an admission of incompetence. Why would a question have been posed if the (prestigious?) individuals who created it did not believe that one ought to have an answer? With surveys, silence may carry the additional burden of disenfranchising oneself by not contributing a vote to public opinion. With psychological experiments, it may be awkward to get out, or to get payment, until one has responded in a way that is acceptable to the experimenter.

One indication of the level of perfunctory responses in surveys may be seen in the repeated finding (Schuman and Presser, 1981) that explicitly offering a *don't know* option greatly increases the likelihood of subjects offering no opinion (e.g., from 5% to 25%). Yet, even that option is a rather crude measure. Respondents must determine how intense a degree of ignorance or indifference *don't know* implies (e.g., Does it mean absolutely, positively having no idea?). Investigators must, then, guess at how respondents have interpreted the option.

Hoping to say something more about the intensity of reported beliefs, survey researchers have conducted a lively debate over alternative statistical analyses of seemingly inconsistent attitudes (e.g., Achen, 1975; Converse, 1964). Its resolution is complicated by the difficulty of simultaneously evaluating questions and answers (Schuman and Presser, 1981; T. Smith, 1984). For example, one potential measure of value articulation is the stability of responses over time. When people say different things at different times, they might just be responding randomly. However, they might also have changed their underlying beliefs or settled on different interpretations of poorly worded questions. Changes in underlying opinions may themselves reflect exogenous changes in the issues addressed by the question (e.g., "My headaches are worse now than the last time I was asked") or endogenous changes in one's thinking (e.g., "I finally came to realize that it's crazy to be squirreling money away in the bank rather than using it to make myself less miserable").

A striking aspect of many contingent valuation studies is the high rate of refusals to provide acceptable responses among individuals who have already agreed to participate in the study (Cummings et al., 1986; Mitchell and Carson, 1989; Tolley et al., 1986). These protest responses take several forms: simply refusing to answer the evaluation question, offering to pay $0 for a good that one has admitted to be worth something, and offering to pay what seems to be an unreasonably high amount (e.g., more than 10% of disposable income for relieving a headache). For investigators under contract to monetize environmental goods, these responses are quite troublesome.[6] For investigators who have the leisure to entertain alternative perspectives, these responses provide some insight into how respondents having only basic values cope with pressure to produce more. It is perhaps a testimony to the coerciveness of interview situations how rarely participants say *don't know*, much less try to bolt (as they have in these contingent valuation studies).

The term *protest response* implies hostility toward the investigator. Some of that emotion may constitute displaced frustration with one's own lack of articulated values. The investigator's "crime" is forcing one to confront not knowing exactly what an important good is worth. Perhaps a more legitimate complaint is that investigators force that confrontation without providing any help in its resolution.

As mentioned, investigators within the articulated values paradigm provide no help as a matter of principle. Elicited values are intended to be entirely those of the respondent, without any hint from the questioner. This stance might also be appropriate to investigators in the basic values paradigm in cases in which they want to know what is in there to begin with when an issue is first raised. However, basic values investigators might also be interested in prompting the inferential process of deriving specific values from general ones. That might be done nondirectively by leaving respondents to their own devices after posing an evaluative question and promising to come back later for an answer. In the interim, respondents can do whatever they usually do, such as ruminate, ask friends, listen to music, review Scripture, or experiment. Such surveys might be thought of as accelerating natural experiences, guided by descriptive research into how people do converge on values in their everyday life.

Alternatively, investigators can adopt a multiply directive approach. They can suggest alternative positions, helping respondents to think through how those positions might or might not be consistent with their basic values. Doing so requires a normative analysis of alternative positions that might merit adoption. That might require adding professions like economics or philosophy to the research team. Surveys that present multiple perspectives are, in effect, respondent centered, more akin to decision analysis than to traditional question-centered social research, with its impassive interviewers bouncing stimuli off objectified respondents. Studies that propose alternative perspectives incur a greater risk of sins of commission, in the sense of inadvertently pushing subjects in one of the suggested directions, and a reduced risk of sins of omission, in the sense of letting respondents mislead themselves by incompletely understanding the implications of the questions that they answer.

As shown in the discussion of the questions opening this article, a clear hint that people have only basic values to offer is when they show undue sensitivity to changes in irrelevant features of a question. It can also be suggested by undue *in*sensitivity to relevant features. Figure 12.2 shows the proportion of women who reported that they expect no additional births, either in all future years (a) or in the next five years (b).[7] In each panel, there was considerable agreement between responses elicited by two respected survey houses. So, here is a case in which all of the irrelevant differences in procedures (e.g., interviewers, sampling, preceding questions) had no aggregate effect on responses. Across panels, however, there is a disturbing lack of difference. If there are women who intend to give birth after the next five years, then the curves should be lower in the left panel than in the right one. Although the investigators took care to specify time period, respondents either did not notice or could not make use of that critical detail.

An analogous result in contingent valuation research was Tolley et al.'s (1986) finding that people were willing to pay as much for 10 days worth as for 180 days worth of a fixed improvement in atmospheric visibility. Even more dramatic is Kahneman and Knetsch's (Kahneman, 1986) finding that respondents to a phone survey were willing to pay equal amounts to preserve the fisheries in one Ontario lake, in several Ontario lakes, and in all of the lakes in Ontario. These results could, of course, reflect articulated values based on utility functions that flattened out abruptly after 10 days and one lake. More likely, they reflect a vague willingness to pay a little money for a little good.

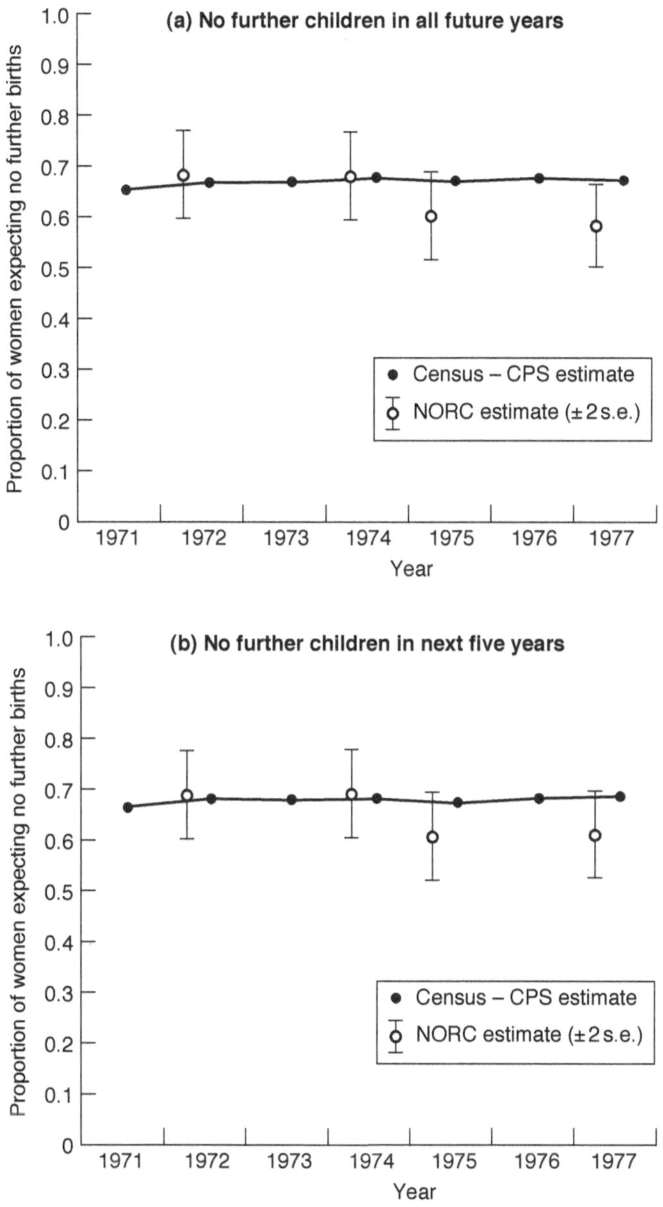

FIGURE 12.2 Estimates of fertility expectations of American women: proportion of women expecting no further children in (a) all future years, and (b) the next five years.

Note. Samples included only married women aged 18–39; sample sizes in each year were approximately 4,000 (Census-CPS) and 220 (NORC). CPS = Current Population Survey; NORC = National Opinion Research Center. From "Why Do Surveys Disagree? Some Preliminary Hypotheses and Some Disagreeable Examples" (p. 192) by C. F. Turner, 1984, in C. F. Turner and E. Martin, *Surveying Subjective Phenomena,* New York: Russell Sage

Philosophy of partial perspectives

By adopting an intermediate position, individuals working within the partial perspectives paradigm must worry about the problems concerning both extremes. On the one hand, they face the risk of inadequately formulated and understood questions, preventing respondents from accessing those partially articulated perspectives that they do have. On the other hand, investigators must worry about reading too much into expressions of value produced under pressure to say something.

These worries may, however, take on a somewhat different face. In particular, the existence of partial perspectives may give a deceptive robustness to expressions of value. Thus, investigators using a single method may routinely elicit similar responses without realizing the extent to which their success depends on the method's ability to evoke a common perspective. That fact may be obscured further when a family of related methods produces similar consistency. It takes considerable self-reflection for investigators to discern the structural communalities in methods that seem to them rather different. Speculative examples might include a tendency for surveys to emphasize hedonic rather than social values by asking respondents for their personal opinions, or for experimental gambles to encourage risk taking because participants cannot leave with less than they went in with,[8] or to discourage emotional involvement because the scientific setting seems to call for a particularly calculating approach. Discovering the perspectives that it inadvertently imposes on itself is part of the continuing renewal process for any scientific discipline. In the social sciences, these perspectives may also be imposed on the people being studied, whose unruly behavior may, in turn, serve as a clue to disciplinary blinders (e.g., Furby, 1986; Gergen, 1973; Gilligan, 1982; Wagenaar, 1989).

Research methods may create consistent response sets, as well as evoke existing ones (Tune, 1964). When asked a series of obviously related questions on a common topic, respondents may devise a response strategy to cope with the experiment. The resulting responses may be consistent with one another, but not with responses in other settings. Indeed, those investigations most concerned about testing for consistency may also be the most vulnerable to generating what they are seeking. Think, for example, of an experiment eliciting evaluations for stimuli representing all cells of a factorial design in which each factor is a different outcome attribute. Why not come up with some simple rule for getting through the task?

For example, Poulton (1968, 1989) has conducted detailed secondary analyses of the quantitative estimates elicited in psychophysics experiments in an effort to capture the subjective intensity of physical stimuli (e.g., sweetness, loudness). He argued that the remarkable internal consistency of estimates across stimulus dimensions (Stevens, 1975) reflects the stability of investigators' conventions in setting up the details of their experiments. Although subjects have no fixed orientation to such unfamiliar forms of evaluation, they do respond similarly to structuring cues such as the kind of numbers to be used (e.g., integers vs. decimals) and the place of the standard stimulus in the range of possibilities.[9] The (nontrivial) antidotes are what might be called *looking for trouble* and *asking for trouble*—eliciting values in significantly different ways and using sufficiently open-ended methods to allow latent incoherence to emerge.

Economists hope to reduce these problems by discerning people's values from the preferences revealed in market behavior. Such actions ought to be relatively free of pressures to respond. After all, you don't have to buy. Or do you? Even if choices are voluntary, they can only be made between options that are on offer and with whatever information respondents happen to have. For example, you may hate ranch style homes but have little choice other than to choose one that makes the best of a bad situation in some locales. In that case, the preferences thereby revealed are highly conditional. Furthermore, even if the choice sets are relatively open and well understood, they may be presented in ways that evoke only a limited subset of people's values. By some accounts, evoking partial perspectives is the main mission of advertising (by other accounts, it is just to provide information). Some critics have argued that some perspectives (e.g., the value of possessing material goods) are emphasized so effectively that they change from being imposed perspectives to becoming endorsed ones.[10]

If one wants to predict how people will behave in situations presenting a particular perspective, then one should elicit their values in ways evoking that perspective.[11] If one wants to get at all their potentially relevant perspectives, then more diverse probing is needed. This is the work of many counselors and consultants. Although some try to construct their clients' subjective problem representation from basic values (along the lines of decision analysis), others try to match clients with general diagnostic categories. Each category then carries prognoses and recommendations. As mentioned, the counselor stance is unusual in social research. Like any direct interaction, it carries the risk of suggesting and imposing the counselor's favored perspective. Presumably, there is a limit to how quickly people can absorb new outlooks. At some point, they may lose cognitive control of the issue, wondering perhaps, "Whose problem is it, anyway?"

How could they think this way?

Described in its own right, any paradigm sounds like something of a caricature. Could proponents really believe that one size fits all when it comes to methodology? Surely, decision analysts realize that some values are already so well articulated that their decomposition procedures will only induce confusion. Surely, survey researchers realize that some value issues are so important and so unfamiliar in their details that respondents will be unable to resist giving uninformed answers to poorly understood questions. Surely they do. Yet, equally surely, there is strong temptation to stretch the envelope of applications for one's favored tools.

Some reasons for exaggerating the applicability of one's own discipline are common to all disciplines. Anyone can exaggerate the extent to which they are ready for a challenge. Each discipline has an intact critique of its competitors. People who ask questions know what they mean and also know how they would answer. What might be called anthropology's great truth is that we underestimate how and by how much others see the world differently than we do. Paradigms train one to soldier on and solve problems, rather than to reflect on the whole enterprise.

The inconsistent responses opening this article present an interesting challenge for that soldiering. As shown in the discussion of those results, each paradigm has a way to accommodate them. Yet, investigators in the basic values paradigm seem much more comfortable with such accommodation. They seem more ready to accept them as real (i.e., produced from sound, replicable studies) and much more ready to see them as common. Basic values investigators sometimes seem to revel in such discrepancies (e.g., Hogarth, 1982; Nisbett and Ross, 1980), whereas articulated values investigators seem to view them as bona fide, but still sporadic, problems (e.g., Schuman and Presser, 1981).[12] Insight into these discrepant views about discrepancies can be gained by examining the institutional and methodological practices of these paradigms.

Interest in discrepancies

Basic values investigators would like to believe that there are many robust discrepancies "out there in the world" because they serve a vital purpose for this kind of science. Discovering a peculiar pattern of unexpected results has been the starting point for many theories (Kahneman and Tversky, 1982). McGuire (1969) has gone so far as to describe the history of experimental psychology as the history of turning artifacts into main effects. For example, increased awareness of experimenter effects (Rosenthal, 1967) stimulated studies of nonverbal communication (e.g., Ekman, 1985). In fact, some critics have argued that psychology is so much driven by anomalies that it tends to exaggerate their importance and generality (Berkeley and Humphreys, 1982). Anomalies make such a good story that it is hard to keep them in focus, relative to the sometimes unquirky processes that produce them (Fischhoff, 1988).

Interest in order

On the other hand, articulated values investigators are more interested in what people think than in how they think. For those purposes, all these quirks are a major headache. They mean that every question may require a substantial development effort before it can be asked responsibly, with elaborate pretesting of alternative presentations. The possibility of anomalies also raises the risk that respondents cannot answer the questions that interest the investigators—at least without the sort of interactive or directive elicitation that is an anathema within this paradigm.

Of course, investigators in this paradigm are concerned about these issues. Some of the most careful studies of artifacts have come from survey researchers (e.g., Schuman and Presser, 1981). Classic examples of the effort needed to tie down the subjective interpretation of even seemingly simple questions may be found in the U.S. Department of Commerce studies of how to ask about employment status (Bailar and Rothwell, 1984). However, every research program with resource constraints is limited in its ability to pursue methodological nuances. When those nuances could represent fatal problems, then it is natural to want to believe that they are rare.

For survey research houses, these constraints are magnified by the commercial pressures to keep the shop open and running at a reasonable price. To some extent, clients go to quality and will pay for it. However, there is a limit to the methodological skepticism that even sophisticated clients will tolerate. They need assurance that investigators have the general skill needed to create workable items out of their questions. Clients might know, at some level, that "different questions might have produced different answers" (according to the strange wording that quality newspapers sometimes append to survey results). However, they still need some fiction of tractability.

Ability to experiment

A further constraint on articulated values scientists is their theoretical commitment to representative sampling. The expense of such samples means that very few tests of alternative wording can be conducted. Conversely, it means that many discrepancies (like the happiness questions) are only discovered in secondary analyses of studies conducted for other purposes. As a result, there are typically confounding differences in method that blur the comparison between questions.

By contrast, basic values scientists are typically willing to work with convenience samples of subjects. As a result, they can run many tightly controlled experiments, increasing their chances of finding discrepancies. Multiple testing also increases the chances of finding differences by chance. If they are conscientious, these scientists should be able to deal with this risk through replications (which are, in turn, relatively easy to conduct). This indifference to sampling might reflect a self-serving and cavalier attitude. On the other hand, it may be the case that *how* people think might be relatively invariant with respect to demographic features that are known to make a big difference in *what* they think.

Precision of search

The theories that basic values scientists derive to account for discrepancies are not always correct. When they are, however, they allow investigators to produce inconsistent responses almost at will. Much of experimental psychology is directed at determining the precise operation of known effects. For example, at the core of prospect theory is a set of framing operations designed to produce inconsistencies. The prevalence of phenomena under laboratory conditions has, of course, no necessary relationship to their prevalence elsewhere. Some extrapolation of prevalence rates from the lab to the world would, however, be only natural (Tversky and Kahneman, 1973). Furthermore, continuing absorption with a phenomenon should sharpen one's eagerness and ability to spot examples. Investigators who want and expect to see a phenomenon are likely to find it more often than investigators who do not. It would be only natural if the confirmation offered by such anecdotal evidence were overestimated (Chapman and Chapman, 1969).

The theoretical tools for seeking nuisance effects in an articulated values study would likely be more poorly defined. For example, the question might be posed as generally as "How common are order effects?" Given the enormous diversity of questions whose order might be reversed, the answer is, doubtless, "very low" in the domain of all possible questions. However, with questions of related content, order effects might be much more common (Poulton and Freeman, 1966). Moreover, questions are more likely to appear in surveys with somewhat related ones, rather than with completely related ones—even in amalgam surveys pooling items from different customers. Without a theory of relatedness, researchers are in a bind. Failure to find an order effect can just be taken as proof that the items were not related.[13]

Criterion of interest

Surveys are often conducted in order to resolve practical questions, such as which candidate to support in an election or which product to introduce on the market. As a result, the magnitude of an effect provides the critical test of whether it is worthy of notice. Unless it can be shown to make a difference, who cares? Laboratory results come from out of this world. If they cannot be mapped clearly onto practical problems, then they are likely to seem like curiosities. The psychologists' criterion of statistical significance carries little weight here. Survey researchers, with their large samples, know that even small absolute differences can reach statistical significance.

On the other hand, not all survey questions have that direct a relationship to action. One can assess the effects of being off by 5% in a pre-election poll or a product evaluation, as a result of phrasing differences. However, in many other cases, surveys solicit general attitudes and beliefs. These are widely known to be weak predictors of behavior (Ajzen and Fishbein, 1980). As a result, it may be relatively easy to shrug off occasional anomalies as tolerable. Discrepancies should become more important and, perhaps, seem more common as the questions driving research become sharper. The discrepancies associated with contingent valuation studies have come under great scrutiny recently because of their enormous, economic consequences. Changes in wording can, in principle, mean the difference between success and failure for entire companies or industries.

Thinking about lability

How common are artifacts? is an ill-formed question, insofar as there is no clear universe over which the relative frequency of instances can be defined. Nonetheless, investigators' intuitive feeling for overall frequency must determine their commitment to their paradigms and their ability to soldier on in the absence of definitive data. Understanding the nature and source of one's own disciplinary prejudices is essential for paradigms to be used wisely and to evolve. Understanding other disciplines' (more and less legitimate) prejudices is necessary for collaboration. Implicit assumptions about the nature of human values seem to create a substantial divide among the social sciences. If they were to work together, the focal question might shift from how well articulated are values to where are they well articulated. Table 12.3 offers one possible set of conditions favorable to articulated values. Turner (1981) offered another. It might be informative to review the evidentiary record of discrepancies and nondiscrepancies in the light of such schemes.

TABLE 12.3 Conditions favorable to articulated values

Personally familiar (time to think)
Personally consequential (motivation to think)
Publicly discussed (opportunity to hear, share views)
Uncontroversial (stable tastes, no need to justify)
Few consequences (simplicity)
Similar consequences (commensurability)
Experienced consequences (meaningfulness)
Certain consequences (comprehensibility)
Single or compatible roles (absence of conflict)
Diverse appearances (multiple perspectives)
Direct relation to action (concreteness)
Unbundled topic (considered in isolation)
Familiar formulation

Acknowledgements

Preparation of this article was supported by National Science Foundation Grant No. SES-8175564 and by the Carnegie Corporation of New York, Council on Adolescent Development. The views expressed are those of the author.

My special thanks go to Robert Abelson, who suggested the juxtaposition that is explored here, and to Charles Turner, who has stimulated concern for non-sampling error for many years. My thinking on these issues has benefited from discussions with many people, including Lita Furby, Robyn Dawes, Paul Slovic, Sarah Lichtenstein, Amos Tversky, Daniel Kahneman, Alan Randall, and Robin Gregory.[1] have also received valuable comments from participants in the National Research Council Panel on Survey Measure of Subjective Phenomena; the Russell Sage Foundation Conference, "Towards a Scientific Analysis of Values"; and the U.S. Forest Service Conference on Amenity Resource Valuation.

Notes

1 The two questions did differ slightly in their introductory phrase. One began "taken all together," the other "taking all things together." Only the bravest of theoretician would try to trace the pattern in Figure 12.1 to this difference.
2 Subsequent research (Turner, 1984; Turner and Martin, 1984) has shown a somewhat more complicated set of affairs—which may have changed further by the time this article is printed and read. Incorporating the most recent twists in this research would change the details but not the thrust of the discussion in the text.
3 Thus, these results would lead one to expect lower renewal rates on insurance policies were subscribers to receive periodic bills for sure losses, rather than for premiums.
4 One is reminded of the finding that undetected computational errors tend to favor investigators' hypotheses. A nonmotivational explanation of this trend is that one is more likely to double-check all aspects of procedure, including calculations, when results are surprising (Rosenthal and Rosnow, 1969).
5 The processes by which these fears were allayed might be usefully compared with the processes by which psychology convinced itself that it knew how to manage the effects of experimenter expectations (Rosenthal, 1967).
6 In actual studies, investigators sometimes just throw out protest responses. At times, they adjust them to more reasonable values (e.g., reducing high values to 10% of disposable income).
7 This is a question of prediction, rather than of evaluation, except in the sense that intentions to have children reflect the perceived value of having them.
8 The need to protect human subjects poses this constraint. Even without it, there would be problems getting people to risk their own money in a gamble contrived by some, possibly mistrusted, scientist.
9 Many contingent valuation studies have elicited values by asking subjects questions such as "Would you pay $l, $2, $3, …?" until they say no. One might compare the implicit structuring of this series of questions with that achieved by "Would you pay $10, $20, $30, …?" or by moving down from $100 in $1 increments.
10 This is just the tip of the iceberg regarding the methodological difficulties of inferring values from observed market behavior (Campen, 1986; Fischhoff and Cox, 1985; Peterson, Driver, and Gregory, 1988). In many cases, technical difficulties make inferring values from behavior an engaging fiction.
11 Fischhoff (1983) considered some of the difficulties of predicting which frames are evoked by naturally occurring situations.

12 This observation was sharply drawn by Professor Robert Abelson at a meeting of the National Research Council Panel on Survey Measure of Subjective Phenomena (Turner and Martin, 1981). This section of my article is, in large part, an attempt to work up the pattern that he highlighted.

13 A related example—for which I have unfortunately misplaced the reference and must rely on memory—is the finding that people respond more consistently to items on a common topic when those are grouped in a survey than when the questions are scattered. An (expensive) attempt to replicate this finding took as its common topic attitudes toward shop stewards, and found nothing. That could mean that the first result was a fluke or that *shop stewards* is not a meaningful concept of the sort that could induce consistent attitudes when brought to people's attention.

References

Achen, C. H. 1975. Mass political attitudes and the survey response. *American Political Science Review, 69*, 1218–1231.

Ajzen, I., and Fishbein, M. 1980. *Understanding attitudes and predicting social behavior*. Englewood Cliffs, NJ: Prentice-Hall.

Bailar, B. A., and Rothwell, N. D. 1984. Measuring employment and unemployment. In C. F. Turner and E. Martin (Eds.) *Survey measure of subjective phenomena* (pp. 129–142). New York: Russell Sage Foundation.

Bentkover, J., Covello, V., and Mumpower, J. (Eds.) 1985. *Benefits assessment: The state of the art*. Amsterdam: Reidel.

Berkeley, D., and Humphreys, P. 1982. Structuring decision problems and the "bias" heuristic. *Acta Psychologica, 50*, 201–250.

Brookshire, D. S., Ives, C. C, and Schulze, W. D. 1976. The valuation of aesthetic preferences. *Journal of Environmental Economics and Management, 3*, 325–346.

Campbell, A., Converse, P., and Rodgers, W. 1976. *The quality of American life: Perceptions, evaluations, and satisfaction*. New York: Russell Sage Foundation.

Campen, J. T. 1986. *Benefit, cost, and beyond*. Cambridge, MA: Ballinger.

Chapman, L. J., and Chapman, J. P. 1969. Genesis of popular but erroneous psychodiagnostic observations. *Journal of Abnormal Psychology, 74*, 271–280.

Converse, P. E. 1964. The nature of belief systems in mass politics. In D. E. Apter (Ed.) *Ideology and discontent*. Glencoe, NY: Free Press.

Coombs, C. H. 1964. *A theory of data*. New York: Wiley.

Cummings, R. D., Brookshire, D. S., and Schulze, W. D. (Eds.) 1986. *Valuing environmental goods: An assessment of the Contingent Valuation Method*. Totowa, NJ: Rowman and Allenheld.

DeMaio, T. J. 1984. Social desirability and survey measurement: A review. In C. F. Turner and E. Martin (Eds.) *Survey measure of subjective phenomena* (pp. 257–282). New York: Russell Sage Foundation.

Dickie, M., Gerking, S., McClelland, G., and Schulze, W. 1987. *Improving accuracy and reducing costs of environmental benefit assessments: Vol. 1. Valuing morbidity: An overview and state of the art assessment* (USEPA Cooperative Agreement No. CR812954-01-2). Washington, DC: U.S. Environmental Protection Agency.

Ekman, P. 1985. *Telling lies*. New York; Norton.

Ellul, J. 1963. *Propaganda*. New York: Knopf.

Fischhoff, B. 1983. Predicting frames. *Journal of Experimental Psychology: Learning. Memory, and Cognition, 9*, 103–116.

Fischhoff, B. 1988. Judgment and decision making. In R. J. Slernberg and E. E. Smith (Eds.), *The psychology of human thought* (pp. 153–187). New York: Wiley.

Fischhoff; B., and Cox, L. A., Jr. 1985. Conceptual foundation for benefit assessment. In J. D. Bentkover, V. T. Covello, and J. Mumpower (Eds.), *Benefits assessment: The state of the art* (pp. 51–84). Amsterdam: Reidel.

Fischhoff, B., and. Furby, L. 1988. Measuring values: A conceptual framework for interpreting transactions with special reference to contingent valuation of visibility. *Journal of Risk and Uncertainty, I,* 147–184.

Fischhoff, B., Slovic, P., and Lichtenstein, S. 1980. Knowing what you want: Measuring labile values. In T. Wallsten (Ed.) *Cognitive processes in choice and decision behavior* (pp. 117–141). Hillsdale, NJ: Erlbaum.

Furby, L. 1986. Psychology and justice. In R. L. Cohen (Ed.) *Justice: Views from the social sciences* (pp. 153–203). New York: Plenum.

Furby, L., and Fischhoff, B. 1989. *Specifying subjective evaluations: A critique of Dickie et al.'s interpretation of their contingent valuation results for reduced minor health symptoms* (U.S. Environmental Protection Agency Cooperative Agreement No. CR814655-01-0). Eugene, OR: Eugene Research Institute.

Gergen, K. J. 1973. Social psychology as history. *Journal of Personality and Social Psychology, 26,* 309–320.

Gilligan, C. 1982. *In a different voice: Psychological theory and women's development.* Cambridge, MA: Harvard University Press.

Hershey, J. R., and Schoemaker, P. J. H. 1980. Risk taking and problem context in the domain of losses: An expected utility analysis. *Journal of Risk and Insurance, 47,* 111–132.

Hogarth, R. M. (Ed.) 1982. *New directions for methodology of the social sciences: Question framing and response consistency.* San Francisco: Jossey-Bass.

Jabine, T. B., Straf, M. L., Tanur, J. M., and Tourangeau, R. (Eds.) 1984. *Cognitive aspects of survey methodology: Building a bridge between disciplines.* Washington, DC: National Academy Press.

Kahneman. D. 1986. Comment. In R. D. Cummings, D. S. Brookshire, and W. D. Schulze (Eds.) *Valuing environmental goods: An assessment of the Contingent Valuation Method.* Totowa, NJ: Rowman and Allenheld.

Kahneman, D., and Tversky, A. 1979. Prospect theory. *Econometrica, 47,* 263–292.

Kahneman, D., and Tversky, A. 1982. On the study of statistical intuitions. *Cognition, II,* 123–141.

Kuhn, T. S. 1962. *The structure of scientific revolution.* Chicago: University of Chicago Press.

Kunreuther, H., Ginsberg, R., Miller, L., Sagi, P., Slovic, P., Borkin, B., and Katz, N. 1978. *Disaster insurance protection: Public policy lessons.* New York: Wiley.

Lakatos, I., and Musgrave, A. (Eds.) 1970. *Criticism and the growth of scientific knowledge.* Cambridge, England: Cambridge University Press.

McGuire, W. J. 1969. Suspiciousness of experimenter's intent. In R. Rosenthal and R. L. Rosnow (Eds.) *Artifact in behavioral research.* San Diego, CA: Academic Press.

Mitchell, R. C. and Carson, R. T. 1989. *Using surveys to value public goods: The Contingent Valuation Method.* Washington, DC: Resources for the Future.

National Opinion Research Center. 1978. *General Social Surveys, 1972–1978: Cumulative codebook.* Chicago: Author.

Nisbett, R. E., and Ross, L. 1980. *Human inference: Strategies and shortcomings of social judgment.* Englewood Cliffs, NJ: Prentice-Hall.

Nunnally, J. C. 1968. *Psychometric theory* (2nd ed.). New York: McGraw-Hill.

Peterson, G. L., Driver, B. L., and Gregory, R. (Eds.) 1988. *Amenity resource valuation: Integrating economics with other disciplines*. State College, PA: Venture.

Poulton, E. C. 1968. The new psychophysics: Six models for magnitude estimation. *Psychological Bulletin, 69*, 1–19.

Poulton, E. C. 1989. *Bias in quantifying judgments*. London: Erlbaum.

Poulton, E. C. and Freeman, P. R. 1966. Unwanted asymmetrical transfer effects with balanced experimental designs. *Psychological Bulletin, 66*, 1–8.

Raiffa, H. 1968. *Decision analysis*. Reading, MA: Addison-Wesley.

Rokeach, M. 1973. *The nature of human values*. New York: Free Press.

Rosenthal, R. 1967. Covert communication in the psychological experiment. *Psychological Bulletin, 67*, 356–367.

Rosenthal, R., and Rosnow, R, L. (Eds.) 1969. *Artifact in behavioral research*. San Diego, CA: Academic Press.

Samuelson, P. 1954. The pure theory of public expenditure. *Review of Economics and Statistics. 36*, 387–389.

Schuman, H., and Presser, S. 1981. *Questions and answers*. San Diego, CA: Academic Press.

Smith, T. 1984. Nonattitudes: A review and evaluation. In C. F. Turner and E. Martin (Eds.) *Survey measure of subjective phenomena* (pp. 215–256). New York: Russell Sage Foundation.

Smith, V. K., and Desvousges, W. H. 1988. *Measuring water quality benefits*. Boston: Kluwer-Nijhoff.

Stevens, S. S. 1975. *Psychophysics: Introduction to its perceptual, neural, and social prospects*. New York: Wiley.

Sudman, S., and Bradburn, N. M. 1982. *Asking questions: A practical guide to questionnaire design*. San Francisco: Jossey-Bass.

Tollcy, G. *et al.* 1986. *Establishing and valuing the effects of improved visibility in the eastern United States* (USEPA Grant No. 807768-01-0). Washington, DC: U.S. Environmental Protection Agency.

Tune, G. S. 1964. Response preferences: A review of some relevant literature. *Psychological Bulletin, 61*, 286–302.

Turner, C. F. 1981. Surveys of subjective phenomena: A working paper. In D. Johnson (Ed.) *Measurement of subjective phenomena*. Washington, DC: U.S. Government Printing Office.

Turner, C. F. 1984. Why do surveys disagree? Some preliminary hypotheses and some disagreeable examples. In C. F. Turner and E. Martin (Eds.) *Surveying subjective phenomena* (pp. 159–214). New York: Russell Sage Foundation.

Turner, C. F., and Krauss, E. 1978. Fallible indicators of the subjective state of the nation. *American Psychologist, 33*, 456–470.

Turner, C. F., and Martin, E. (Eds.) 1981. *Surveys of subjective phenomena*. Washington, DC: National Academy Press.

Turner, C. F., and Martin, E. (Eds.) 1984. *Surveying subjective phenomena*. New York: Russell Sage Foundation.

Tversky, A., and Kahneman, D. 1973. Availability: A heuristic for judging frequency and probability. *Cognitive Psychology, 5*, 207–232.

Tversky, A., and Kahneman, D. 1981. The framing of decisions and the psychology of choice. *Science, 211*, 453–458.

von Winterfeldt, D., and Edwards, W. 1986. *Decision analysis and behavioral research*. New York: Cambridge University Press.

Wagenaar, W. A. 1989. *Paradoxes of gambling behavior*. London: Erlbaum.

Watson, S., and Buede, D. 1988. *Decision synthesis*. New York: Cambridge University Press.

13

INFORMED CONSENT FOR ELICITING ENVIRONMENTAL VALUES

Baruch Fischhoff

DEPARTMENT OF SOCIAL AND DECISION SCIENCES AND DEPARTMENT OF ENGINEERING AND PUBLIC POLICY, CARNEGIE MELLON UNIVERSITY, PITTSBURGH, PENNSYLVANIA 15213–3890

Abstract

Agreeing to participate in a study eliciting environmental values means agreeing to abide by the commitment implied by any proposal that one accepts or rejects in it. That might mean anything from addressing the gist of an issue to expressing an explicit willingness to pay for an environmental change. By soliciting such participation, investigators promise to provide the information that participants need in order to evaluate the proposals being presented. This paper proposes a standard for providing such information that must be met in order to conduct valid and ethical value-elicitation studies. Namely, investigators must secure the *informed* consent of participants.

Drawing on research in risk analysis and communication, this approach allows setting priorities among the facts that could conceivably be conveyed. It also leads to two general strategies for increasing the capacity of the communication channel. The paper concludes by discussing how the policy relevance of value-elicitation studies depends on their ability to ensure informed consent.

Introduction

Physicians are often required to secure *informed consent* from the participants in their procedures. Social researchers often must do the same for participants in their studies. Individuals who sign the associated consent forms agree (a) that the benefits of participation outweigh the risks, (b) that they have received the information needed to make that determination, and (c) that they were free to decline before starting—and are free to leave midstream if their expectations are not met. Like their medical counterpart, research informed-consent procedures were designed primarily to protect participants. However, they can also protect the user of a study's results by ensuring the quality of its data. The user can be assured that participants knew what they were talking about and accept the use made of their responses.

This paper applies the informed-consent standard to studies eliciting environmental values. In this context, the standard means creating conditions such that respondents will stand behind researchers' interpretations of their responses—as expressing their informed personal preferences. The next section sets out the importance of providing such quality assurance. The following section operationalizes this principle in decision analytic terms, using existing risk analyses to set priorities in communicating task-related information and to evaluate a study's success. Two complementary strategies are then proposed for using best the limited opportunity for communicating essential information. The final section considers the options facing policy-makers hoping to use value-elicitation studies responsibly.

Risks and benefits

Before deciding to participate in a study, a candidate should receive a candid disclosure of its risks and benefits. Studies eliciting environmental values seldom pose direct risks to participants, except perhaps depressing them about the state of the environment (although, see ref. *1*). For most potential participants, those small direct risks should be outweighed by the (perhaps small) direct benefits of thinking about the issues, along with any payment for responding. Such impacts should be over once the task has been completed. It should pose little problem for investigators to inform potential participants about what they would be getting themselves into. People have thought and talked about the environment before.

However, policy-related research can have *indirect* effects extending long after the data collection is over. Such effects are familiar to many environmental researchers. For example, a study finding pollution in a community can undermine its economic viability and its residents' insurability; it can also get their plight taken seriously. These possibilities have prompted protective procedures, like those used by the Agency for Toxic Substances and Disease Registry, as well as increased calls for *participatory research* with citizens involved in the design, conduct, and reporting of studies (*2, 3*). Without an understanding of how the data will be used, citizens may hesitate to provide their views, much less tissue or soil samples.

With studies eliciting environmental values, the indirect impacts of participation depend on how literally responses are interpreted. At one extreme, lie what might be called *gist* studies: Investigators claim to have received no more than general answers to general questions, reflecting respondents' general attitude toward a general topic (e.g., supporting the environment, worrying about pollution, having confidence in industry, mistrusting government). Gist responses provide only vague direction for policy-makers. As a result, the indirect effects of such studies on respondents *should* be small. They have not said very much; no one should read very much into their responses. Of course, policy-makers may claim strong mandates on the basis of such weak indicators, producing strong indirect effects. However, in that case, they will have misrepresented what respondents said. Researchers who followed the use of their work would then be obligated to deny such misuse, saying in effect, "All they (the public) said was that they wanted cleaner cars; they didn't say that they wanted to mandate that particular fuel system."

At the other extreme, lie *contract* studies: Specific transactions are proposed to respondents, to support claims such as "people are willing to pay 7% for "green' products" or "to incur a 0.2% rise in unemployment in order to meet Kyoto obligations." By participating, respondents will promote those policies that fit the investigator's interpretation of their responses. Respondents will benefit if those are policies that they actually desire. They will lose if the investigators misinterpret them. That could occur malevolently through deliberate misrepresentation. However, it could also occur if the investigator simply understands the contract differently than do the respondents. One form of misinterpretation is mistaking gist responses for contractual ones (e.g., assuming that support for the World Trade Organization implies knowledge and advocacy of all its environmental provisions).

These possibilities can be seen in the familiar context of contingent valuation (CV) studies (*4*), eliciting respondents' willingness to pay (or to accept compensation) for proposed environmental changes. Participation provides a unique opportunity for citizens to evaluate specific environmental issues. That chance presumably encourages many people to accept and stick with CV tasks despite the hard work required. However, unless their responses are interpreted as intended, participants will have signed an IOU that they were not really willing to pay (or an invoice that they did not really intend to deliver—when someone else will pay the contingent value). That could happen if respondents missed a critical task feature (which the investigator assumed was understood), leading them to evaluate a different transaction than the one that was offered. Or it could happen if they simply did not realize how seriously their response would be taken, hence gave but a gist (which the investigator treated as contractual).

For present purposes, I assume that people mean what they say in order to focus on the possibility that they do not understand the task well enough to know what they are talking about. Thus, I will not consider the contentious issue of strategic behavior (deliberately misrepresenting one's true beliefs) and hypotheticality (when the evaluated good may not really be provided and the amount that respondents claim to be willing to pay may not really be extracted). Extensive discussions of each can be found in the general literature.

Explaining the terms of a study

When investigators go to secure informed consent, their first obligation is to explain how responses will be used by placing their study on the gist—contract continuum. Doing so should be simple. Just say (the equivalent of) either "we're interested in your general opinion about [this general topic]: or "your response will be used literally in setting policy, hence will be treated as though you absorbed and acceded to every detail in the description to follow". However, an investigator's aspirations may not go without saying. Table 13.1 suggests what can happen when respondents take their task much more or less seriously than was intended. One of the sustained controversies regarding CV studies focuses on whether they lie in the lower right or lower left quadrant. Critics argue that respondents typically cannot translate their feelings about the goods presented in such studies into hedonically equivalent changes in wealth. When that happens, they provide general expressions of concern, little related to an economic interpretation of WTP (*5, 6*). Were that true, then respondents would be in a gist study that researchers (and policy-makers) have treated as a contractual one, violating their informed consent.

Investigators' next obligation is to provide the conditions that participants need to generate responses of the required depth. For gist studies, that should be easy. Their "come as you are" perspective means that respondents need not (indeed, should not) be provided with additional information about the topic. Moreover, as long as respondents share a common culture with the

TABLE 13.1 Compatibility between respondent and researcher view of study intent

	Respondent's view	
Researcher's view	*Gist*	*Contract*
Gist	casual, general response taken casually and generally; respondent works no harder than is warranted; researcher makes no more of responses than is warranted	serious response taken much more casually than intended; respondents have invested much more effort than was warranted; researcher has ignored the details that respondents considered when formulating their responses
Contract	casual, general response treated much more seriously than intended; researcher assumes that respondents have processed details of contract, when they have just considered general issue; degree of commitment overstated	specific response taken as seriously as it was intended; respondents are rewarded for the effort that they invested in considering the details of the contract

researchers (and policy-makers), the meaning of possible responses should be quickly apparent. Everyone involved should extract a similar message from a general opinion about the environment.

If a contract question is familiar, then the same thinking applies. Conveying its content is easy if the work has already been accomplished by natural processes. For example, as the vote approaches on a widely debated referendum, many citizens should know what commitment it implies and respond reliably to a poll eliciting their values. Many citizens know the verdict that they want from our periodic show trials, as those near their conclusions, and what would happen if their views determined how the law of the land was applied.

Communicating a novel contract can, however, be much more challenging. Many different features could matter to respondents. Table 13.2 shows a framework for specifying contractual tasks eliciting environmental values (7). In this view, a properly informed respondent must understand how the investigator interprets the *good*, the *payment*, and the associated *social context* of a proposed *transaction*. If not, then respondents are answering a different question than the one being asked. An investigator who secures informed consent for a contractual study promises to ensure that level of mutual understanding.

TABLE 13.2 Framework for defining transactions; features that may require specification if an evaluation task is to be understood similarly by respondents, investigators, and policy-makers[a]

The Good (e.g., Visibility)
substantive definition (aspects of proposed change in good that may matter to evaluators)
focal attribute(s)
(e.g., haze intensity, visual range, plume color, light extinction)
context (giving particular value to attribute)
(e.g., natural or built, judged uniqueness, associated activities [such as hiking, viewing, playing], significance [such as religious, cultural, historical])
source of change (in focal attribute)
predominantly natural (e.g., vegetation, forest fires, dust storms, humidity) or human (e.g., power plant, other factory, field burning, slash burning, motor vehicles)
formal definition (specifying extent of change in valued focal attributes)
reference and target levels (of good, before and after change)
magnitude and direction of change, statistical summary, form of representation (mode, richness, organization)
extent of change
geographical, temporal
timing of change (when will it happen)
certainty of provision (will it really happen)

The Value Measure (e.g., Money, Time, Discomfort, Effort)

- *substantive definition* (aspects of proposed change in payment that may matter to evaluators)
  - focal attribute(s)
    - dollars (for money)
    - foregone leisure or work (for time)
    - physical or emotional toll (for discomfort or effort)
  - context
    - electric bill, sales tax, income tax, park entry fee, environmental fund (for money)
    - when convenient, when demanded (for time)
    - when rested, when exhausted (for effort)
  - constituency
- *formal definition* (specifying extent of change in valued focal attributes)
  - reference and target levels
    - magnitude and direction of change, statistical summary, elicitation procedure (response mode, response format, cues, feedback)
  - extent of payment
    - frequency, duration
  - timing of payment (when will it happen?)
  - certainty of payment (will it really happen?)

The Social Context

- *other people involved*
  - provider of the good
  - others present
- *resolution mechanism* (determining whether transaction will actually occur)
  - determining parties
  - iterations, constraints
- *other stakes involved*
  - externalities
  - precedents
  - legitimacy of process

[a] Source: Ref 7.

Doing so could constitute *reactive measurement*, whereby the elicitation experience changes respondents. However, it could just make the task clear enough that respondents can express their unchanged selves, just as would happen with a "plain English" car, apartment, or video rental contract. Nonetheless, seeing the issues laid out in a concentrated form might lead some respondents to think differently about them than they would otherwise. If so, then the researcher would want to treat them as expressing the views of any (similarly selected) citizens who made a similar effort. For some purposes, that might be the appropriate level of involvement for determining citizens' values on important issues. Indeed, policy-makers might want researchers to take the extra step of informing participants about the underlying issues (and not just about the specific question). Those individuals might be treated as a citizens jury, impaneled to examine the issue on behalf of their fellow citizens, or as a projection of what others might think were the issues publicly aired.

The specific examples in Table 13.2 were culled from studies eliciting values for changes in atmospheric visibility. Each represents a feature that might need to be understood to evaluate a proposal. When a contractual response requires understanding many such features, investigators bear a heavy communication burden. The remainder of this paper considers analytical and psychological approaches to meeting that challenge and to assessing investigators' success in doing so.

I like this framework and have used it in various contexts (*8–12*). Its set of potentially relevant details seems initially daunting. As discussed below, not all or even many need be relevant in a given application. However, the communication burden created by the relevant features reflects the complexity of the contractual choice, not the specific framework used to characterize it. Unless investigators make a relevant feature clear, respondents must make it up. If respondents and investigators make the same assumption about an unstated feature, then it might "go without saying." If not, then respondents are answering different questions than the one attributed to them. Respondents who chanced upon the investigators' reports could justifiably ask, "How dare you interpret my responses that way? I meant nothing of the kind. You have violated my informed consent to participate in your study."

A materiality standard for informed consent

The legal doctrine for informed consent arose in medicine. It has two general expressions. About half of the United States have a professional standard: physicians must say what their peers say. Although they pool the judgments of a community, professional standards can also entrench flawed practices. These risks may be particularly large when there are few opportunities for systematic feedback and evaluation.

The remaining states have a *materiality* standard: physicians must say whatever is "material" to their patients' decisions. One way to implement this demand is to provide a laundry list of possible side effects. As in the pasta sauce ad, everything that one could possible want is "in there" somewhere. However, although that form of assurance may satisfy lawyers, it can frustrate patients—whose time, energy, and cognitive capacity are all limited. [The same might be said of Material Safety Data Sheets and many other warnings (*17*).] The same challenge would face respondents trying to master the essential details of a contractual study. In either case, a systematic approach is needed to set priorities so that attention is focused primarily on the features that really matter.

If materiality is interpreted as focusing on what people most need to know, then it can be operationalized in *value of information analysis* terms (*13, 14*). That is, information is material to the extent that receiving it affects the expected utility of recipients' choices (*15, 16*). The next section considers two general strategies for achieving this standard and for assessing the success of attempts.

Two strategies for communicating material facts efficiently

Investigators can approach their communication challenge in a piecemeal or holistic way. That is, they can strive either to convey the most material individual facts or to create a meaningful whole that facilitates recalling and inferring material facts.

Piecemeal strategies: identifying a supply curve for information. Other things being equal, communication should focus on the facts that respondents most need to know but currently do not. Figure 13.1 shows a medical example, setting priorities among information about treatment side effects (whose probability has been estimated, with appropriate uncertainty, by risk assessors). In this case, although many facts might have been material to patients, the communication task turns out to be fairly easy. Only three side effects matter much, in the sense that knowing them has much chance of affecting patients' evaluation of the surgery option. Moreover, the probabilities of these three side effects turn out to be large enough to be readily comprehended (rather than being, say, thousandths of a percent). The nature of the two top side effects (death, stroke) should be familiar, especially to candidates for this procedure who face them already. As a result, not much needs to be said about the events to which those probabilities are attached. Thus, communication might focus on what the third side effect (facial nerve paralysis) is like.

The example in Figure 13.1 assumes patients who know nothing about the probabilities of the possible outcomes. That seems reasonable for an unfamiliar surgery. For patients with prior beliefs, the analysis would simply consider how different facts would update those priors. In either case, the analysis creates a *supply curve*, showing the optimal order for providing new facts. That order would change if some important facts proved too hard to communicate. De-emphasizing them saves recipients' time—while acknowledging that less overall understanding is possible. The order would also change if some facts proved redundant, so that respondents who know one fact can predict the

Carotid endarterechtomy involves scraping out the artery leading to the brain. It can reduce the risk of stroke for patients with arteriosclerosis there. However, having a sharp object in one's neck can also cause problems, as can major surgery itself. Consider a population of patients, for all of whom the surgery would be a rational choice, if there were no risks and were money no object). They vary in their physical condition, represented by probability distributions over possible consequences, and in their values, represented by distributions on tradeoffs across those consequences. Monte Carlo sampling produces a set of hypothetical patients, each characterized by a combination of physical states and personal values. Calculate the expected utility of the surgery decision for each such patient, ignoring all risks. Then recalculate it, including knowledge of the probability of each possible consequence. Information about a risk is material, if it changes the expected utility of surgery from positive to negative.

The table shows the results of such an analysis. About 15% of such patients should decline the surgery, upon learning the probability of immediate death. Another 5% should decline, if told the risk of iatrogenic stroke. An additional 3% should be dissuaded by hearing the risk of facial paralysis. Although many other side effects are possible, few would affect many choices. Thus, although the set of potentially relevant facts is large, the set of critically relevant ones is small.

The percentage of a simulated patient population that would decline carotid endarterechtomy, upon learning of each risk.

Outcome	Percentage decline (standard deviation)
Death	15.0 (0.3)
Stroke and neurological deficit	5.0 (0.2)
Facial Nerve paralysis	3.0 (0.2)
MI (myocardial infarction)	1.1 (0.1)
Lung damage	0.9 (0.06)
Headache	0.8 (0.1)
Resurgery	0.4 (0.03)
Tracheostomy	0.2 (0.03)
Gastrointestinal upset	0.09 (0.1)
Broken teeth	0.01 (0.01)
Liver damage	0.01 (0.01)
Parotiditis	0.01 (0.01)
Kidney dysfunction	0.01 (0.01)

Source: Merz et al. 1993

FIGURE 13.1 Materiality standard for prioritizing information

others. With artificial tasks, any two features can appear together (e.g., the sizes and probabilities of gains and losses in laboratory gambles), allowing no such efficiencies. However, when evaluating real environmental changes, respondents may already have many beliefs and be quick to infer more. If respondents can guess the investigators' intentions, even a short description may communicate a lot. Thus, efficient communication requires a combination of analytical and empirical research.

The supply curve perspective also allows evaluating the adequacy of imperfect communications, as might occur when recipients can only absorb a few facts (*17*). Sometimes (as in Figure 13.1), only a few things really matter; sometimes, many things do. If participants cannot get very far in understanding the value elicitation question (much less their response to it), then investigators have two options. One is using a more intensive and potentially reactive communication procedure (*1, 18, 19*). The other is settling for a gist study, admitting that it is impossible to ensure the understanding needed for a contract study.

Holistic strategies: completing mental models. People absorb information more quickly when they can organize it into *chunks*, clusters of facts cognitively processed as units (*20*). Mnemonists have mastered this skill to a high art (*21*). They can create chunks from diverse sets of facts by integrating their members into highly flexible templates. Ordinary people tend to organize information spontaneously into less coherent, domain-specific *mental models*. When accurate and appropriate, these mental models provide a platform for integrating task features. Invoking a mental model takes advantage of existing beliefs, allowing some task features to go without saying while making others easier to absorb. However, mental models can also encourage unintended inferences, while hindering the processing of unexpected features. For example, "referendum" is a widely used metaphor in studies eliciting environmental values, with respondents being asked to imagine that they are voting on a proposed transaction. How well it works depends on the appropriateness of the associations that it evokes for respondents in a given study. (Imagine that they infer a government sanctioned, legally binding, take-it-or-leave-it choice, decided by majority rule and open to all citizens? How should the data be analyzed, if the researcher knows some of these assumptions to be false?).

Psychology has developed many approaches to studying mental models (*22–27*) suited to particular domains. Our research group has developed a version for use with complex, loosely bounded problems with substantial uncertainty, like many health, safety, and environmental issues. Our approach attempts to accommodate the messiness of lay cognitions while still allowing them to be analyzed in terms of scientific understanding (so as to guide communications). We have applied it to varied topics, including radon, climate change, solvents, EMF, mammography, HIV/AIDS, and nuclear energy in space (*28–31*).

Each application begins by creating an *expert model* of the problem, capturing its formal structure and the uncertainty in the scientific community. To the extent possible, we make these models computationally tractable, even if we lack many of the requisite parameter estimates. Doing so increases the opportunities for formal analyses of information needs as well as sharpening our own thinking. We then conduct semi-structured, open-ended interviews, eliciting respondents' beliefs about the domain, trying to evoke their natural ways of thinking and talking. The interviews begin with a request to "tell me whatever you know or have heard about [the topic]." Follow-up questions ask respondents to elaborate on what they say and then to elaborate on their elaborations. Once they lose momentum, respondents are asked more specific questions about the major domains in the expert model. For example, respondents who have not discussed remediation are asked, "what can be done about …?"

What respondents say is then compared with the expert model. As with the piecemeal strategy, the analysis attempts to identify the critical gaps between what people know and what they need to know. However, the resulting communication attempts to take advantage of the coherence of natural systems and create a picture of the problem as a whole—not just fill in isolated facts. If successful, it allows recipients to integrate their fragmentary beliefs, absorb new information, and make inferences about additional issues. With this base of understanding, they can begin the next task, identifying their value for the environmental good.

Measuring success

Whether investigators use these methods or others, their success must still be assessed. One standard method for determining what respondents take away from task descriptions is the *think-aloud protocol* (*32–34*). Respondents describe what comes into their minds as they read the task. The interviewer asks them to elaborate on what they say as a check on their intended meaning. People often use words in ways that differ from their official definitions—for example, "safe sex" (*35*), "climate" (*36*), "employed," or "room" (*37*).

"Manipulation checks" provide a second and less costly way to measure success. Placed at the end of a task, they ask respondents to report their understanding of its critical elements. Figure 13.2 shows the results of three such checks, administered after a short environmental evaluation task. Only the first check revealed a majority of respondents reporting the value actually used in the description. Despite the brevity of this task, most respondents did not hear, remember, or believe these details. They were, in effect, answering a different question than what the researchers intended to ask. In such cases, their answers would have to be discarded or reinterpreted (*11*).

When think-aloud protocols and manipulation checks are performed on pretests, they allow investigators to tell potential participants in the actual study how far a good-faith effort should get them. Leveling with them fulfills an ethical obligation (when requesting informed consent). It might also encourage them to try harder, by enlisting them as collaborators. When performed on

Offered Values of Stimulus Features

Condition (= length of proposed cleanup)	*Miles of River in Proposed Cleanup*			
	0–100 miles	101–1000 miles	1001–10000 miles	Don't Know
30 miles	**62**	9	1	29
1000 miles	4	**34**	32	31
	Feasibility of Proposed Cleanup			
	Eliminated Completely	Good Headway	Not Much_ Progress	Did not Think
30 miles	**35**	46	4	13
1000 miles	**18**	52	10	18
	Payment Vehicle of Proposed Cleanup			
	Taxes	Higher Prices	Donations	Other
30 miles	63	**25**	22	19
1000 miles	65	**31**	28	16

In a phone interview, Pittsburgh-area respondents were asked about their willingness to pay in **higher prices** for goods and services) for completely cleaning up the following problem: "Presently, a large number of the rivers in Pennsylvania are seriously polluted. These rivers include the Delaware, Susquehanna, Monongahela, Allegheny, Ohio, Clarion, Schuylkill and Lehigh. All together, there are more than 3,000 miles of rivers, of which more than **1,000 miles** are polluted. Authorities caution against swimming in or eating fish caught from these polluted portions of the rivers." The **30-mile** condition mentioned only the Susquehanna River. (Source: Fischhoff et al., 1993)

FIGURE 13.2 Manipulation checks. Percent of respondents choosing each option; boldface type indicates the value actually given in stimulus.

actual respondents, such measures of success help the users of a study to decide what conclusions it can support—by determining where respondents fall relative to the acceptable level of misunderstanding.

Discussion

Before answering an evaluation question (in a study or in life), one needs to understand it. Before agreeing to try, one should understand how tractable that task is. Unless respondents understand their task, investigators should have little to report (although, unless they measure their success, they may not know what questions respondents are really answering). Unless potential respondents understand how far they will get, investigators should not engage their service or report their answers.

As a result, environmental risk communication is an essential part of environmental change evaluation. Most researchers take these obligations seriously, subjecting their designs to hard thought and pretests. However, they typically rely on their own intuitions to get things right and to measure their success. With basic research, investigators pay the price if these intuitions are wrong. If a task confuses respondents, then interesting signals can be lost in the resultant noise. If a task misleads respondents, then their evaluations will mislead the investigator, producing conclusions that cannot be replicated. However, when research has policy implications, the respondents themselves are at risk. Their views could be missed or misinterpreted. The claim made in this paper is that the front end of environmental value elicitation requires explicit empirical research. Trusting intuition places respondent, investigator, and policy-maker at risk.

In the short run, the additional design work will increase costs—even if it takes advantage of cognate literatures (e.g., refs *37–39*). In the long run, though, deliberate design research should reduce costs by creating a cumulative empirical record, focused on issues specific to environmental value elicitation (e.g., how to convey the magnitude of an environmental change or the time period for a payment, how to convince respondents that a change will really happen). In some cases, this research will show clever ways to simplify tasks. However, it may also show us, as investigators, to be victims of the *curse of cleverness*: We prize novel tasks whose formulation captures nuances that eluded our colleagues or which address emerging policy concerns. Yet, such tasks are even more novel for respondents. As a result, the tasks may lie beyond respondents' grasp. At least within the constraints of conventional interview and experimental settings, respondents may not be able to understand the task well enough to proceed (even if many will accede to the obligations and pressures of the interview situation and produce some kind of response). Nor may they be able to understand what they want, given the complex tradeoffs involved with real environmental choices. In such cases, respondents may need a more interactive and instructional setting—just to understand what we want from them—before starting to figure out what they want for the environment (*18, 40*).

Acknowledgments

Originally prepared for the "Workshop on Alternatives to Traditional CVM in Environmental Valuation," Vanderbilt University, October 15–16, 1998. The comments of the participants are greatly appreciated as is the support of the Environmental Protection Agency and National Science Foundation. The opinions expressed are those of the author.

Literature cited

1 Whittington, D. 1988. *World Dev. 26*, 21.
2 Institute of Medicine. 1998. *Scientific opportunities and public needs*; National Academy Press: Washington, DC.
3 Institute of Medicine. 1999. *Toward environmental justice*; National Academy Press: Washington, DC.
4 Mitchell, R. C.; Carson, R. T. 1989. *Using surveys to value public goods: The contingent valuation method*; Resources for the Future: Washington, DC.
5 Arrow, K.; Solow, R.; Portney, P.; Leamer, E.; Radner, R.; Schuman, H. 1993. *Fed. Regist. 58*, 4601.
6 Kahneman, D.; Knetsch, J. L. 1992. *J. Environ. Econ. Manage. 22*, 57.
7 Fischhoff, B.; Furby, L. 1988. *J. Risk Uncertainty 1*, 147.
8 Fischhoff, B. 1993. *J. Risk Uncertainty* 7, 53.
9 Fischhoff, B.; Furby, L. 1986. *A review and critique of Tolley, Randall et al. "Establishing and valuing the effects of improved visibility in the Eastern United States*; ERI Technical Report 86–8; Eugene Research Institute: Eugene, OR.
10 Fischhoff, B.; Quadrel, M. J.; Kamlet, M.; Loewenstein, G.; Dawes, R.; Fischbeck, P.; Klepper, S.; Leland, J.; Stroh, P. 1993. *J. Risk Uncertainty 6*, 211.
11 Fischhoff, B.; Welch, N.; Frederick, S. 1999. *J. Risk Uncertainty 19*, 139–164.
12 Furby, L.; Fischhoff, B. 1988. *Specifying subjective evaluations. A critique of Dickie et al.'s interpretation of their contingent valuation results for reduced minor health symptoms*; U.S. EPA Cooperative Agreement CR814655-01-0; Eugene Research Institute: Eugene, OR.
13 Fischhoff, B. 1985. In *To Breathe Freely: Risk, Consent, and Air*; Gibson, M., Ed., Rowman and Allanheld: Totowa, NJ; pp 169–185.
14 Merz, J.; Fischhoff, B.; Mazur, D. J.; Fischbeck, P. S. 1993. *J. Toxics Liab. 15*, 191.
15 Clemen, R. 1991. *Making Hard Decisions*; PWS-Kent: Boston.
16 Raiffa, H. 1968. *Decision Analysis*; Addison-Wesley: Reading, MA.
17 Fischhoff, B.; Riley, D.; Kovacs, D.; Small, M. 1998. *Psychol. Marketing 15*, 663.
18 Fischhoff, B. 1997. In *Determining the Value of Nonmarketed Goods*; Kopp, R. J., Pommerehne, W. W., Schwarz, N., Eds.; Plenum: New York; pp 189–217.
19 Gregory, R.; Lichtenstein, S.; Slovic, P. 1993. *J. Risk Uncertainty* 7, 139.
20 Miller, G. A. 1956. *Psychol. Rev. 63*, 81–97.
21 Luria, A. R. 1968. *The Mind of a Mnemonist*; Basic Books: New York.
22 Bartlett, R. C. 1932. *Remembering*; Cambridge: Cambridge, MA.
23 Gentner, D.; Stevens, A. L. 1983. *Mental Models*; Erlbaum: Hillsdale, NJ.
24 Hirschfeld, L. A.; Gelman, S. A. 1994. *Mapping the Mind*; Cambridge University Press: New York.
25 Johnson-Laird, P. N. 1983. *Mental Models*; Cambridge University Press: New York.
26 Newell, A.; Simon, H. A. 1972. *Human Problem Solving*; Prentice Hall: Englewood Cliffs, NJ.
27 Rouse, W. B.; Morris N. M. 1986. *Psychcol. Bull. 100*, 349.
28 Bostrom, A.; Fischhoff, B.; Morgan, M. G. 1992. *J. Social Issues 48*, 85.
29 Morgan, M. G.; Fischhoff, B.; Bostrom, A.; Lave, L.; Atman, C. J. 1992. *Environ. Sci. Technol. 26*, 2048.
30 Fischhoff, B. 1998. *Reliab. Eng. Syst. Saf. 59*, 63.
31 Fischhoff, B. 1999. *J. Natl. Cancer Inst. Monogr. 25*, 7.
32 Ericsson, K. A.; Simon, H. A. 1994. *Protocol Analysis: Verbal Reports as Data,* 2nd ed.; MIT Press: Cambridge, MA.
33 Schkade, D.; Payne, J. W. 1994. *J. Environ. Econ. Manage. 26*, 88.
34 Schriver, K. A. 1989. *IEEE Trans. Profess. Commun. 32*, 238.
35 McIntyre, S.; West, P. 1992. *AIDS* 7, 121.

36 Read, D.; Bostrom, A.; Morgan, M. G.; Fischhoff, B.; Smuts, T. 1994. *Risk Anal. 14*, 971–982.
37 Turner, C. F.; Martin, E. 1984. *Surveying Subjective Phenomena*; Russell Sage Foundation: New York.
38 Garner, G. T.; Stern, P. C. 1996. *Environmental Problems and Human Behavior*; Allyn and Bacon: Needham Heights, MA.
39 Schwarz, N. 1999. *Am. Psychol. 54*, 93.
40 Fischhoff, B. 1991. *Am. Psychol. 46*, 835.

PART V

Deciding

Utility theory provides a normative standard for evaluating individuals' choices: do they choose the options with the greatest expected utility? However, applying that standard requires knowing how a decision should be structured. That is, which options, outcomes, values, and uncertainties should it consider? Utility theory is mute on these questions, whose answers require substantive knowledge about people and their circumstances. Utility theory also says nothing about the possibility that its formulae are less relevant because people have developed efficient heuristic rules, learned what they like through trial and error, or willfully chosen to make non-rational choices (e.g., based on passion or uncompromising values). My research here has sought to bridge the reasoned abstractions of normative analyses and the messy worlds in which decisions are made.

"Giving advice: Decision theory perspectives on sexual assault" (Chapter 14) applies behavioral decision research to providing sound advice, exemplified by this pervasive risk. It arose from Lita Furby's observation that, "The powerful can afford decision analysis, what about the vulnerable?" It benefited from my membership on the Eugene (Oregon) Commission on the Rights of Women, which addressed policies that affected the decisions that women faced (e.g., were counseling services readily available? would Lane County maintain its Interagency Rape Team? how well would the riverside running path be lit?). Our normative analyses summarized the (disappointingly limited) evidence regarding the effectiveness of self-defense strategies. Our descriptive research elicited women's (surprisingly diverse) goals and (overly optimistic) judgments of strategy effectiveness. We had planned to develop and test a prescriptive intervention building on these results. However, NIMH had eliminated its Center for the Prevention and Treatment of Rape before we could complete the project.

"The real world: What good is it?" (Chapter 15) contrasts the pictures of decision making that emerge from the highly structured tasks favored by experimental research and from semi-structured tasks that allow people to express themselves more freely, while still allowing comparison with normative accounts. One study in the article analyzes teens' accounts of their everyday decisions, described in their own terms. Some of its results replicate ones found with structured tasks. For example, teens seem to focus naturally on one choice option, overlooking the opportunity costs of not choosing another option (as has been found with structured tasks drawing their attention to one option). Some results, though, are different. For example, teens seem to pay more attention to uncertainty about what outcomes will be like (as personal experiences) than to uncertainty about whether they will happen (the focus of most research). Other studies ask whether people interpret structured tasks as researchers intended them. If not, then reasoned responses may be mistaken for biased ones and vice versa. The article concludes that semi-structured tasks are needed to reveal both what people make of structured tasks and what decision-making processes those tasks obscure.

"Assessing adolescent decision-making competence" (Chapter 16) asks what behavioral decision research has to say about adolescents' ability to make autonomous decisions. Although many people have a dismal view of adolescent decision making, the research finds that teens' cognitive abilities resemble those of adults. That (perhaps surprising) conclusion partly reflects the research's "inside" view, asking how teens think about their choices, rather than whether they do what adults want. It also reflects the field's view of adults as imperfect decision makers, lowering the bar for comparisons. The chapter also provides a partial answer to the question of how much these decision-making abilities matter. It summarizes a study finding that teens who do better on behavioral decision research tasks also tend to do better in life, even after holding scores on intelligence tests constant. The chapter concludes by discussing the relationship between decision making and emotional self-regulation, arguing that people who cannot control their emotions may make poor decisions, whereas people who cannot make good decisions may be left to their emotions.

"Questions of competence: The duty to inform and the limits to choice" (Chapter 17) asks about adults' ability to make autonomous decisions. Underestimating those abilities can mean needlessly denying people the right to manage their own affairs; overestimating them can mean needlessly denying people protection, such as disclosures for financial products and medical procedures. Using case studies involving health and safety, the chapter shows how normative research can assess what constitutes decision-making competence, descriptive studies can establish where people need help, and prescriptive interventions can see how far the envelope of competence can be extended. The case studies find no single answer. Sometimes, people are fine with no help at all; sometimes, interventions can provide the help that they need; sometimes, they need paternalistic institutions to protect them when they cannot make decisions on their own.

14

GIVING ADVICE

Decision theory perspectives on sexual assault

Baruch Fischhoff

DEPARTMENT OF SOCIAL AND DECISION SCIENCES, DEPARTMENT OF ENGINEERING AND PUBLIC POLICY, CARNEGIE MELLON UNIVERSITY

Abstract

What entitles us to give general advice? This general question is explored here in the specific context of providing responsible advice to women on how to make decisions about possible ways of reducing their risk of sexual assault. An approach is advanced that is a combination of decision analysis, used to provide a formal characterization of decision situations, and behavioral decision theory, used to provide a descriptive characterization of how people perceive those situations. The approach is illustrated with a set of studies using three diverse groups of women; a group of men, paralleling one of the groups of women; and a national sample of sexual assault experts. The approach is evaluated in terms of its feasibility, its strengths and weaknesses relative to alternative approaches, and its implicit position on broader political and philosophical issues.

Introduction

Psychologists are often called upon to give general advice to people on how they should manage their lives. At times, they provide such advice without being asked. The ostensible basis of this advice is psychologists' access to the relevant research literature, their training in analytical thinking, and their familiarity with individuals who have taken various steps and had to live with the consequences.

One intrinsic motivation for providing general advice is the chance to help many people at once. One extrinsic motivation is being recognized for one's expertise. It is even possible to be paid for the wisdom, either directly (e.g., for books, talks, or individual counseling) or indirectly (e.g., through the greater access to resources that recognition can bring).

Knowing the most about a topic need not, however, mean being in a position to direct other people's actions. General advice cannot be equally applicable to all individuals. Even personalized advice can mislead if undue confidence is placed in it. In some cases, the research base may be so slim that the expert is just one more person guessing.

One systematic approach to advice giving is offered by decision analysis, a family of techniques for applying the abstract principles of decision theory to the practicalities of everyday problems (Fischhoff, 1980, 1988; Raiffa, 1968; von Winterfeldt and Edwards, 1986; Watson and Buede, 1987). In these applications, analysts sit with their clients and go through the following steps: Identify the options for action, identify the consequences that might follow those actions, weight them according to their relative importance, assess the probability that each will be incurred, and combine these considerations in order to reach a decision.

Decision analysts have a sophisticated set of techniques for accomplishing these tasks. This set includes ways to express complex personal value structures, ways to accommodate subjective beliefs and statistical information, and ways to estimate the definitiveness of recommendations (Keeney and Raiffa, 1976; Morgan and Henrion, 1990).

Unfortunately, applying any but the most rudimentary versions of these techniques requires considerable training. Until that training is more widely available (Baron and Brown, 1991), decision analysis will be a labor intensive, and expensive, enterprise. In addition, good analysis is time consuming. It would be impossible to analyze every decision thoroughly, even if every individual were trained in analysis or provided with a personal analyst. How often do we hear (or make) complaints about the burden of having to show street smarts on topics as diverse as choosing a long-distance telephone carrier, picking foods that are safe and healthy, or selecting an employee benefits package—not to mention such traditional problems as choosing a career, a partner, or a parenting strategy? That frustration would be reduced somewhat if we could be reassured that we were using the best available data and decision-making methods—but only somewhat—there are just too many decisions of too great complexity to analyze each in detail. Thus, there is a need for general advice. But one needs assurance that it is trustworthy and personally relevant.

The remainder of this article offers an approach to generating such general advice, applied to the specific problem of helping women decide what, if anything, to do to reduce their risk of sexual assault. Simply speaking, the approach is to push decision analysis as far as possible under the circumstances. That requires at least two compromises from the ideal of detailed, personalized analyses. One is analyzing the situations of classes of potential clients, rather than those of individuals. Thus, the product of an analysis is not, "Here is what you should do," but "Here are some ways of thinking about the problem. Each is adapted to the circumstances of a hypothetical individual, whose situation might be quite different than your own. Consider where you might fit in." The second compromise is being satisfied with clarifying components of a decision problem (e.g., what the options are), without producing summary advice at all. These components may aid the thinking of many people, whereas summary recommendations may suit few or none.

Conflicting advice about sexual assaults

Aside from their inherent importance, decisions about sexual assault demonstrate the difficulties and dangers of producing general advice. Women receive a wide variety of advice about what to do if they are assaulted by a man intent on rape. Some experts tell women to fight back. Others say that women should do anything but defend themselves physically. Some experts recommend appealing to the rapist's "human" side by telling him how traumatic rape is for the victim. Others advise against doing anything that shows weakness. Still other experts recommend bizarre behavior, such as throwing up; their critics contend that those strategies are useless. Similar disagreement surrounds advice about how to prevent assaults from ever beginning (Morgan, 1986).

A woman who attempted to rely on this conflicting advice would be sorely confused. If the experts cannot agree among themselves, then how could she know what is right for her? The confident but contradictory pronouncements of these "experts" may make a difficult situation seem even less manageable.

Experts reach different conclusions about which strategies to use, in part, because they disagree about what effects each strategy will have. For example, some think that kicking and biting increase a woman's chances of escaping from an assailant. Others believe that such fighting is likely to make an assailant madder, leading to even greater violence.

There is plenty of room for disagreement about these effects. No one really knows the effectiveness of any strategy. It is very hard to study what happens in rape assaults and relatively few studies have been conducted (Furby and Fischhoff, 1988). Even if there were many studies, they would be unlikely to show any simple picture. The consequences of using a given strategy can depend on the particulars of the woman, the man, and the setting. A well-trained woman might fight effectively against an unarmed man who tries to pull her into a car in a public place. She might not do as well if he had a knife or if he grabbed her from behind in an office building after hours.

Faced with this uncertainty about what works, women are forced to gamble, trying to choose the strategies that give them the best chance in particular situations. Faced with this uncertainty, experts should be forced to mumble, giving no more definitive advice than the data can support. If they say more, then they risk compounding women's problems by creating the illusion that women should be able to pick strategies that will protect them. That could both reduce the pressure on society to reduce the overall threat of assaults and increase women's exposure to criticism (and guilt) after assaults in which they have been unable to defend themselves. Whatever such women have done, some "experts" have advised against it.

Experts also reach different conclusions, in part, because they make different assumptions about what is important to women. For one woman, putting up a fight may be so important to her feeling of self-worth that she would be willing to risk an increased chance of being killed in the process (if that trade-off were necessary). Another woman may care solely about getting out of the assault alive. As a result, different gambles may be right for different women. Experts who fail to make their assumptions explicit when they give advice may be projecting their personal values onto any women who listen to them.

Generalized decision analysis

Effective advice-giving must begin with communication from clients. Experts need to know what their clients believe already, so they can focus on the most critical missing facts or misconceptions. Experts also need to know what their clients want, so they can devise ways to achieve those goals, rather than impose their own. In both cases, experts must be prepared for surprises. Their clients may know and value things that they do not know or value.

The training of good decision analysts, like that of good clinical psychologists, emphasizes the ability both to listen to clients' unique problems and to look up generally applicable facts. Where the clients are a population, the way to listen is to interview individuals like those targeted for the advice. The place to look for facts is in the relevant scientific literature.

In the research described here, we applied these tools to each stage of the decision-making process. For example, in studying *options* for reducing the chances of being raped, we collected potential strategies from the research and advice literatures, then asked people which they had adopted or considered. In studying the *importance* of consequences, we asked people what concerned them and how willing they were to make various trade-offs, as well consulting the same literature. In studying the *probabilities* of consequences, we performed a secondary analysis of all the empirical studies that examined the effectiveness of different strategies for reducing the chances of completed rape, then asked people what they believed those probabilities to be. The following sections review portions of this research effort, selected to illustrate the overall approach. Readers, particularly those interested in sexual assault, are strongly urged to refer to the technical articles whose methods, results, and qualifications are necessarily cited very briefly here. Two concluding sections consider what general advice, if any, might be synthesized from such research and what legitimacy that advice would have.

The "people" in our studies included women drawn from three populations: students at the University of Oregon; alumnae of the university (all over 40 years of age); and mothers belonging to a parenting program, largely from low-income homes. We also had samples of male students, paralleling the female ones, and samples of sexual assault experts, drawn nationally from private consultants, criminal justice personnel, counselors, and researchers. Further details about the samples can be found in Furby, Fischhoff, and Morgan (1990).

The studies consider both self-defense decisions, concerning how to deal with assaults once they have begun, and prevention decisions, concerning how to prevent assaults from ever beginning. In some cases, we distinguished between self-defense strategies undertaken during an assault and those undertaken in preparing for an imminent assault. Like many other decisions for dealing with acute situations, these self-defense decisions might be contemplated both at moments of crisis and in anticipation of such moments. That anticipation could involve both organizing one's thoughts for situations in which there will be little time to think and improving one's repertoire of possible responses (e.g., buying Mace and taking self-defense classes). Both prevention and self-defense decisions might be thought of as examples of portfolio selection, attempting to identify a set of strategies that provide the greatest protection at the least cost in whatever one values (e.g., time, money, or personal freedom).

Prevention decisions can be taken by society as a whole, as well as by individual women. Women must be occupied with these decisions only because their society has failed to provide its citizens with adequate protection. The focus on individual decisions in the present research was intended to improve the advice given to women faced with this unacceptable situation. Were there a societal commitment to addressing these problems, then the techniques of decision theory, and even some of the present results, might make a contribution to making that commitment as effective as possible.

Identify options

When faced with the threat of sexual assault (either long-term or immediate), a woman must choose a strategy (or set of strategies). Even "doing nothing" is a strategy, and sometimes might be the most appropriate one (e.g., for escaping an assailant who is primarily after the "satisfaction" of subduing a woman who has actively resisted him). Doing nothing about personal protection might be the appropriate strategy for an individual woman who felt that her personal energies were best invested in working for societal action.

If a woman does not consider a strategy, then she cannot choose it. As a group, women have thought of a very large number of possibilities. Unfortunately, only a few of those strategies may be available to individual women at the time that they are needed. We posed a series of open-ended questions asking what a woman could do to prevent a sexual assault from ever occurring or to respond to an actual assault (Furby et al., 1990). Groups drawn from our three populations of women (N = 43–45) each produced at least 300 different strategies (in which strategies were held to be different if we could see plausible reasons why women would view them as differing in their probability of producing any significant consequence). On average, respondents produced about 26 different strategies (13.6 ways to prevent an assault from ever occurring, 4.0 ways to prepare for reacting to an assault, and 8.8 ways to defend oneself during an assault). Although this is a large number of strategies for an individual to write on an open-ended questionnaire, it is still less than one tenth of the options produced by each group as a whole.[1]

We supplemented the list of strategies produced by our respondents with others culled from a sample of 50 books, articles, and pamphlets on sexual assault written for lay or professional audiences, producing a total of more than 1,100 strategies. This is a bewildering number of possibilities for women, or even researchers, to consider in any detail. To bring some order to this welter of possibilities, we created a "strategy grammar." In it, each strategy was described in the form, "Doing *action* X in order to achieve *intended effect* Y [which is believed to reduce the risk of rape]." The typology distinguishes further between the *stage* of an assault at which a strategy is directed (preventing an assault from occurring, preparing to react to an assault, or defending oneself during an assault) and the *level of action* involved (individual or societal). Appendix A shows the typology, with examples produced by our respondents (Fischhoff, Furby, and Morgan, 1987).

In this view, a strategy may be ineffective either because it fails to produce the intended effect or because that effect does not deter sexual assaults. For example, threatening to report an assailant might not work either because it does not make punishment seem more likely or because fear of punishment does not reduce his intent to rape. The grammar recognizes that the same action (e.g., shouting "Jim, come quick") might be done for different reasons (getting help from an actual Jim or creating the illusion that help is around). Conversely, two very different strategies (e.g., screaming and turning up the radio) might have the same intended effect (getting help).

For researchers, such a grouping allows them to study the effectiveness of different intended effects on preventing rape or the effectiveness of different actions for achieving an intended effect. In a review of empirical studies of the effectiveness of self-defense strategies (described later), we (Furby and Fischhoff, 1988) found that there were seldom enough data to make strong statements about individual actions. However, at the level of intended effects, some moderately strong patterns emerged.

For women, thinking about intended effects may help focus the consideration of rape prevention measures. Goal-directed prompts ("How can I achieve [intended effect X]?" may be a good way to stimulate option production; Pitz, Sachs, and Heerboth, 1980). Knowing that other women have produced many possible ways to achieve every intended effect might be a good stimulus to "keep thinking," even when all seems hopeless and some strategies have failed already.

For would-be advisors, realizing that there are so many strategies in each category should increase the need for precision. Kicking, throwing things, and using karate are all ways to impede a rapist physically. However, they might have quite different effects on the chances of escaping. So, when experts advocate "fighting back," it is essential for them to say just what they mean.

Identify and weigh consequences

In an open-ended questionnaire, we (Furby, Fischhoff, and Morgan, 1991) asked people to list the positive and negative consequences (other than reducing the risk of rape) that were possible for the rape-prevention and self-defense strategies that they had listed as being ones that they, or other women, used. We identified 162 different consequences in their responses. They included consequences for the woman using the strategy (71% of all those mentioned), for specific others (e.g., her children), for the assailant, for women in general, and for society as a whole. The consequences were remarkably varied, including changes in housing options, sense of security, relations with other men, sleep, personal appearance, and public commitment to the issue.

For strategies designed to prevent assaults from ever occurring, members of the lay groups produced between 6 and 8 consequences on average; the experts produced about 11. The two consequences mentioned most frequently by both women and experts were the positive consequence of promoting good mental health and self-esteem and the negative consequence of restricting a woman's freedom and mobility. All but 1 of the 19 consequences mentioned by at least 20% of one respondent group dealt with an effect on the woman undertaking the strategy.[2] Across these 19 consequences, there was a rank correlation of .53 between frequency of mention by women and by experts. Experts were significantly ($p < .05$) more likely to mention six consequences: promoting good mental health and self-esteem, limiting a woman's entertainment and recreation activities, costing money, causing a woman to be less social, limiting a woman's job options, and improving assertiveness, honesty, and clarity in communications.

The experts also mentioned roughly twice as many self-defense strategies as did the female respondents (5.7 vs. 3.3). The three consequences mentioned most frequently (by women, men, and experts) were increasing the woman's chance of illness or injury, making the assailant madder or more violent, and promoting mental health and self-esteem. The last was mentioned significantly more frequently by the experts ($p < .05$). They were also somewhat (although not significantly) more likely to mention such aftermath consequences as psychological problems, feeling fearful or stressed, and legal repercussions. These differences may have reflected some mixture of experts knowing things that women do not, experts failing to recognize the concerns that actually occupy women, experts working harder on this pencil-and-paper task, and experts having provided more options in the preceding task (giving them more prompts for thinking about possible consequences).

A second study (Furby et al., 199I) asked respondents to judge the importance of the 22 most frequently produced consequences (14 for preventing assaults from occurring, 8 for self-defense) in terms of "how much it might influence you in deciding whether to use a strategy which has that consequence." The 5-point rating scale ranged from *would not affect my willingness to use a strategy at all* (0) to *would greatly affect my willingness* (4). The words *It would somewhat affect my willingness to use a strategy* appeared under (2). There were no significant correlations between mean ratings here and frequency of mention in the study just cited, for either the rape-prevention or the self-defense strategies. Thus, it may be that important consequences need not come to mind spontaneously. For the female respondents, three of the four most important consequences of strategies to prevent assaults from occurring (in addition to reducing the chances of being raped) were positive ones: reducing the woman's chances of being a victim of other crimes, reducing her chances of other injury or illness, and promoting her self-esteem. The three least important ones were the costs in money, in inconvenience, and in time, energy, and attention. Possibly, these consequences pale before threats to one's person; possibly, these consequences have become so routine in our society that they are taken for granted (Hindelang, Gottfredson, and Garofalo, 1978; Riger and Gordon, 1981), like a silent tax.

There was a strong rank correlation between the importance ratings of women and experts ($r = .71$). However, there were some significant differences in the ratings of individual consequences. Women attached greater importance to changes in their social relations, in the chances of injury or illness to other people with or near the woman, in the chances of catching the assailant, and in the costs in their personal time, energy, and attention.

Across female respondents, mean importance ratings for the different consequences ranged from 2.0 to 3.7, on the 0–4 scale. Although there were no significant differences between the groups of women, there were substantial differences within groups. On this four-point scale, the mean standard deviation across the 22 rated consequences was about 1.0. These disagreements highlight the disservice that might be done by providing universal advice, intended to fit all women.

To clarify the trade-offs implied by these levels of concerns, subjects were asked to evaluate the nine most frequently mentioned negative consequences of prevention strategies on a second scale. It asked, for example, "How much would a strategy have to reduce your risk of being assaulted before you would use it if it also *restricts your job options?*" The options were 1% (it would only have to reduce my chances by 1/100th), 10%, 20%, …, 100% (it would have to eliminate completely the chance that I would be assaulted). Responses to the two scales were highly correlated (e.g., for women, $r = .74$). For women, the mean reduction in the probability of an assault needed to tolerate each negative consequence ranged from 47% to 85%, with a mean of 65%. For men and experts, the means were lower (about 50%) and the ranges greater (about 30% to 80%). If these numbers are to be taken literally, then women care a great deal about the price exacted by efforts to prevent assaults from ever occurring, and even more than men and experts imagine. As before, there were substantial within-group differences (for women, the mean of the 9 standard deviations was 22).

Probability of consequences

When considering a strategy, women need to know not just if it "works," but how well it works. For women concerned about consequences other than just their chances of not being raped, detailed information is needed about the other effects of adopting a strategy. It is not enough to hear qualitative statements, such as "fighting decreases your chances of being raped, but increases your chances of being otherwise injured." A woman might be willing to risk a 5% greater chance of other injury in return for a 25% greater chance of avoiding rape, but not for a 10% chance. A woman might accept the threat to her career due to refusing to work late at the office if it reduced her chances of being assaulted by 25%, but not for a 10% reduction.

If experts fail to say what they know in quantitative terms, then they leave women to guess what they mean. If women guess wrong, then they have been misinformed, even if inadvertently. For example, imagine a magazine article saying that "physically resisting makes rape less likely." Unfortunately, verbal quantifiers, such as "likely," can be interpreted in different ways by different people, or by the same person in different circumstances (e.g., Budescu and Wallsten, 1985; Merz, Druzdzel, and Mazur, 1991; Poulton, 1989). A reader might assume that physical resistance makes a substantial difference; otherwise, why would it be newsworthy? She may assume that it must mean at least a doubling of her chances of escape without being raped. The writer of the article might have had a much smaller increase in mind—thinking that any increase, however small, is worth publicizing. The article might be based on research reporting a statistically significant effect, without mention of its practical significance (i.e., the magnitude of the difference).

Experts need to be precise, not only about what they believe but also about why they believe it. An estimated 25% increase in the chances of escaping without being raped is most impressive if based on a meta-analysis of many relevant studies. It is less interesting, but still useful, if taken from one detailed study of hundreds of women who have been assaulted. It is of questionable value, possibly even dangerous, if it is based on casual discussion with but a few women.

In addition to revealing the precision of its estimates, responsible advice must describe any systematic biases in those estimates. For example, a study of strategy effectiveness might, reasonably, be based on sexual assault victims seen by the police. However, relatively few victims go to the police. Those who do may include a disproportionately large number of women who were unable to escape (hence, have a more serious crime to report) and, among them, a disproportionately large number of women who have physically defended themselves (hence, can testify more effectively that they did not give consent). That combination might lead to underestimating the chances of escape in general and the effectiveness of physical self-defense in particular.

Other sources of information on strategy effectiveness are workers in rape crisis centers, self-defense instructors, hospital emergency room personnel, crime reporters, sex offender therapists, and national victimization surveys.

Each samples different cases and hears different things about them (e.g., women might be more open with sympathetic counselors than with impersonal interviewers; victims might tell different stories than [their] assailants). Each source may also define *rape* and *consent* differently. It is difficult for experts to identify the biases in their information. However, unless they provide appropriate cues, they may mislead those who rely on their advice. At the time of our survey, there were no studies of the effectiveness of strategies either for preventing an assault from ever occurring or for preparing to react to an assault.

Not finding a thorough summary, we examined every available study of the effectiveness of strategies for defense during an assault, some two dozen in all (Furby and Fischhoff, 1988). Wherever possible, we calculated the difference in the probability of avoiding a so-called *completed* rape for women who did and did not use each strategy.[3] Then, we grouped strategies according to their intended effects and looked for patterns of results that were repeated in multiple studies. We also grouped studies by data source, in order to assess systematic biases.

It is a tribute to the diligence and idealism of the individuals who conducted these difficult studies that some identifiable patterns did emerge. It is particularly difficult to summarize the findings from such an involved review in the limited space available here. Interested readers are strongly urged to consult the review itself. It provides quantitative estimates of effect size as well as essential details of method and interpretation.

One general result was that women who used strategies intended to reduce an assailant's propensity to rape (do crude, unfeminine things; make him see you as a human) were no less likely to be raped than were women who used other strategies. Indeed, what evidence there is suggests that they are somewhat more likely to be raped. Women who attempted to physically impede or incapacitate their assailants had a smaller chance of being raped than did women who used other strategies. Generally speaking, there seemed to be a smaller chance of being raped as strategies became more forceful and physical (Furby and Fischhoff, 1988).

There are at least two possible interpretations of these results, depending on the inferred direction of causation. Strategies that try to change an assailant's propensity to rape may have no effect (or even a slightly negative one), or it may be that women try those strategies when there is nothing else that they can do. Physically impeding an assailant may decrease the chances of being raped, or it may be that women are most likely to resist physically in those situations where that strategy is most likely to work.

Only further research will either clarify this question or allow more precise effectiveness estimates.[4] At the moment, what one can say might be,

> Women who try to work on an assailant's psychology should not expect too much of it. Those strategies might best be used to buy time to try more effective things. Physical resistance has increased some women's chances of escaping assaults. It might, therefore, pay for women to prepare themselves for exercising strategies from that category.

These are awfully broad statements, referring to effectiveness in all assaults. Unfortunately, there is too little research to refine them very much. In our review, this overall pattern was stronger when the assailant was a stranger. Perhaps women are more able to resist strangers physically; perhaps they are more able to work an acquaintance's psychology.

Other than the chances of avoiding rape, the only other consequence of self-defense strategies that had been studied extensively was the chance of physical injury other than rape. Our (Furby and Fischhoff, 1988) review of those 13 studies suggested that women who use more forceful strategies are more likely to suffer other injury. Here, too, the direction of causation must be questioned. Several studies examining the temporal order of events in assaults suggest that greater violence by the man tends to produce more physical resistance by the woman, rather than vice versa (e.g., Griffin and Griffin, 1981; Quinsey and Upford, Î985). That is, women fight more when they are being hurt, rather than being hurt more when they fight.[5]

Applied research is, of course, most valuable when it tells people things that they do not already know (although confirming beliefs can also be useful). In order to understand people's beliefs about strategy effectiveness, we (Furby, Fischhoff, and Morgan, 1989) asked subjects drawn from our five groups to evaluate the 14 self-defense strategies mentioned most commonly in Furby, Fischhoff, and Morgan (1989). They were asked how each strategy would affect the chance that an assailant would rape them (once an assault had begun) compared with "if you did nothing to stop him." The response scale was anchored at 0% ("It would not affect the chances that he would rape me") and 100% ("It would be absolutely certain to keep him from raping me"); 50% was labeled "It would reduce the chances he would rape me by about half." They were also offered the option of "harmful—It would *increase* the chances he would rape me." They were also asked to evaluate the effectiveness of the 16 strategies for preventing sexual assault from ever occurring, mentioned most commonly in Furby et al. (1990).[6]

For the self-defense strategies, mean effectiveness judgments for the three groups of women were very similar to one another, as well as to the groups of men and experts. Moreover, their judgments of *relative* effectiveness roughly corresponded to those revealed in our review. Their judgments of *absolute* effectiveness seemed, however, much too high. Over all strategies, the mean expected increase in the chances of escape was about 40%. No strategy emerged from our review (Furby and Fischhoff, 1988) as having anywhere near that impact on the probability of avoiding rape. Assuming that these estimates do not reflect some measurement artifact, advice givers should focus on communicating how much the strategies can do. Women who know which strategies are best can still make poor decisions if they expect too much of them.

For strategies intended to prevent assaults from ever occurring, the picture was somewhat different. The three groups of women largely agreed with one another and with the men, but not with the experts, regarding the relative effectiveness of different strategies. Again, absolute estimates of effectiveness seemed much too high (averaging, over all 19 strategies, 62% for women, 52% for men, and 47% for experts).[7] The biggest group differences lay with strategies for increasing the chances of outside intervention (e.g., parking in well-lit places), for which the experts had much lower expectations. In order to be useful here, experts face three tasks. The first is to understand the reasons why these lay beliefs seem out of line. The second is to figure out what the experts can justifiably say—in the absence of empirical studies of the effectiveness of such strategies. The third is to communicate their best guesses, in a comprehensible way that is candid about the limits of their knowledge.

Synthesis

Decision theory offers explicit and elaborate ways of putting all of these considerations together in order to identify the best possible strategy for a particular individual in a particular situation. Basically, they involve weighting the possible consequences of a strategy by how important and likely they are. It is, however, hard to imagine women performing such calculations on their own or having the services of a professional consultant.[8]

If experts believe that laypeople are ill informed about the components of a decision, then the treatment is conceptually straightforward. The experts can disseminate information about neglected options, misestimated risks, or consequences that are afforded too much or too little attention. Accomplishing this task is, however, quite complicated. One must identify just what these gaps are, establish that the experts' knowledge really is superior, and then convey the missing information in a comprehensible fashion that will be accepted and incorporated into people's belief systems.

It is quite a different enterprise to provide summary advice to people who have different circumstances, capabilities, and values. One compromise approach to this problem is to provide "clients" with simple rules, capturing some of the wisdom of decision theory, which they might apply to their personal situations. A second compromise approach is to apply the full apparatus of decision theory to archetypal women, then show how the advice computed for them might be adapted to someone with differing values, threats, capabilities, and so on.[9]

Simple decision rules might include the following: (a) Eliminate from consideration any option that does not have any feature that you consider essential (e.g., preserving your self-esteem), (b) eliminate any option that has any totally unacceptable consequence (e.g., increasing your chance of death), and (c) eliminate any option where another option is at least as good in all respects.[10] If such rules do not point to a single strategy (or set of strategies) as being clearly best, then one faces trade-offs such as just how much of a decreased chance of being raped would it take to offset a 10% increase in the chances of alienating other men?

Providing full decision analyses of the situations of several women in archetypal situations would demonstrate the principles of both individual differences and individual sovereignty; people face different situations and have the right to choose what seems best to them. Such detailed analyses might also be an effective way to introduce the logic of decision theory and pertinent results from the scientific literature, through a sort of tutorial by concrete example. Extracting personal lessons from the analysis of the circumstance of others should be fairly natural for people; it underlies such diverse enterprises as reading advice columns, listening to talk shows, comparing notes with friends, and following moral tales.

Yet, one should no more expose people to untested advice than to untested medicine (Fischhoff, 1987). People could read too much or too little into such indirect advice. They may make too much of analogies, failing to see critical differences from their own circumstances. Or, they could dismiss lessons too quickly, perhaps hastening to distinguish between themselves and others so as to avoid facing threatening situations. We know too little about what people extract from the advice that others receive.

Conclusions

This sort of calm, deliberate, calculating analysis is far removed from the emotional intensity of an actual assault. No one could envision women getting out their calculators and clipboards when they hear a suspicious noise outside or find themselves alone in an elevator with a strange man. Doing so would represent a parody of the decision-making perspective and extreme insensitivity to women's predicament.

A more legitimate role for this perspective is to aid women who are worrying about such contingencies before they arise. The hope is that it will help them to consider more options, to concentrate on the options that are most relevant to them, to recognize the full set of possible consequences, to know which consequences are more likely to occur, and to read the advice of experts with a more critical eye.

Just being able to organize one's thinking on a stressful topic might aid in coping with it. Anything that reduced women's general level of anxiety, without compromising their safety, would have something to say for it. If a woman was actually confronted by an assault situation like one she had imagined, then her theoretical analysis might help her decide what to do and improve her resolve to carry it out. Even if the assault situation were quite different, then all that prior thinking might help her to think more effectively on the spot. Just being able to keep calm might have a salutary effect.

Decisions regarding strategies to prevent assaults from ever happening are made under less stressed circumstances. Yet, with more than 900 possible strategies, 150 possible consequences, and large uncertainties, comprehensive analysis is infeasible. A full-blown decision-theoretic perspective might at best be useful in evaluating a few select choices. More general uses of the perspective may be in organizing one's thinking, having a framework to accommodate new evidence, and scrutinizing the recommendations provided by experts.

Indeed, the greatest help to women from the decision-theoretic perspective might be from imposing it on would-be advisors. It should force them to recognize individual women's sovereignty over their decisions and the uniqueness of their situations. It should compel empirical studies to determine what women believe and desire, the results of which might help prevent experts from projecting their own values on others—not only acting as though "one size fits all" when it comes to advice but making that size fit the expert's predispositions. Identifying the information most critical to women's decisions should focus research efforts—showing, for example, the glaring absence of studies concerning the effectiveness of strategies for preventing assaults from ever occurring. It should also encourage experts to report results in a relevant format (e.g., changes in the probability of preventing rape, rather than just the statistical significance of effects).

Finally, a decision-analytic perspective provides a constant reminder that dealing with sexual assault is a gamble. No one knows what will happen. Nor does anyone know what would have happened had, say, an individual woman pleaded, rather than fought, or vice versa. Even choosing the best strategy for a particular situation provides no guarantee of having the best possible outcome. Most men may back off from a physically assertive woman, but not every man will, and it may be impossible for a woman to tell what kind of man she is facing. Grappling with this unpredictability is a miserable business. However, it is essential if women are to choose wisely and to be judged fairly in hindsight, by others and themselves, after the outcome of an assault is known. It is all too easy to ignore all the uncertainty present at the time of a decision and to judge the quality of a decision by its outcome (Baron and Hershey, 1988; Fischhoff, 1975; Hawkins and Hastie, 1990).

Unfortunately, this promise of helping women is just a promise. The empirical studies reported here were the first of their kind and will, we hope, be superseded by future research, using better methods and more diverse subject populations. Some of our best guesses may eventually be shown to have pointed in inappropriate directions. There are also clear omissions from the research done to date, such as exploring the role of emotion or social interaction in how women manage these risks. Nor have we resolved the sources of the differences between the perceptions of women and experts regarding the magnitude of risks, the effectiveness of measures, and the importance of consequences. We never produced, much less tested, the sort of conditional summary advice described in the previous section. There is always the chance that our attempts to clarify the structure and components of the problem will confuse or mislead. That risk may be particularly great with a summary description such as the present one, which necessarily omits details of methods, results, and interpretation.

Some of these lacunae are the sort of "technical" problems that affect any new area of research and will gradually be reduced over time. What could not be overcome is any fundamental flaw in the philosophy underlying this approach. In its defense, one can point to the fact that the approach recognizes people's right to make their own decisions, with experts relegated to the role of aiding in those decisions, entrusted with a deep obligation to be frank about the limits to their own knowledge and about the values underlying their advice. The approach also recognizes the possibility that laypeople may be reservoirs of knowledge about their situations; thus, every time the public and the experts disagree need not mean that the experts are right (Fischhoff, 1989, 1990). The approach emphasizes—and might help to justify—the research needed to serve women's decisions. The parallel structure for individual and societal strategies, in the typology of Appendix A, may help maintain awareness of the need for action on both levels.

There seem, however, to be several risks. The greatest of these is that, despite our protestations and cautions, the approach may be seen as effectively holding women responsible for their own safety. As a result, the approach may do nothing to move society toward protecting women from this intolerable threat. A second risk is that the hyperrationality of the decision-analytic approach may anesthetize feeling about this topic (Tribe, 1973). It might not only reduce the urgency of the problem, but muffle the emotions of those who must deal with the threat and pay the price of coping. A third risk is the introduction of yet another group of experts, decision theorists, into women's lives. Although those theorists might claim to be playing no more than a coordinating and structuring role, they provide a kind of expertise that is hard to scrutinize, hence, that cannot be empowering women.

Variants of these risks and promises face any attempt to provide general advice. Whether an approach merits support requires a detailed analysis of how it rates in these regards. Finally, it requires the metadecision of how good it is, relative to the alternatives.

Editor's note

Articles based on APA award addresses that appear in the *American Psychologist* are scholarly articles by distinguished contributors to the field. As such, they are given special consideration in the *American Psychologist's* editorial selection process.

This article was originally presented as a Distinguished Early Career Contribution to Psychology in the Public Interest award address at the 99th Annual Convention of the American Psychological Association in San Francisco in August 1991.

Author's note

The research reported here was supported by the National Institute of Mental Health, Center for the Prevention and Control of Rape (now disbanded), under Grant MH40481-02 to the Eugene Research Institute. Preparation of this article was supported by the National Science Foundation, under Grant SES-8715564 to Carnegie Mellon University. The views expressed are those of the author.

All of the research was conducted in collaboration with Lita Furby. This project never would have happened without her. Her diligence, commitment, and imagination are gratefully acknowledged. My thanks also go to Marcia Morgan, Nancy Collins, Mark Layman, Cecilia Hagen, Leisha Sanders, Geri Hanson, Ann Curran, Beth Fischhoff Melina, Robyn Dawes, Marilyn Jacobs Quadrel, Ann Bostrom, Claire Palmgren, and Rosa Lio Stipanovic for their valuable contributions to the project and this article.

Notes

1 There were relatively few differences in the judgments of the three groups of women, despite their differences in age, income, and education. Thus, in this sense, the threat of sexual assault posed a common reality for these women. The male students often responded similarly to their female counterparts, suggesting some understanding of the situation that women face. By contrast, the sexual assault experts often responded rather differently. On the one hand, that means that they may have something to teach women. On the other hand, that suggests the existence of barriers to making themselves understood, as well as the possibility that they may have something to learn about what women see and want. Unfortunately, there is no opportunity to go into these issues here.

2 The one exception was increasing the chance of injury or illness to others with or near the woman, mentioned by 22% of women, 16% of men, and 9% of experts.

3 In some discussions, we (Furby and Fischhoff, 1988) found a confusion between these conditional probabilities and the "inverse" probabilities, namely, the proportion of women using a strategy among those who escaped. For example, it would not be an endorsement of "verbal resistance" relative to "physical resistance" if it were found that 50% of women who escaped had used verbal resistance, whereas only 20% had used physical resistance. It may be true that all of the, relatively few, women who resisted physically succeeded in escaping, making this a very effective strategy. Conversely, women who used verbal resistance may still have been less likely to escape than those who did not. This form of confusion is known as the *base-rate fallacy* (Bar Hillel, 1990; Eddy, 1982; Tversky and Kahneman, 1974).

4 We (Furby and Fischhoff, 1988) closed the books on our review in 1986. A computer search yielded two subsequent studies of strategy effectiveness (a remarkably small number considering the importance of the problem). Both found that more forceful resistance reduced the chances of rape completion, consistent with the studies that we had surveyed (Kleck and Sayles, 1990; Levine-MacCombie and Koss, 1986).

5 Two studies subsequent to our (Furby and Fischhoff, 1988) review reported related results. In a correlational analysis, Ruback and Ivie (1988) found a positive relationship between physical resistance and injury, which was stronger when the assailant was a stranger rather than an acquaintance. Kleck and Sayles (1990) found little evidence that resistance increased injury.

6 For the male student and expert samples, "you" was replaced by "a woman." The students were asked to think about the risks faced by women at their university. The experts were asked to think about a woman in an area such as Eugene-Springfield.

7 If a strategy was 100% effective in preventing assaults when it was used, in order to effect a 62% reduction in the overall chance of an assault occurring, it would have to apply in 62% of assault situations. Although, as mentioned, there were no studies of the effectiveness of these strategies, it seems very unlikely that any would be so effective and usable in so large a set of circumstances. The article discusses these issues in some detail. One clue to the meaning of these estimates may be found in the similarity between them and those elicited in a study (Fischhoff, Furby, and Morgan, 1988) asking about the relative riskiness of situations differing in whether a protective strategy had been taken. For example, on average, respondents there estimated that a woman at home was 2.1 times safer if her doors and windows were locked; that is equivalent to a 52% reduction in her risk. In Furby, Fischhoff, and Morgan (1989), always locking was judged to reduce risk by 68%. This similarity (68% vs 52%) suggests that respondents in Furby et al. (1989) judged the effect of strategies on risk levels in the specific situations in which they were used, rather than on overall risk levels (as they were asked). Although this interpretation provides an excuse for the apparently exaggerated effectiveness judgments, it may create a problem for communication, namely ensuring that respondents have understood the risk levels that were intended (Fischhoff and Quadrel, 1991).

8 However much we might anguish over our profession's success in following through on its commitment to "give psychology away," the thought of giving consulting services away hardly arises.

9 A high-tech third option would be to develop interactive computer programs for decision analysis. They could, in principle, provide people with the needed decision analytic perspective, latest scientific evidence, and computational capacity. Some fascinating examples are under development (e.g., BARN for helping adolescents make decisions about their sexual behavior, Gustafson, Bosworth, and Hawkins, 1988). However, it will be some time before they can be developed, tested, and disseminated.

10 For example, if you think that all the negative consequences of biting are as bad or worse than those of kicking, and all the positive consequences of kicking are as good or better than those of biting, forget about biting.

References

Bar Hillel, M. 1990. Back to base rates. In R. M. Hogarth Ed.), *Insights in decision making* pp. 200–216). Chicago: University of Chicago Press.

Baron, J., and Brown, R. V. (Eds.). 1991. *Teaching adolescents decision making* Hillsdale, NJ: Erlbaum.

Baron, J., and Hershey, J., 1988. Outcome bias in decision evaluation. *Journal of Personality and Social Psychology, 54*, 569–579.

Budescu, D. V., and Wallsten, T. S. 1985. Consistency in the interpretation of probabilistic phrases. *Organizational Behavior and Human Decision Processes, 36*, 391–405.

Eddy, D. M., 1982. Probabilistic reasoning in clinical medicine: Problems arid opportunities. In D. Kahneman, P. Slovic, and A. Tversky (Eds.), *Judgment under uncertainty: Heuristics and biases* (pp. 249–267). New York; Cambridge University Press.

Fischhoff, B. 1975. Hindsight ≠ foresight: The effect of outcome knowledge on judgment under uncertainty. *Journal of Experimental Psychology: Human Perception and Performance. 104*, 288–299.

Fischhoff, B. 1980. Clinical decision analysis. *Operations Research. 28*, 28–41.

Fischhoff, B. 1987. Treating the public with risk communications: A public health perspective. *Science, Technology, and Human Values, 12* (3, 4), 13–19.

Fischhoff, B. 1988. Judgment and decision making. In R. Sternberg and E. E. Smith (Eds.), *The psychology of thinking processes* (pp. 153–187). New York: Cambridge University Press.

Fischhoff, B. 1989. Risk: A guide to controversy (Appendix to National Research Council). *Improving risk communications* (pp. 211–319). Washington, DC: National Academy Press.

Fischhoff, B. 1990. Psychology: Tool or toolmaker? *American Psychologist, 45*, 647–653.

Fischhoff, B., Furby, L., and Morgan, M. 1987. Rape prevention: A typology of strategies. *Journal of Interpersonal Violence, 2*, 292–308.

Fischhoff, B., Furby, L., and Morgan, M. 1988. *Judging the risks of rape*. Eugene, OR: Eugene Research Institute.

Fischhoff, B., and Quadrel, M. J. 1991. Adolescent alcohol decisions. *Alcohol Health and Research World, 15*, 43–51.

Furby, L., and Fischhoff, B. 1988. Rape self-defense strategies: A review of their effectiveness. *Victimology*. Eugene: Eugene Research Institute.

Furby, L., Fischhoff, B., and Morgan, M. 1989. Judged effectiveness of common rape prevention and self-defense strategies *Journal of Interpersonal Violence, 44*, 44–64.

Furby, L., Fischhoff, B., and Morgan, M. 1990. Preventing rape: How people perceive the options for assault prevention. In E. Viano (Ed.), *The victimology research handbook* (pp. 227–259). New York: Garland.

Furby, L., Fischhoff, B., and Morgan, M. 1991. Rape prevention and self-defense: At what price? *Women's Studies International Forum, 14*, 49–62.

Furby, L., Fischhoff, B., and Morgan, M. 1989. Judged effectiveness of common rape prevention and self-defense strategies. *Journal of Interpersonal Violence, 4*, 44–64.

Griffin, B. S., and Griffin, C. T. 1981. Victims in rape confrontation. *Victimology, 6*, 59–75.

Gustafson, D., Bosworth, K., and Hawkins, R. P. 1988. *Body awareness resource network.* Madison: University of Wisconsin—Madison, Center for Health Systems Research and Analysis.

Hawkins, S., and Hastie, R. 1990. Hindsight: Biased judgments of past events after the outcomes are known. *Psychological Bulletin, 107*, 311–327.

Hindelang, M. J., Gottfredson, M. R., and Garofalo, J. 1978. *Victims of personal crime: An empirical foundation for a theory of personal victimization.* Cambridge, MA: Ballinger.

Keeney, R., and Raiffa, H. 1976. *Decision making with multiple objectives.* New York: Wiley.

Kleck, G., and Sayles, S. 1990. Rape and resistance. *Social Problems, 37*, 149–162.

Levine-MacCombie, J., and Koss, M. P. 1986. Acquaintance rape: Effective avoidance strategies. *Psychology of Women Quarterly, 10*, 311–319.

Merz, J. F., Druzdzel, M. J., and Mazur, D. J. 1991. Verbal expressions of probability in informed consent litigation. *Journal of Medical Decision Making, 11*, 273–281.

Morgan, M. 1986. Conflict and confusion: What rape prevention experts are telling women. *Sexual Coercion and Assault, 1*, 160–168.

Morgan, M. G., and Henrion, M. 1990. *Uncertainty: A guide to dealing with uncertainty in quantitative risk and policy analysis.* New York: Cambridge University Press.

Pitz, G. F., Sachs, N. J., and Heerboth, J. 1980. Procedures for eliciting choices in the analysis of individual decisions. *Organizational Behavior and Human Performance, 26*, 396–408.

Poulton, E. C. 1989. *Bias in quantitative judgment.* Hillsdale, NJ: Erlbaum.

Quinsey, V. L., and Upford, D. 1985. Rape completion and victim injury as a function of female resistance strategy. *Canadian Journal of Behavioral Research, 17*, 40–50.

Raiffa, H. 1968. *Decision analysis.* Reading, MA: Addison-Wesley.

Riger, S., and Gordon, M. T. 1981. The fear of rape: A study in social control. *Journal of Social Issues, 37*, 71–92.

Ruback, R. B., and Ivie, D. L. 1988. Prior relationship, resistance, and injury in rapes: An analysis of crisis center records. *Violence and Victims. 3*, 99–111.

Tribe, L. 1973. Technology assessment and the fourth continuity: The limits of instrumental rationality. *Southern California Law Review, 46*, 617–660.

Tversky, A., and Kahneman, D. 1974. Judgment under uncertainty: Heuristics and biases. *Science, 185*, 1124–1131.

von Winterfeldt, D., and Edwards, W. 1986. *Decision analysis and behavioral research.* New York: Cambridge University Press.

Watson, S., and Buede, D. 1987. *Decision synthesis.* New York: Cambridge University Press.

Appendix A

Rape prevention strategy typology (with verbatim examples from respondents)

Prevent assault from ever occurring

Societal action

1. Reduce visibility of women to potential assailant (measures intended to prevent him from ever seeing a woman or knowing of her existence)
2. Reduce accessibility of women to potential assailant (measures intended to maintain a physical barrier or distance between woman and assailant, given that he knows of her existence)
    - Provide safe houses
    - Law forbidding hitchhiking
3. Increase women's perceived ability to cope with potential assailant if assault were to take place
    - Set up dog escort services
    - Publicize that there are rape resistance study groups and self-defense courses for women (so rapists will know)
4. Increase perceived chances of outside intervention if assault were to take place
    - Put emergency sounding devices on every block
    - Set up neighborhood watch programs
5. Increase perceived chances of punishment if assault were to take place
    - Make laws against marital rape
    - Encourage your community to set up a rape crisis center
6. Reduce men's propensity to rape
    - Have complete psychological rehabilitation for rapists
    - Therapy for men who are lacking in self-confidence
7. Increase women's ability to implement prevention measures successfully
    - Public school education programs for females on rape prevention
    - Post notices on bulletin boards and in newspapers of rape prevention clinics

8. Alter societal beliefs and attitudes that promote rape
   Portray rape as a violent crime, not as a crime of passion
   Change attitude that sex is a commodity
9. Alter structural characteristics of society that promote rape
   Eliminate poverty
   Full employment in the country

Individual action

1. Reduce visibility of women to potential assailant
   Avoid dangerous neighborhoods
   Do not hang around bus terminals
2. Reduce accessibility of women to potential assailant
   Do not hitchhike
   Move to a place with a doorman
3. Increase perceived ability to cope with assailant if assault were to take place
   When approached by a stranger, make direct eye contact
   When entering house, let dog in first to scare person
4. Increase perceived chances of outside intervention if assault were to take place
   Do not drive alone
   Fake presence of others
5. Increase perceived chances of punishment if assault were to take place
   Wear identifying armband to designate membership in rape prevention group
   Report known rapists/press charges
6. Reduce potential assailant's propensity to rape
   Don't wear tight or revealing clothes
   Women should be available for proper relationships and willing to go on dates
7. Manage yourself in ways that increase ability to implement prevention measures successfully
   Get educated about high risk situations
   Notice other people's behavior
8. Contribute to societal action
   Be involved in political action
   Encourage setting up or participate in rape crisis center

Prepare for reacting to an assault

Societal action

1. Increase women's ability to cope with assailant in the event of an assault
   Inform women about common elements in rape assaults
   Whistles, batons, and other defensive weapons provided by police departments

2. Increase chances of outside intervention in the event of an assault
   - Set up escort services
   - Public transportation buses should run later in the evening

Individual action

1. Increase ability to cope with assailant in the event of an assault
   - Own a dog
   - Learn self-defense
2. Increase chances of outside intervention in the event of an assault
   - Install burglar alarm system
   - Carry noisemaker

Defend yourself during an assault

Societal action (not applicable)

Individual action

1. Manage yourself in ways that maximize your ability to implement self-defense measures successfully
   - Do not faint or pass out
   - Assess attacker's personality
2. Reduce/minimize assailant's propensity to rape
   - Do crude, unfeminine things
   - Make him see you as a human
3. Increase perceived ability to cope with assailant
   - Make it known you have a weapon
   - Clear verbal resistance
4. Increase perceived chances of outside intervention
   - Fake arrival of others
5. Increase actual chances of outside intervention
   - Yell "fire"
   - Summon nearest male
6. Increase perceived chances of punishment
   - State you will press charges against attacker
7. Establish distance or barrier between self and assailant
   - Get out of house
   - Run away
8. Physically impede or incapacitate assailant
   - Incapacitate him with drugs or alcohol

Note. From "Rape Prevention: A Typology of Strategies" by B., Fischhoff, L. Furby, and M. Morgan, 1987, *Journal of Interpersonal Violence. 2.* pp. 303–305. © 1987 by Sage Publications, Inc. Adapted by permission.

Appendix B

Possible bottom lines

Premise

The clearest message of the research literature is that these are very complex decisions. There are many possible options, with difficult tradeoffs, and great uncertainty about the outcomes. Although complexity is an inherent part of these situations, decision theory would be a limited tool if it could do no more than describe what is there. It might even have negative value, if it promoted paralysis (rather than just caution).

What follows is an attempt to derive responsible advice, consistent with a decision-theory perspective. It is organized into "Living With the Situation," suggesting ways to think about this ugly reality, and "Dealing With the Situation," suggesting ways to narrow the set of possible options.

WARNING. One should no more expose people to untested advice than to untested medicine. People may take it too seriously or too casually. These suggestions are but points of departure, to be used and disseminated with caution.

Living with the situation

Do not judge the quality of a decision by how it turns out. These decisions are inevitably gambles. No one can think of all contingencies or predict consequences with certainty. Good decisions may be followed by bad outcomes. As a result, all one can hope to do is think as clearly and comprehensively as possible.

Question confident recommendations. The relative effectiveness of different strategies is only partially understood. "Experts" who speak too confidently misrepresent the degree of control that women can exert over these situations, perhaps exaggerating it, perhaps understating it.

Question universal recommendations. At times, achieving one goal comes at the price of a reduced ability to achieve other goals. In that case, trade-offs must be made. It is an individual's right to decide what those should be. "Experts" who provide the same advice for everyone violate this principle, and risk imposing their own values on others.

Recognize lay expertise. Ordinary citizens may have unique information about the risks and resources in their own community (e.g., which places to avoid, which public services to trust). They may also have a uniquely relevant understanding of what it means to experience different actions and consequences (e.g., how it feels to have taken a self-defense course, to have been stalked).

Remember that sexual assault is the community's problem. Although these problems must be faced at the individual level, they cannot be solved there. Decisions about individual actions should be viewed as a stopgap, until an acceptable environment can be created (as should research and advice intended to aid those decisions).

Dealing with the situation

Keep thinking. One cannot choose the best option unless one has at least thought of it. People who think hard can generate many possible options. When trying to produce options, it may help to consult with others and think about how to achieve specific goals.

Evaluate options by category. Although there are a great many possible options, most express one of a fairly small set of theories (e.g., assailants have a human side, assailants fear punishment). One should focus on options that embody the most credible theories. Once a category seems promising, one may be able to increase the set of ways to implement that theory by preparatory actions (e.g., installing locks, taking self-defense classes, traveling with others).

Consider base rates (the overall frequency of events). Focus on strategies to be used in those situations in which assaults are most likely. Unless you have good evidence to the contrary, assume that your risk is as large as that faced by others in your general circumstances, assume that your date is as likely as the average man to assault you, assume that your assailant will be deterred by the same things as other assailants, and so on.

Follow the scientific literature. Good studies can collect large samples of data with relatively clear interpretation. They provide the best hope for sorting out the conflicting claims that arise from anecdotal evidence. Existing studies have found more assertive self-defense strategies to be more effective. That should shift the burden of proof onto those doubting the efficacy of such resistance.

Analyze the sources of experts' claims. What experts believe depends on what they have seen, in their studies, in their personal experience, and in their reading. Identifying the sources of bias in that evidence may make it possible to use experts' claims more critically (e.g., have they disproportionately seen cases of successful escapes, or of fatalities?).

Analyze the assumptions underlying experts' recommendations. Do they share your values? Are they talking about people in your situation? If they are, then you can treat their recommendations as their best guess as to what you should do. If they are not, then knowing the experts' assumptions is the starting point for adjusting their advice to your own circumstances.

Try simple decision rules to reduce the set of possibilities. For example: (a) reject any options lacking features you consider essential (e.g., preserving your self-esteem); (b) reject any option with any totally unacceptable consequence (e.g., increasing the chance of death); (c) eliminate any option where another option is at least as good in all respects.

15

THE REAL WORLD

What good is it?

Baruch Fischhoff

Abstract

Any experimental science runs the risk of implosion. The tasks that were created as surrogates for complicated real-world situations can take on lives of their own. When that happens, scientists become fascinated with the nuances of variations within that little world. Their theoretical accounts end up with little place for phenomena that could not be observed there. Extrapolations to other settings may require large doses of conjecture (or leaps of faith). This essay examines some possibilities for understanding the world of the lab by grappling with applied problems in the world outside and perhaps doing a little good along the way.

Choice A

> In this task, you will be asked to choose between a certain loss and a gamble that exposes you to some chance of loss. Specifically, you must choose either: Situation A. One chance in 4 to lose $200 (and 3 chances in 4 to lose nothing). OR Situation B. A certain loss of $50. Of course, you'd probably prefer not to be in either of these situations, but, if forced to either play the gamble (A) or accept the certain loss (B), which would you prefer to do? (Fischhoff, Slovic, and Lichtenstein, 1980)

Choice B

> My cousins … ordinarily, I'm like really close with my cousins and everything. My cousin was having this big graduation party, but my friend—she used to live here and we went to … like started preschool together, you

> know. And then in 7th grade her stepdad got a job in Ohio, so she had to move there. So she was in Ohio and she invited me up for a weekend. And I've always had so much fun when I'd go up there for a weekend. But, it was like my cousin's graduation party was then, too—like on that same weekend. And I was just like I wanted to go to like both things so bad, you know. I think I wanted to go more to like up to Ohio, you know, to have this great time and everything, but I knew my cousin—I mean, it would be kind of rude to say, "Well, my friend invited me up, you know for the weekend." And my cousins from out of town were coming in and everything. So I didn't know what to do. And I wanted my mom to say, "Well, you have to stay home," so then I wouldn't have to make the decision. But she said, "I'm not going to tell you, you have to stay home. You decide what you want to do." And I hate when she does that because it's just so much easier if she just tells you what you have to do. So I decided to stay home basically because I would feel really stupid and rude telling my cousin, well, I'm not going to be there. And I did have a really good time at her graduation party, but I was kind of thinking I could be in Ohio right now. (Fischhoff, Furby, Quadrel, and Richardson, 1991)

The first of these choices looks like the kind of tidy stimulus found in many studies of behavioral decision making. The second looks like the kind of messy process found in many people's minds, when they try to grapple with what life sends their way. Our goal in studying choices like A is to illuminate situations like B. A nagging worry has to be whether you can get there from here.

There are good reasons for studying A, rather than B. So much is going on in B that it is hard to tell which factors influence behavior. Choice B is so unique that one cannot say how people generally behave in such situations, at least without grappling with what "such situations" means. The richness of Choice B comes through the filter of verbal reports, which can add both random and systematic error (e.g., Ericsson and Simon, 1984; Nisbett and Wilson, 1977). Observing Choice B *in vivo*, and even eliciting concurrent verbal protocols, might reduce these problems at the price of incurring others (e.g., influencing decision makers' behavior).

Other professions (e.g., history, ethnography) have complementary reasons for avoiding the risks that come with the standardization and experimentation that we prefer. Some are so comfortable with studying the particular that they deny the wisdom (or even possibility) of generalization (e.g., deriving laws of history) (e.g., Gallie, 1964; Hexter, 1971). They might be skeptical of our hope of eventually informing specific decisions with general insights (Kren and Rapoport, 1976). Even psychologists committed to generalization have cautioned against the premature closure that can come with neglecting detail (Frisch, 1993; Gergen, 1994; Gilligan, 1982)

This essay considers some ways to bridge the gap between the general and the specific in decision making, driven by the demands of trying to be useful in real-world settings. Responses to these demands have been divided into *applied basic* psychology and *basic applied* psychology (Baddeley, 1979). The former uses basic research tools to solve applied problems, possibly learning something general from the obstacles encountered in the process. The latter looks for unsolved basic issues embedded in applied problems. The former may be more generous (giving psychology away), while the latter may be more modest (acknowledging our need to learn). Both can provide our science with exogenous sources of change, complementing the endogenous ones that arise from self-generated problems (Kuhn, 1962; Lakatos, 1970).

The set of studies described below begins and ends with attempts to look simultaneously at many aspects of the same decision. In the middle are studies focused on components of decisions (options, consequences, uncertainties). In each case, the research attempts to learn something general by wrestling with the particulars of specific decisions. A recurrent concern in these studies is whether investigators' conceptualizations of decisions match those of the people making them. At the level of individual studies, identifying discrepancies can clarify the interpretation of results. Looking across studies, the pattern of discrepancies (and convergences) might contribute to a theory of context, showing how people impose structure on the choices that life presents them.

These studies are chosen from my own work, conducted with several valued colleagues. This restriction allows me to speak from personal knowledge regarding the practical concerns that motivated each study, as well as how we tried to "make science" out of the issues that it raised. I try to be candid about how far these studies have made either basic or applied contributions. The exposition presents enough results to give some feeling for whether this level of detail contains a message or just a mess. Some related efforts by other investigators are noted, although space constraints prohibit doing them justice.

Whole decision strategies I: teen decisions

Some years ago, we received support, from the Carnegie Council on Adolescent Development, to study adolescent risk taking. Part of our response was to replicate, with teens, structured experiments that had previously been conducted with adults, in which subjects judge the magnitude of risks and the extent of their knowledge about them (Quadrel, Fischhoff, and Davis, 1993; Quadrel, Fischhoff, Fischhoff, and Halpern, 1994a). Teens' responses were similar to those observed with adults, a mixed pattern of strengths and weaknesses. Indeed, they were so similar as to raise the suspicion that the method had shaped the message. Especially given the deep concern over adolescents' "reckless" behavior (e.g., Arnett, 1993; Elkind, 1967), we worried that our methods had prompted "adult-like" modes of thought, unlike adolescents' normal way of thinking, out there in the world.

As a result, we supplemented these structured studies with more open-ended ones, allowing a wider range of thought processes to emerge.[1]

Fortunately, our funding carried relatively weak strings, allowing us to attempt much more speculative studies than were likely to have survived ordinary proposal review.

The most ambitious of these studies involved 1- to 2-h interviews with 105 teenage girls, drawn from disparate social backgrounds, in Eugene and Pittsburgh. After a brief introduction, these teens were asked to describe briefly recent decisions in several specific domains (e.g., about school, parents, clothes). Then, they described more thoroughly three difficult decisions of their own choosing, two made recently and one still facing them. They used their own words, shaped only by our prompts to consider specific issues (after their spontaneous descriptions had ebbed). Choice B is the initial description of one of these decisions. Our coding of the transcripts considered the content, structure, and process of these decisions. What follows is a sampling of results from each of these three areas. These results are from 66 younger teens (age: 12–14; grade: 6–8) from varied socioeconomic backgrounds, with the median being lower-middle class. Fischhoff *et al.* (1991) provide a full report.

Content

Those who study decision making in general often have relatively little interest in the topics of everyday decisions. Their primary concern is processes that emerge across domains. It would be a diversion for them to develop a theory of context, characterizing decision topics and examining their effects on thought processes.[2] One aspect of content that sometimes does appear in theoretical discussions is familiarity. Performance may improve if familiarity provides fluency with the issues, making it easier to focus on task structure. Or, familiarity may reduce performance, if it encourages habitual responses, leading people to ignore task structure. Familiarity may also affect the vocabulary for discussing a domain, perhaps stabilizing usage, perhaps leading to the divergence found in a living language. Unrecognized differences in usage can produce tasks that mislead subjects and responses that mislead investigators. People should have better elaborated beliefs about topics that are on their minds. They may also have more accurate beliefs, if they have had the opportunity to learn. Thus, how people speak can illuminate the experiences that they have had.

In order to find out what topics were on these teens' minds, we asked them to describe briefly recent decisions in each of seven domains (school, free time, clothing, friendships, health, money, and parents). They had no trouble doing so, producing 17.4 decisions per subject. The largest category of these decisions dealt with friendships (4.6), while very few concerned the sort of fateful issues that concern parents and educators (e.g., drugs, sex, violence, alcohol). Such "risk behaviors" were, however, somewhat more common when the teens later described three hard decisions in detail. The most common topic of hard decisions was openness with friends, followed, at some distance, by openness with parents, drugs, career plans, and school performance.

Of course, these domains are not independent. For example, relations with friends, parents, and teachers shape and are shaped by decisions about drugs. Our subjects' reports suggest that health-risk issues emerge sequentially, after other decisions have been made. Experimental research has demonstrated predictable differences in people's choices when decisions are framed in simultaneous or sequential form; moreover, people are unlikely to generate alternative frames independently (Hogarth, 1982; Kahneman and Tversky, 1984; Thaler, 1985). More open-ended procedures may clarify which frames people adopt spontaneously (Fischhoff, 1983; Frisch, 1993). Thus, teens concerned about relations with friends may not even think about how those choices affect subsequent decisions about smoking or sex.

Some of the most effective programs for reducing teen risk behaviors might be seen as attempts to eliminate such framing effects. These programs train teens in the cognitive skills needed to understand decisions and in the social skills needed to implement choices, by withstanding pressure to take unwise actions (Baron and Brown, 1991; Botvin, 1983; Dryfoos, 1990). Unfortunately, the empirical evaluations of these complex interventions reveal little about how they affect teens' decision-making processes en route to influencing their behaviors (Beyth-Marom *et al.*, 1991).

Structure

We used several coding schemes to ask whether the decisions described by teens had the structure assumed by decision theory: distinct options, clear consequences, and intervening uncertainties—as well as what forms these components take.

Options. The introductory segments of the interview defined decisions as choices among two or more options and saw that subjects could produce suitable examples. The decisions that subjects subsequently produced, however, typically focused on a single option. Indeed, the largest category of decisions (about 45%) involved statements of resolve, like "eat more healthfully" or "stop blaming others." In effect, they expressed the decision to stick with a previous choice, which might or might not have been implemented. These descriptions leave the alternative options unstated. Explicit alternatives were also absent from the next most common category (20% of descriptions), involving decisions about whether to do something (e.g., smoke cigarettes). Roughly equal numbers of decisions (about 15%) described two distinct options (e.g., whether to go to school or hang out with friends) or a set of options that seemed to have been identified, even if it was not described completely (e.g., which class to take, what to wear, with whom to have lunch). Very few decisions (5%) involved seeking or "designing" options (e.g., how to spend my free time, what to do about having fought with a friend). The three hard decisions that subjects described in detail were similarly focused on single options. There was, however, a much greater tendency to formulate them in terms of "whether to do X" (especially with the pending decisions, which were seldom described in terms of decisions "to do X"). The hard decisions were also more likely to involve two distinct options, rather than a set of possibilities. These results, and others, appear in Table 15.1.

These descriptions suggested some additional underlying mechanisms: (a) These teens rarely described their past decisions as having offered the possibility of creating new options, which might have gotten them out of difficult spots. (b) The chance to create options was more common with pending hard choices (suggesting that behavior looks more constrained in hindsight than in foresight). (c) Subjects' hard decisions (like Choice B) were somewhat more likely to involve two clear options. (d) The more decisions that a subject described, the more likely those decisions were to be statements of resolve.

If these descriptions and our coding capture subjects' everyday thinking, then their decision making focuses on the acceptance or rejection of a single option, or even just the reiteration of a previous choice. Paying undue attention to a focal option is seen in several phenomena demonstrated in laboratory experiments. These include neglect of opportunity costs, overemphasis of sunk costs, and various confirmation biases (e.g., Dawes, 1988; Fischhoff and Beyth-Marom, 1983; Klayman and Ha, 1987; Thaler, 1980, 1985). If people think the way they talk, then our subjects' descriptions suggest that such phenomena are not only possible, but common.

TABLE 15.1 Option structure

	Structure					
Decision	*To do X*	*Whether to do*	*Two choices*	*Finite choices*	*What to do about*	*Missing*[a]
Recent						
School	36.2%	29.0%	13.0%	13.0%	2.9%	5.8%
Free time	31.9	18.8	17.3	20.3	4.3	7.2
Clothing	30.4	10.1	11.5	31.9	2.9	13.0
Peers	47.8	24.6	10.1	5.8	4.3	7.2
Health	55.0	20.2	1.4	2.9	2.9	17.4
Money	52.2	11.6	10.1	4.3	10.1	11.6
Parents	30.4	23.2	18.9	4.3	8.7	14.5
Hard						
First past	39.1	33.3	20.2	—	5.8	1.4
Second past	20.2	44.9	18.8	5.8	4.3	5.8
Current	2.9	44.9	23.2	13.0	13.0	2.9

[a] Includes cases where respondents produced no answer or an uncodable one, or where the question was not asked due to a procedural error. These three cases constituted 27, 51, and 21% of missing responses, respectively.

Source: Fischhoff, Furby, Quadrel and Richardson (1991).

Breadth. Decisions might be arranged along a continuum from *tactical*, one-time decisions regarding specific limited choices to *strategic* decisions, setting long-term, relatively irreversible policies. In order to see whether teens address policy issues spontaneously, we categorized their descriptions into (a) concrete choices, regarding specific situations; (b) limited policies, regarding a single repeated decision (e.g., ride my horse every day); (c) limited policies, regarding multiple decisions (e.g., save money for horse equipment); (d) general policies, with broad principles governing diverse circumstances (e.g., take better care of myself).

The vast majority of all decisions involved concrete, one-time choices. This was even more true for the detailed hard decisions than for the briefly described recent decisions (about 85% vs. 75%). The clearest exception was decisions about health and money, 48 and 36% of which involved limited policies (e.g., what sort of diet or spending pattern to adopt). It is not hard to imagine general policy choices in the other areas (e.g., how to spend free time, what to do about homework, how to keep parents off one's back). However, they were rarely produced. Although hard decisions provide an opportunity to reflect about the long term, it was not one that these teens seized.

In such situations, tactical situations add up to general policies. For example, each decision about doing homework (or hanging out with the crowd) partially defines a teen's policies regarding schoolwork and responsibility (among other things). If one-time and repeated choices are made differently (Keren and Wagenaar, 1987), then results from single-choice experiments may be most pertinent to predicting real-world behavior, in situations that could be construed either way.

Consequences and uncertainties. Subjects chose one of their two past hard decisions for a fuller and more directed description. Among other things, they were asked to list six kinds of consequences. The first was "good things that you were certain would happen to you if you made that choice." Subsequent requests dealt, in order, with uncertain good things, certain bad things, uncertain bad things, good things given up by making that choice, and bad things avoided by making that choice. The last two categories were called *indirect* consequences. Such open-ended, but directive questions show how subjects can think about these issues, an indirect indicator of how they actually do.

On average, subjects produced two good consequences of their chosen option, and one consequence in each other category. Three times as many subjects produced no certain bad consequence as produced no certain good ones (27% vs. 9%). Most subjects produced at least one good and one bad *uncertain* consequence (84%, 80%), indicating that they could think about uncertainty — if asked explicitly. Despite the direct query, quite a few subjects produced no indirect consequences, either good ones given up by the choice or bad ones avoided by it (29%, 30%). Here, too, there is confirmation of experimental results, showing the difficulty of thinking up opportunity costs, even when asked explicitly. This reduced sensitivity to indirect consequences applied equally to avoided costs and forgone benefits.

Process

We did not observe these decisions being made. Even for the pending hard decisions, we got only a midstream selection of issues. As a result, we also asked subjects to describe their decision-making processes. These descriptions are vulnerable to various biases (e.g., flawed introspection, self-serving biases). Even so, they may have a life of their own, if they reflect the stories that people tell themselves, about their decision making.

What made this a hard decision? Subjects produced two reasons, on average, for each of their three hard decisions. For the two past decisions, the most common concern was negative aspects of the chosen option, followed by negative features of the rejected option. They seldom mentioned positive features of either the chosen option or a rejected one (i.e., opportunity costs). Thus, they emphasized loss avoidance over benefit maximization. Similarly, the difficulty of their current hard decisions was attributed much more often to both options having negative features than to both having positive features.

Subjects seldom mentioned uncertainty as a major source of difficulty, even though they had produced uncertain consequences in response to a previous question. Subjects never mentioned some potential reasons that we had included in our coding scheme because they are often mentioned as sources of decision-making difficulty: complexity (many options to choose from, many consequences to consider), time pressure, and personal inability to make decisions or make changes. In this light, either these are not the factors that make decisions difficult or people do not recognize their influence.

A subsequent question asked "what did you think about while making the decision?" It produced three issues on average. Similar patterns emerged (e.g., a focus on negative and concrete consequences). One difference was greater reference to uncertainty. However, it typically concerned how an event would be experienced (e.g., "what would it be like if I went to live with my mother" or "went to the other school"), rather than whether something would happen.

What did you do to help you think about your decision? For about one third of their past decisions, subjects reported doing nothing. Most of the remaining cases involved various ways of focusing their thinking or seeking information and support. Those who sought help went to friends and parents. A few subjects reported trying not to think. Even fewer reported reading something.

Subjects were much more likely to report active responses (beyond just thinking) for their current hard decisions, with a particular tendency to experimentation. Trying out different options (e.g., ways of dealing with parents) can reduce cognitive load (by acting out possibilities) and reduce uncertainty (by seeing how actions work and feel). Current experiments may be more visible than past ones because many plans do not actually get carried out. Thus, past decisions may provide better indicators of which experiments they get around to trying.

Do you wish that you had made a different choice? Only 15–20% of subjects said that they did. Their primary reason was not liking the option that they had selected, rather than wishing that they had taken the rejected option. Those who expressed no regret were also twice as likely to mention good aspects of what they had chosen, rather than bad aspects of what they had rejected. The lack of regret may reflect which hard decisions subjects chose to describe, how well those decisions were made, or how they were reinterpreted in retrospect. These descriptions suggest a less balanced analysis of competing alternatives than that appearing in regret theories (Loomes and Sugden, 1983).

In response to other questions: (a) Subjects saw themselves as working about as hard as their peers at decision making, (b) Some 39% saw themselves as equally good decision makers. Those who made a distinction were twice as likely to describe themselves as better than average decision makers than as worse (42% vs. 18%). (c) Individual differences in the difficulty of making decisions were attributed to the circumstances that teens faced and the social support that they enjoyed, rather than to personal characteristics. These beliefs contrast with the popular notion that teens view themselves as unique and invulnerable (e.g., Elkind, 1967; Quadrel *et al.*, 1993). They suggest resistance to the fundamental attribution error (Nisbett and Ross, 1980).

Summary

Taken at their word, these teens emerge as active, absorbed decision makers. After minimal instruction, they produced many examples of decisions in their lives. Their descriptions were sufficiently clear that we could code them from a variety of decision theoretic perspectives. The teens seemed eager to discuss their decisions and, as in Choice B, recognized their complexity. Nonetheless, their descriptions showed some potential problems. The focus on single options is reminiscent of several phenomena in the literature (e.g., neglect of opportunity costs), while the focus on negative consequences might be a corollary of loss aversion. The limited references to uncertainty about what will happen contrast with the focus on probability assessment in decision-making research. It might also provide weak general support for research showing problems with probability judgments, assuming that people do poorly at tasks that attract little of their attention. References to uncertainty centered on what (relatively certain) outcomes would be like. Understanding those uncertainties requires studying how people predict their tastes, how confident they are in those predictions, and how they resolve uncertain preferences (Fischhoff, 1991; Frisch, Jones and O'Brien, 1994; Loewenstein, 1996).

Components of decisions

Letting people discuss whatever decisions are on their minds allows them to speak about relatively well-worked topics. However, it produces such diverse decisions that only general, structural analyses are possible. It is difficult to study how people deal with the substance of particular domains. The following sections consider some possibilities for an open-ended focus on specific components of decisions.

Uncertainties

A natural question in considering teens' (or anyone else's) decisions about risks is how well they understand the size of those risks. Our literature review found many surveys of adolescents' risk perceptions. However, few had questions formulated precisely enough for the accuracy of the answers to be evaluated. As might be expected, most tasks used verbal quantifiers (e.g., likely, rarely), whose ambiguity is well known (Lichtenstein and Newman, 1967; Merz, Druzdzel, and Mazur, 1991; Wallsten, Budescu, Rapoport, Zwick, and Forsyth, 1986). Thus, these investigators had not yet begun to deal with the difficulties of eliciting quantitative estimates in unfamiliar units (Fischhoff and MacGregor, 1983; Lichtenstein, Slovic, Fischhoff, Layman, and Combs, 1978; Poulton, 1989).

Less expected was the discovery of similar ambiguity in the descriptions of the risks being judged. Even had a precise response mode been used, the accuracy of those answers still could hardly be evaluated. Table 15.2 shows two questions from a large-sample survey, sponsored by the National Center for Health Statistics of the Centers for Disease Control. Following each question are alternative definitions of the quantity and intensity of that kind of risk behavior. If one believed that these exposures create any risk at all, then the magnitude of that risk would depend on what one inferred about these unspecified elements of the definition (among others). Subjects' risk judgments could be interpreted only if all subjects made the inferences — which investigators were then able to guess. The numbers in parentheses indicate the percentage of subjects who endorsed each alternative interpretation, after having answered the question on a previous page. These subjects were drawn from a relatively homogeneous population (juniors at an Ivy League college). The range of opinion shows some of the ambiguity in these questions and some of these students' linguistic norms and mental models.

The options in Table 15.2 were the investigators' own invention. Suggesting them might have artificially put some ideas in subjects' minds, while neglecting other ideas that were already there (thereby exaggerating or understating the degree of ambiguity). Quadrel (1990; Quadrel, Fischhoff, and Palmgren, 1994) adopted a more open-ended strategy: having teens think aloud as they answered risk-estimation questions that were deliberately ambiguous (at about the level of typical surveys). For example, "What is the probability that a person will have an accident while drinking and driving?" or "What is the probability that a person will get AIDS from having sex with someone who has AIDS?" Subjects were drawn from both low-risk settings (e.g., suburban high school teams and clubs) and high-risk ones (e.g., treatment homes).

TABLE 15.2 Interpretations of AIDS risk questions

How likely do you think it is that a person would get AIDS or the AIDS virus from sharing plates, forks or glasses with someone who has AIDS?

How did you interpret "sharing plates, forks, or glasses with someone who has AIDS?"

—— Sharing utensils during a meal (e.g., passing them around, eating off one another's plates). (81.8%)

—— Using the same utensils after they have been washed. (10.9%)

—— I was uncertain about the interpretation. (5.8%)

—— Multiple interpretations (1.5%)

Did you interpret "sharing plates, forks, or glasses with someone who has AIDS" as

—— Occurring on a single occasion? (39.1%)

—— Occurring on several occasions? (19.6%)

—— Occurring routinely? (27.5%)

—— I was uncertain about the interpretation. (12.3%)

—— Multiple interpretations (1.4%)

How likely do you think it is that a person would get AIDS or the AIDS virus from having sex with a person who has AIDS?

How did you interpret "having sex with a person who has AIDS?"

—— Having vaginal intercourse without a condom. (72.5%)

—— Having vaginal intercourse with a condom. (4.3%)

—— Having other kinds of sex. (6.5%)

—— I was uncertain about the interpretation. (6.7%)

—— Multiple interpretations. (8.0%)

Did you interpret "having sex with a person who has AIDS" as

—— Occurring on a single occasion? (61.6%)

—— Occurring on several occasions? (22.5%)

—— Occurring on many occasions? (7.2%)

—— I was uncertain about the interpretation. (5.1%)

—— Multiple interpretations. (3.6%)

Note. Entries are the percentage of subjects (in a sample of 135 students at an Ivy League college) who reported having inferred each definition of the phrase when they had answered a question about the risk that it entailed. The response mode comprised five verbal quantifiers: definitely not possible, very unlikely, somewhat unlikely, somewhat likely, and very likely.

Source: Fischhoff (1994).

Table 15.3 shows the coding framework for the risk factors that subjects produced, illustrated with examples from the drinking-and-driving question. Data analyses showed great variation in subjects' assumptions about the unstated elements of these risks (e.g., how much driving, what kind of driving, how serious an accident). As a result, these subjects were effectively answering different questions from one another and from any investigator's expectations. In these domains at least, it may be hard to learn very much about risk judgments without much sharper questions and answers. A follow-up experiment varied individual risk factors within highly specified questions. Subjects there were most responsive to those factors that had been cited most frequently as relevant in the open-ended study.

Methodologically, these results suggest that well-specified questions are both needed and possible. Substantively, they suggest that teens have fairly robust intuitive theories regarding these risks, which emerge similarly in different tasks. The content of those beliefs suggest how they have processed past experiences with these risks and how they might make future decisions about them. For example, most subjects spontaneously discussed the frequency of most risk behaviors, indicating the salience of dose-response relationships. However, they did not ask about the frequency of sex as a determinant of the risks of pregnancy and of AIDS.

These particular results provide an additional perspective on a recurrent topic in experimental studies: how people estimate the cumulative probability of compound events (e.g., Bar-Hillel, 1974; Cohen, Chesnick and Haran, 1971; Keren, 1990). In structured tasks, we have observed particularly large underestimation of how quickly risks mount up through repeated sexual exposures (Linville, Fischer, and Fischhoff, 1993; Shaklee and Fischhoff, 1990). The present, open-ended results suggest that the idea of cumulative risk does not even occur to many people in this domain (Luker, 1975). If an issue does not arise naturally, then people may do particularly poorly when asked to address it. There are many studies of intuitive beliefs in specific domains (e.g., Chi, Glaser and Farr, 1988; Furnham, 1988; Leventhal and Cameron, 1987; Morgan, 1993). Behavioral decision making might exploit them to identify and to understand the properties of the topics it happens to use for stimuli.

Consequences

A focal argument in youth policy is whether information "works" with adolescents. If education does not influence teen decision making, then there is a stronger case for implementing coercive policies. One approach to determining the role of information in teen decisions has been to predict those decisions as a linear function of the judged attractiveness of a preselected set of consequences, weighted by their judged likelihood. The moderate success of such models (e.g., Bauman, 1980; Bauman, Fisher and Koch, 1988) could provide some weak support for the claim that teens' choices are under cognitive control.

TABLE 15.3 Coding framework

Framework element	*Risk factor categories*	*Example variables drinking-and-driving question*
Behavior	Dose	Amount of alcohol consumed
	Amount	
	Potency	
	Method	
Other behaviors	Risk buffers	Amount of food eaten
	Risk amplifiers	Other drugs consumed
	Time-related	Night or day; day of the week
	Place-related	Where alcohol was consumed
Actor	Physical	Tolerance to alcohol
	Cognitive	Awareness of effects of alcohol
	Social – psychological	Mood
	Material	Wealth
	Spiritual	Faith
	Skill	Driving skill
	Character	Responsible; mature
	Age	
	Gender	
	Genetic history	
	Status	
	Luck	
	Motivation	
	Self other	
Context	Social	
	General, cultural	Drinking norms
	Family	Family approval
	Peers, others	Peer approval
	Environmental	Road conditions
Risk outcome	Social reactions	Get in trouble
	Personal effects	
	Physical	Injury
	Psychological	Worry, guilt
	Cognitive – physiological	Kill brain cells
	Cognitive – psychological	Can't think
	Material	Lose car, lose license
	Accidents	Get in a wreck while high
	Lifestyle	Become a bum
	Complex	Get high
	Effects on others	Hurt your friends, family
	Behaviors	Use more, do heavier drugs
	Severity, type when measured	

Source: Quadrel, Fischhoff and Palmgren (1996)

However, it has long been known, in behavioral decision-making research, that many weighting schemes will provide somewhat successful predictions, as long as the consequences span the space of relevant concerns—or are correlated to ones that do (Camerer, 1981; Dawes and Corrigan, 1974; Goldberg, 1968). The price for these practical successes is difficulty in distinguishing alternative specifications and, hence, in gaining much insight into cognitive processes (Hoffman, 1960).

In order to gain some additional purchase on which consequences matter to teens, we asked some teens to list the "things that might happen" if they accepted (or declined) opportunities to engage in six potentially risky behaviors (e.g., driving with a friend who has been drinking, skipping school to go to a mall) (Beyth-Marom, Austin, Fischhoff, Quadrel and Palmgren, 1993). We found very similar responses in matched groups of adults and adolescents, drawn from low-risk settings. In addition to sharing beliefs about these possibilities, these teens and adults were equally sensitive to how their task was formulated. Although accepting and declining a risky opportunity are logically complementary actions, they were not psychologically complementary. For example, engaging in a risk behavior generated more consequences overall than did avoiding it, but fewer indirect consequences (i.e., ones that do not occur as a result of the choice). Accepting a risky option produced four times as many bad consequences as good ones; however, rejecting it produced equal numbers of good and bad consequences. Subjects saw rejecting a risk as eroding their social standing much more than they saw accepting a risk as securing social approval. Another kind of sensitivity to question framing was that the prospect of repeatedly engaging in a risk behavior evoked many more bad consequences than did doing it just once.

These results suggest general decision-making processes. For example, one might infer that action and inaction evoke quite different perspectives, as do short-term and long-term time perspectives. Without explicit prompts, thinking about one perspective does not invoke the other. These results also elaborate or qualify conclusions reached in studies using structured methods. For example, they show another context within which opportunity costs are neglected, and that such neglect is less when rejecting risky options than when accepting them. Methodologically, these results suggests some possibilities for increasing the prediction ceiling on linear models (e.g., distinguishing between accepting and rejecting an action, using teens' terminology for formulating consequences, adopting the appropriate time perspective).

In terms of the policy debates that motivated them, these studies show teens as having a higher capacity for cognitive performance than one might have expected. They also suggest some specifics on how to communicate best with teens. These two issues are interrelated. Poorly designed communications erode faith in teens, by making them responsible for our failures to speak clearly (we knew what we meant, why didn't they listen?). If teens are seen as cognitively competent, then there is more place for providing them with the sophisticated information needed for effective decision making, addressing the values that make them willfully choose behaviors that adults deplore, and alleviating circumstances that restrict their ability to act maturely.

Options

If only one thing is standardized in a structured study, it is likely to be the options. Life, however, does not always provide ordered sets of options. Indeed, some of the few studies of option generation have shown sensitivity to just how the tasks are posed by investigators (Gettys *et al.*, 1987). The differences between the perceived consequences of accepting and rejecting a risk behavior (Beyth-Marom *et al.*, 1993) suggest that an option need not even evoke its complement.

Some years ago, Lita Furby and I became interested in the confident, universal, and contradictory advice being offered to women, regarding how to reduce the risk of sexual assault (Morgan, 1986). Even if well intended, such advice can do women a serious disservice. Not only can it lead to ill-advised actions, but unfounded advice can increase the blame placed on those women whose chosen options do not succeed—because whatever they did was against the advice of some self-styled experts. Universal advice can misdirect women who do not accept its underlying trade-offs.

As part of a systematic look at this advice and the evidence supporting it (Fischhoff, 1992), we examined the scope, organization, and formulation of options among both experts and laypeople. In the advice literature, we found that options are often lumped into coarse categories with loaded labels, like active and passive.[3] In order to understand lay perceptions, we posed a series of open-ended questions asking what a woman could do to protect herself in various circumstances (Furby, Fischhoff, and Morgan, 1990). Our subjects were 43–45 women drawn from each of three populations, students, middle-aged alumna of the same university, and working-class young mothers. Each woman produced 26 options on average. Although this is a large number of strategies for an individual to write on an open-ended questionnaire, it is still less than a tenth of the 300-plus different options produced by each group as a whole.[4]

We supplemented this list of options with ones produced by a group of men and a national sample of sexual assault experts, and additional ones found in a sample of 50 publications written for lay or professional audiences. In total, we found over 1100 different options. This is a bewildering number of possibilities for women or researchers to consider in any detail. In order to bring some order to this welter of options, we created a "strategy grammar." In it, each strategy was described in the form, "Doing *action* X in order to achieve *intended effect* Y [which is believed to reduce the risk of rape]." The typology distinguishes further between the *stage* of an assault at which a strategy is directed (preventing an assault from occurring, preparing to react to an assault, or defending oneself during an assault) and the *level of action* involved (individual or societal).

In this view, a strategy may be ineffective either because it fails to produce the intended effect or because that effect does not deter sexual assaults. Making that distinction might help researchers to understand the effects of strategies (by looking separately at the two contingencies). Indeed, this categorization revealed a moderately consistent pattern in the otherwise confusing results from studies of the effectiveness of assault prevention strategies (Furby and Fischhoff, 1988). Making that distinction might help women to generate options by prompting the search for ways to achieve these ends (Pitz, Sachs and Heerboth, 1980). Table 15.4 shows the most commonly mentioned strategies, for what women can do if assaulted (Stage III). All are individual level actions for situations where societal action has failed. The details in the table suggest the differing experiences of laymen, women, and experts. Responses of the three groups of women were remarkably similar and are combined here.

The messiness of these options contrasts with the apparent simplicity of the few (perhaps two) well-defined options offered in many lab experiments. However, the simplicity of the options in experiments belies the complexity of the underlying option space. The options are often selected from an uncountable space of possibilities by a deliberately inscrutable procedure. (If subjects knew how stimuli were derived, then they might discern the purpose of the study and provide artifactually orderly responses). There may be value to studying how the nature of outcome spaces affects people's thinking. For example, do people draw on different skills for arbitrarily created tasks and for messily specified ones? If so, then the generality of behavior observed with artificial stimuli might depend on their underlying outcome space (Evans, 1986; Hammond, 1966; Hogarth, 1981).

Whole decision strategies II: transaction analysis

Contingent valuation

According to Executive Order 12291 (Bentkover, Covello, and Mumpower, 1985), cost–benefit analyses must be conducted for all significant federal actions. Where those actions affect the environment, that often means putting a price tag on goods not traded in any marketplace. These goods include such "intangible" effects as changes in genetic diversity, visibility at national parks, and discomfort (due to increased tropospheric ozone).

Fearing the neglect of these effects, some resource economists have resorted to asking respondents for dollar equivalents. They call their family of procedures "contingent valuation" mechanisms, requiring participants to act *as if* a market existed for the environmental change. These investigators have soldiered on for 15 or more years, despite membership in a profession with little faith in "expressed preferences" (Mitchell and Carson, 1989; Portney, 1994). Recently, they have achieved some notable institutional successes: a panel of (more and less) disinterested scholars, convened by NOAA (1994), has issued

TABLE 15.4 What people think women are likely to do if assaulted[a]

		% Who mentioned strategy[c] (rank order in parentheses)		
Stage-level-intended effect[b]	*Strategy*	*Females*	*Males*	*Experts*
III-i-outside intervention	Scream or yell	66.7	56.8 (1)	69.8 (1/2)
III-i-establish distance or barrier	Run away	57.6	52.3 (2)	69.8 (1/2)
III-i-reduce assailant's propensity to rape	Talk way out of situation, reasoning	44.7	18.2 (8/10)	41.9 (4)
III-i-physically impede assailant	Fighting/physical self-defense techniques (e.g., martial arts or specific blows/hits)	43.9	38.6 (4)	53.5 (3)
III-i-physically impede assailant	Knee/kick him in the groin	42.4	47.7 (3)	11.6 (28/32)
II-i-reduce assailant's propensity to rape	Make yourself seem less attractive; do crude "unfeminine," gross, dirty things (e.g., vomit, urinate, defecate, pick nose, fake menstrual cramp, pass gas)	30.3	11.4 (17/18)	27.9 (8/9)
III-i-minimize assailant's propensity to rape	Cooperate with rapist/submit	28.8	15.9 (11/3)	34.9 (5)
III-i-physically impede assailant	Poke, jab, or claw rapist's eyes	26.5	27.3 (5)	20.9 (12/15)
III-i-physically impede assailant	Kick	25.8	18.2 (8/10)	25.6 (10)
III-i-reduce assailant's propensity to rape	State you have VD, are a virgin, menstruating, recovering from operation, diseased	22.7	13.6 (14/16)	32.6 (6)
III-i-physically impede assailant	Bite attacker, draw blood (e.g., bite his tongue or lips hard)	21.2	20.5 (6/7)	23.3 (11)
III-i-physically impede assailant	Use Mace, tear gas and chemical sprays	16.7	15.9 (11/13)	11.6 (28/32)

[a] Strategies mentioned by at least 10% of respondents in each group, in answer to questions R and S (females) or R (males and experts). See Furby, Fischhoff, and Morgan (1990) for verbatim questions. A difference of approximately 20% between females and either of the other groups is significant at $p < .01$.
[b] Stage codes: I, prevent assault from occurring; II, prepare for reacting to an assault; III, defend yourself during a assault. Action level codes: s, societal action; i, individual action. Intended effect subscripts: p, perceived; a, actual. See Fischhoff, Furby, and Morgan (1990).
[c] Tied ranks are indicated by giving the range of ranks that are tied (e.g., 8/10 means that there were three strategies all mentioned with the same frequency and thus tied for the 8th, 9th, and 10th slots).

Source: Furby, Fischhoff, and Morgan (1992).

a cautious endorsement. Contingent valuation results influenced the monetary value of the Exxon Valdez settlement (to the extent that science played any role at all).[5]

A great attraction of contingent valuation is that it allows investigators to pose questions about anything that interests them. However, this flexibility exacts a price. Because they can be asked anything, respondents can take nothing for granted. As a result, investigators must first convey the complex details of a CV question and, then, allow respondents to articulate stable preferences. Many critics have worried about the cognitive demands imposed on CV respondents (e.g., Baron, 1994; Diamond and Hausman, 1994; Fischhoff, 1988, 1991; Kahneman and Knetsch, 1992; Schkade and Payne, 1994).

After reviewing the contingent valuation studies in a single domain (atmospheric visibility), Lita Furby and I settled on a rather different concern: that the intellectual demands of these tasks had overwhelmed the theoretical capacity of economics. We found enormous variability in task specification across studies, without clear guidelines for its resolution. Details emphasized by one investigator were often ignored entirely by another. Where one scientist feared misinterpretation, another felt that the same detail went without saying. It seemed as though the analysis of existing markets, the customary task of economics, had not shown how to create new (contingent) ones.

In response, we created a framework, identifying the details required in a well-specified "transaction" (Fischhoff and Furby, 1988). It distinguished the *good* being offered, the *payment* being requested, and the *social context* within which the proposed transaction would be conducted. The good and payment have *substantive* properties, representing features that one might care about, and *formal* properties, defining the magnitude of those features (which should matter if one cares about them). For example, a substantive property of visibility is haze; its formal properties include the area and time over which a change would be experienced, as well as its probability of being achieved as promised. The social context includes what precedents would be set, who legitimates the proposed transaction, and who else is being asked.

CV investigators typically have little interest in most of these details, caring only about a few focal features. However, there is no guarantee that omitted details will not be imputed by respondents, who need to assume something in order to make the task meaningful. Of course, respondents may be insensitive to these omissions and to much of what is included. Indeed, they may be content to give a general response to a general question. Such "gist" responses may be the best predictors of behavior in situations with low involvement. Unfortunately, CV studies require a much deeper level of processing: monetary evaluations for precisely described goods.

One obvious way to examine depth of processing is with manipulation checks, asking respondents how they interpreted their task. Recently, we quizzed people about the details of a brief evaluation task, asking about a river cleanup, which we had just read them over the phone (Fischhoff *et al.*, 1993).

Many had forgotten (or never believed) essential details in our description, such as the extent of the cleanup, the likelihood of it actually being provided, and who else might pay. Their evaluations were more consistent when interpreted in terms of the task that they reported having answered, rather than the one that we had posed.

Insurance decisions

Our constructive criticism has had some effect on CV practice (Pommerehne and Schwartz, 1995). It has also forced us to look again at the missing details in our own experimental tasks. Ones like Choice A require conveying many fewer details. On the other hand, such tasks leave subjects much greater latitude to read between the lines, because there are so many more details whose omission they might notice. This is, in fact, the same freedom of interpretation that we exploit when extrapolating lab results to the real world. Because we have said so little, we have great freedom in making claims about what in the world looks like our tasks.

Table 15.5 characterizes Choice A in terms of some elements in the framework that we developed for contingent valuation. It performs the same analysis for a second choice problem, identical to A, except that "certain loss" is replaced by "insurance premium." Although the stakes are the same in the two problems, respondents are much more risk averse with the insurance formulation (Fischhoff, Slovic, and Lichtenstein, 1980; Hershey and Schoemaker, 1980).

We attributed this difference to one feature in the framework, the *payment context* (i.e., how the money is extracted). This is a substantive issue, which might be expressed along the lines of "people don't like certain losses, but do protect themselves when insurance is offered." If true, this would mean that people use propositional rules, about how to behave in particular contexts, in addition to analytical rules, for aggregating vectors of dollars and probabilities.

Although brief, these choice descriptions do make several other features explicit. For example, both the good and the payment are in dollars; the choices are single-play, allowing no "iterations" (with opportunities to learn from experience or to have outcomes even out); the probabilities of the risk (.25) and the promised protection (1.0) are given. A single question mark indicates a missing feature that subjects might have imputed, but probably did so similarly for both choices. For example, the hypotheticality might have influenced subjects' interpretations of these probabilities. Our interpretation of the different preferences for the two choices need not change if subjects made similar guesses about these features, even if we do not know what those guesses are.

Life is more complicated if subjects made different assumptions in the two conditions. In that case, we do not know what questions subjects answered nor why they answered differently. Double question marks indicate such cases. For example, the source of the threatened $200 loss may seem different when insurance is available

TABLE 15.5 Transaction analysis insurance/preference

	Preference[a]	*Insurance*[b]
Good		
Attribute	$	$
Source of change	??	??
Extent	one play?	one play?
Certainty of provision (probability)		
Receiving needed compensation	1.00?	1.00?
Needing compensation	.25?	.25?
Payment		
Attribute	$	$
Context	premium	sure loss
Constituency	??	??
Certainty of provision (probability)	1.00	1.00
Social context		
Providers	??	??
Others involved	Exp. group, family, friends?	Exp. group, family, friends?
Determined by	self?	self?
Iterations	none	none
Constraints	??	??
Precedents	??	??
Legitimacy	??	??

Note. ?, uncertain value; ??, uncertain, possibly different value in two conditions.

[a] Choice A with "certain loss" replaced by "insurance premium."

[b] Choice A at the beginning of the article.

Source: Fischhoff (1993)

than when a certain loss will avert it. If the source of a loss matters, then inferences about this missing detail may influence choices. While the context (premium/sure loss) may have been the cue to these inferences, it would be an incomplete explanation of subjects' differential desire for these two kinds of protection.

The same could be said for possibly different interpretations of the unstated social context features: who provides the coverage (and is party to the contract), what constraints are there on possible offers (and opportunities to take advantage of subjects), what precedents are being set (relative to subjects' self-concept or their relationship with the coverage provider), and what social norms might legitimate (or proscribe) such a deal. If they affect choices, these interpretations would elaborate the psychology of "certain loss" and "premium." They would provide a basis for linking this preference reversal to the complex settings within which decisions about protective behavior are actually made.

A framework is not a theory. However, it can identify issues that may warrant theorizing and might be studied across contexts. Thus, one might examine people's sensitivity to the probability of actually receiving a promised good, to which other people observe their choice, or to the schedule for making payments. Insofar as these issues have been studied in their own right, one could then exploit the relevant chapters in psychology, economics, or sociology. Doing so might inform a general theory of context and limit speculations about the boundary conditions on individual studies.

Conclusion

Is this science good for the real world?

Although designed with real-world problems in mind, these studies are still quite a way from improving people's lives. Creating workable interventions requires detailed design work, followed by rigorous empirical evaluation. Baron and Brown (1991) show the work needed for one form of intervention, training in decision making (see also Beyth-Marom *et al.*, 1985). Once proven, an intervention must be sold to those controlling access to applied settings (e.g., school administrators, federal regulators, police officials).

These activities require a focus, a work rhythm, and a skill set quite different than those for managing basic research programs. Rather than retooling themselves, basic researchers might adopt the more modest objective of making their results available to those better situated to acting on them. Unfortunately, providing that access requires considerable repetition. The same message needs to be delivered in person and in writing to a variety of practitioner audiences, tailored to the circumstances and language of each. They may even want to meet and query the individual behind the words. Considerable tending is needed to ensure that the message is heard consistently, in a world where it is convenient to believe one thing or another about human behavior (Fischhoff, 1990).

A more modest interim aspiration is to clarify bits of the long path between the abstractions of the lab and the messiness of the world. Doing so might help basic researchers take responsible steps toward practicality, without promoting lab results too quickly and confidently. It might help practitioners refine existing programs whose cognitive components are underdeveloped. As mentioned, adults often despair of teens (just as technical experts often despair of laypeople) for doing the wrong things "despite being told the facts." Yet, the telling is often conducted with little attention to what people need to know, what they know already, how they interpret messages, and how they process information—not to mention how they might legitimately extract different conclusions from the same set of facts (Fischhoff, Bostrom, and Quadrel, 1993; Merz *et al.*, 1993). The research record shows how we have discovered these issues and partial solutions, in our communications with subjects.

If I were to speculate about the greatest real-world impact of this research, I would place it in this area. Beliefs about competence are important insofar as more competent people are afforded greater autonomy. Decision-making research has sometimes been funded by authorities hoping that psychologists could "do something" about a problematic group of people (e.g., citizens who dislike a technology, adolescents who take health risks). Two early, and oft-repeated, research messages have been: (a) it takes systematic evidence (and not just anecdotal observation) to make responsible statements about others' behavior; and (b) both experts and laypeople show complex patterns of strengths and weaknesses (Fischhoff, 1977, 1990; Fischhoff, Slovic, and Lichtenstein, 1983; Quadrel *et al.*, 1993). I think that these efforts have contributed to a more mutually respectful atmosphere, in arenas where there are strong incentives to bash the public (and to distrust the experts) (National Research Council, 1989).

It is, of course, hard to evaluate a claim of having provided insight to a wide audience. Even invitations to speak and citations in the national press are but circumstantial evidence. Psychology may be no more than a convenient place holder, showing that the public has been represented, even if the details of its concerns are ignored. A somewhat more direct, but still ambiguous contribution is to provide components of larger programs. For example, we developed a set of open-ended procedures to describe and improve people's "mental models" of environmental hazards (Bostrom, Fischhoff, and Morgan, 1992; Morgan *et al.*, 1992). It has influenced the design of two science museum exhibits on climate change and the re-design of EPA's *Citizen's Guide to Radon*. We used it to design a brochure on the health effects of electromagnetic fields, over 100,000 copies of which we have sold (at cost) (Morgan, 1989). A large chemical company has incorporated elements of this approach in its communications, extending it to topics like corporate restructuring. However, even though we designed our procedures for practicality, our colleagues in the company had to improvise extensively in order to use them. What they could use fairly directly was our articulation of a philosophy promoting an open attitude toward their audiences (Fischhoff, 1995). In each case, though, our research was but one piece of a larger process, whose overall impact is poorly documented.

Is the real world good for this science?

It is difficult to study simultaneously how people perceive the world and what they do with those perceptions. A standard strategy in cognitive psychology is to standardize the stimuli, so as to gain access to the process. In this light, subjects become something like battery-raised hens, placed in very similar environments, in hopes of achieving very similar outcomes. The complex intervening processes are clarified whenever changes in individual input factors (e.g., temperature, crowding, diet) produce predictable changes in outputs.

With poultry, this strategy leads to harsh practices. For example, the destabilization of normal social relations may prompt trimming chicken's beaks or putting on little red spectacles, in order to keep them from pecking one another. Although psychologists are prevented (by ethics and regulations) from abusing subjects with strange procedures, we are not protected from deluding ourselves. Live with a task long enough and it may become more real than the setting from which it was abstracted.

Tasks like Choice A run the risk of what might be called the "curse of context": We would like to interpret responses as reflecting deep-seated values, of the sort that come from a lifetime of intense involvement with real-world decisions. Yet, we set minimalist problems before subjects and expect them to resist the temptation to impute a context. We may also be victims of a "curse of cleverness": We pride ourselves on devising just the right tasks for evaluating competing theories of human behavior, tasks that our predecessors were unable to concoct. Then, we expect subjects immediately to discern the structure of these tasks and decide what response is right for them.

The studies reported here "turn subjects loose" within (something like) the domains that more structured studies hope to illuminate. If subjects behave similarly when less constrained, there is added reason for confidence in existing results. If subjects behave differently or interpret tasks in unexpected ways, additional experimental controls may be needed in order to obtain the desired standardization. Although troubling in the short run, such problems may open opportunities for new structured tasks, studying the unexpected behaviors systematically. Assuming that these phenomena can be domesticated as part of experimental science, open-ended approaches can be one element in the anarchic process by which its intellectual capital is created (Feyerabend, 1975). Subjects in these tasks might be compared to free-range chickens, relative to the battery-raised hens of experimental tasks. Nonetheless, even these studies are a long way from real-life decisions. They provide, for example, little in the way of social and emotional pressure. The free-range chicken is still a poor excuse for a bird.

Choosing research strategies is a gamble with a poorly structured option space. Narrowing the focus of tasks reduces one's ability to identify surprising phenomena, while increasing one's ability to make science out of what is found. Pressing harder for applied impact increases the chances of getting practitioners' attention, while decreasing the chance of having something to say—if it reduces the freedom of inquiry needed for basic applied research. Pressing too feebly for applications risks forfeiting the intellectual challenges thrown off by real-world problems. Although practitioners may misdiagnose their situation, there is often something to be understood when they perceive a problem. Doing a little applied basic research along the way may be a way of returning the favor.

Acknowledgements

Talks developing the ideas in this article were given as colloquia at Princeton, Rice, Carnegie Mellon, and New York Universities, and as a Presidential Address to the Society for Judgment and Decision Making. The comments of those audiences are gratefully appreciated, as are the contributions of the co-authors and investigators cited here. Shane Frederick, Gideon Keren, Ola Svenson, and an anonymous reviewer provided valuable comments on a draft of this manuscript. Its preparation was supported by the National Institute on Alcohol Abuse and Alcoholism and the National Science Foundation. The opinions expressed are, of course, those of the author. Address reprint requests to Baruch Fischhoff, Department of Social and Decision Sciences, Carnegie Mellon University, Pittsburgh, PA 15213.

Notes

1 Of course, there is more to the behavior of adolescents, or others, than cognition (Fischhoff, 1992). Although our studies allowed teens to discuss issues like emotions and social pressure, it is still just talk and far removed from observing real-life decision making in its full complexity.
2 One exception is the psychometric studies of risk (Fischhoff, Slovic, Lichtenstein, Read, and Combs, 1978; Slovic, 1987; Vlek and Stallen, 1980), looking for common dimensions of diverse risks. Although widely cited in debates over risk policy, these studies have had little impact on basic behavioral decision-making research.
3 For example, "doing nothing" is a strategy, perhaps the appropriate one (e.g., for escaping an assailant who is primarily after the "satisfaction" of subduing a woman who has actively resisted him). Doing little about personal protection might be the right strategy for a woman who felt that her energies were best invested in social action.
4 Options were coded as "different" if we could see how women might view them as having different probabilities of producing a set of focal consequences. Those consequences were the ones mentioned most frequently in another study, asking similar women to list important ones (Furby *et al.*, 1991).
5 In the interests of disclosure, I was a consultant to the Department of Justice in this case.

References

Arnett, J. 1993. Reckless behavior in adolescence: A developmental perspective. *Developmental Review*, 12, 339–373.

Baddeley, A. 1979. Applied cognitive and cognitive applied research. In L. G. Nilsson (Ed.), *Perspectives on memory research*. Hillsdale, NJ: Erlbaum.

Bar-Hillel, M. 1974. Similarity and probability. *Organizational Behavior and Human Performance*, 11, 277–282.

Baron, J. 1994. *Rationality and invariance: Response to Schuman*. Unpublished manuscript, presented at DOE/EPA Workshop on Using Contingent Valuation to Measure Non-Market Values.

Baron, J. and Brown, R. (Eds.). 1991. *Teaching decision making to adolescents*. Hillsdale, NJ: Erlbaum.

Bauman, K. E. 1980. *Predicting adolescent drug use: Utility structure and marijuana*. New York: Praeger.

Bauman, K. E., Fisher, L. A., and Koch, G. G. 1988. External variables, subjective expected utility and adolescent behavior with alcohol and cigarettes. *Journal of Applied Social Psychology*, 19, 789–804.

Bentkover, J., Covello, V. T., and Mumpower, J. (Eds.). 1985. *Benefits assessment: The state of the art*, Amsterdam: Reidel.

Beyth-Marom, R., Austin, L., Fischhoff, B., Palmgren, C., and Quadrel, M. J. 1993. Perceived consequences of risky behaviors. *Developmental Psychology*, 29, 549–563.

Beyth-Marom, R., Fischhoff, B., Quadrel, M. J., and Furby, L. 1991. Teaching adolescents decision making. In J. Baron and R. Brown (Eds.). *Teaching decision making to adolescents* (pp. 19–60). Hillsdale, NJ: Erlbaum.

Beyth-Marom, R., Dekel, S., R. Gombo, R., and M. Shaked, M. 1985. *An elementary approach to thinking under uncertainty*. Hillsdale, NJ: Erlbaum.

Bostrom, A., Fischhoff, B., and Morgan, M. G. 1992. Characterizing mental models of hazardous processes: A methodology and an application to radon. *Journal of Social Issues*, 48(4), 85–100.

Botvin, G. J. 1983. *Life skills training*. New York: Smithfield Press.

Camerer, C. 1981. General conditions for the success of bootstrapping models. *Organizational Behavior and Human Performance*, 27, 411–422.

Chi, M. T. H., Glaser, R., and Farr, M. J. (Eds.). 1988. *The nature of expertise*. Hillsdale, NJ: Erlbaum.

Cohen, J., Chesnick, E. I., and Haran, D. 1971. Evaluation of compound probabilities in sequential choice. *Nature*, 32, 414–416.

Dawes, R. M. 1988. *Rational choice in an uncertain world*. San Diego, CA: Harcourt Brace Jovanovich.

Dawes, R. M. and Corrigan, B. 1974. Linear models in decision making. *Psychological Bulletin*, 81, 95–106.

Diamond, P. A. and Hausman, J. R. 1994. Contingent valuation: Is some number better than no number? *Journal of Economic Perspectives*, 8(4), 45–64.

Dryfoos, J. G. 1990. *Adolescents at risk: Prevalence and prevention*. New York: Oxford Univ. Press.

Elkind, D. 1967. Egocentrism in adolescence. *Child Development*, 38, 1025–1034.

Ericsson, A. and Simon, H. A. 1984. *Verbal reports as data*. Cambridge, MA: MIT Press.

Evans, J. St. B. T. 1986. *Deduction*. Hillsdale, NJ: Erlbaum.

Feyerabend, P. 1975. *Against method*. New York: NLB Schocken.

Fischhoff, B. 1995. Risk perception and communication unplugged: Twenty years of process. *Risk Analysis*, 15(2), 137–145.

Fischhoff, B. 1994. What forecasts (seem to) mean. *International Journal of Forecasting*, 10, 387–403.

Fischhoff, B. 1993. Transaction analysis: A framework and an application to insurance decisions. *Journal of Risk and Uncertainty*, 7, 53–69.

Fischhoff, B. 1992. Giving advice: Decision theory perspectives on sexual assault. *American Psychologist*, 47, 577–588.

Fischhoff, B. 1991. Value elicitation: Is there anything in there? *American Psychologist*, 46, 835–847.

Fischhoff, B. 1990. Psychology and public policy: Tool or tool maker? *American Psychologist*, 45, 57–63.

Fischhoff, B. 1988. Specifying value measurements. In B. Driver, G. Peterson, and R. Gregory (Eds.), *Evaluating amenity resources* pp. 107–116. New York: Venture.

Fischhoff, B. 1983. Predicting frames. *Journal of Experimental Psychology: Learning Memory & Cognition*, 9, 113–116.

Fischhoff, B. 1977. Cost–benefit analysis and the art of motorcycle maintenance. *Policy Sciences*, 8, 177–202.

Fischhoff, B. and Beyth-Marom, R. 1983. Hypothesis evaluation from a Bayesian perspective. *Psychological Review*, 90, 239–260.

Fischhoff, B., Bostrom, A., and Quadrel, M. J. 1993. Risk perception and communication. *Annual Review of Public Health*, 14, 183–203.

Fischhoff, B. and Furby, L. 1988. Measuring values: A conceptual framework for interpreting transactions. *Journal of Risk and Uncertainty* 1, 147–184.

Fischhoff, B., Furby, L., Quadrel, M. J., and Richardson, E. 1991. Adolescents' construal of choice situations: Are their decisions our "decisions"? Unpublished manuscript.

Fischhoff, B. and MacGregor, D. 1983. Judged lethality: How much people seem to know depends upon how they are asked. *Risk Analysis*, 3, 229–236.

Fischhoff, B., Quadrel, M. J., Kamlet, M., Loewenstein, G., Dawes, R., Fischbeck, P., Klepper, S., Leland, J., and Stroh, P. 1993. Embedding effects: Stimulus representation and response modes. *Journal of Risk and Uncertainty*, 6, 211–234.

Fischhoff, B., Slovic, P., and Lichtenstein, S. 1980. Knowing what you want: Measuring labile values. In T. Wallsten (Ed.), *Cognitive processes in choice and decision behavior* (pp. 117–141). Hillsdale, NJ: Erlbaum.

Fischhoff, B., Slovic, P., and Lichtenstein, S. 1983. The "public" vs. the "experts": Perceived vs. actual disagreement about the risks of nuclear power. In V. Covello, G. Flamm, J. Rodericks, and R. Tardiff (Eds.), *Analysis of actual vs. perceived risks* (pp. 235–249). New York: Plenum.

Fischhoff, B., Slovic, P., Lichtenstein, S., Read, S., and Combs, B. 1978. How safe is safe enough? A psychometric study of attitudes towards technological risks and benefits. *Policy Sciences*, 8, 127–152.

Frisch, D. 1993. Reasons for framing. *Organizational Behavior and Human Decision Processes*, 54, 399–429.

Frisch, D., Jones, K. T., and O'Brien, M. 1994. *Sources of difficulty in real-life decisions*. Unpublished manuscript, University of Oregon.

Furby, L. and Fischhoff, B. 1988. Rape self-defense strategies: A review of their effectiveness. *Victimology*. Eugene: Eugene Research Institute.

Furby, L., Fischhoff, B., and Morgan, M. 1992. Preventing rape: How people perceive the options of defending oneself during an assault. In E. C. Viano (Ed.), *Critical issues in victimology: International perspectives* (pp. 174–189). New York: Springer Publishers.

Furby, L., Fischhoff, B., and Morgan, M. 1990. Preventing rape: How people perceive the options for assault prevention. In E. Viano (Ed.), *The victimology research handbook* (pp 227–259). New York: Garland Publishing, Inc.

Furby, L., Fischhoff, B., and Morgan, M. 1991. Rape prevention and self-defense: At what price? *Women's Studies International Forum*, 14(1/2), 49–62.

Furnham, A. F. 1988. *Lay theories*. London: Pergamon Press.

Gallie, W. B. 1964. *Philosophy and the historical understanding*. London: Chatto & Windus.

Gergen, K. J. 1994. *Realities and relationships*. Cambridge, MA: Harvard Univ. Press.

Gettys, C. F., Pliske, R. M., Manning, C., and Casey, J. T. 1987. An evaluation of human act-generation performance. *Organizational Behavior and Human Decision Processes*, 39, 23–51.

Gilligan, C. 1982. *In a different voice: Psychological theory and women's development*. Cambridge, MA: Harvard Univ. Press.

Goldberg, L. R. 1968. Simple models or simple processes? *American Psychologist*, 23, 483–496.

Hammond, K. R. 1966. *The psychology of Egon Brunswick*. New York: Holt, Rinehart & Winston.

Hershey, J. R. and Schoemaker, P. J. H. 1980. Risk taking and problem context in the domain of losses: An expected utility analysis. *Journal of Risk and Insurance*, 47, 111–132.

Hexter, J. H. 1971. *The history primer*. New York: Basic Books.

Hoffman, P. J. 1960. The paramorphic representation of clinical judgment. *Psychological Bulletin*, 57, 116–131.

Hogarth, R. M. (Ed.) 1982. *New directions for methodology of social and behavioral science: Question framing and response consistency*. San Francisco: Jossey-Bass.

Hogarth, R. M. 1981. Beyond discrete biases. *Psychological Bulletin*, 90, 197–217.

Kahneman, D. and Knetsch, J. 1992. Valuing public goods: The purchase of moral satisfaction. *Journal of Environmental Economics & Management*, 22, 57–70.

Kahneman, D. and Tversky, A. 1984. Choices, values, and frames. *American Psychologist*, 39, 341 –350.

Keren, G. and Wagenaar, W. A. 1987. Violations of utility theory in unique and repeated gambles. *Journal of Experimental Psychology: Learning, Memory & Cognition*, 13, 387–391.

Klayman, J. and Ha, Y-W. 1987. Confirmation, disconfirmation, and information in hypothesis testing. *Psychological Review*, 94, 211–228.

Kren, G. H. and Rapoport, L. H. (Eds.). 1976.. *Varieties of psychohistory*. New York: Springer.

Kuhn, T. S. 1962. *The structure of scientific revolution*. Chicago: University of Chicago Press.

Lakatos, I. 1970. Falsification and scientific research programmes. In I. Lakatos and A. Musgrave (Eds.), *Criticism and the growth of scientific knowledge*. Cambridge: Cambridge Univ. Press.

Leventhal, H. and Cameron, L. 1987. Behavioral theories and the problem of compliance. *Patient Education and Counseling*, 10, 117–138.

Lichtenstein, S. and Newman, J. R. 1967. Empirical scaling of common verbal phrases associated with numerical probabilities. *Psychonomic Science*, 9, 563–564.

Lichtenstein, S., Slovic, P., Fischhoff, B, Layman, M., and Combs, B. 1978. Judged frequency of lethal events. *Journal of Experimental Psychology: Human Learning and Memory*, 4, 551–578.

Linville, P. W., Fischer, G. W., and Fischhoff, B. 1993. AIDS risk perceptions and decision biases. In J. B. Pryor and G. D. Reeder (Eds.), *The social psychology of HIV infection* (pp. 5–38). Hillsdale, NJ: Erlbaum.

Loewenstein, G. (1996). Out of control: Visceral influences on behavior. *Organizational Behavioral and Human Decision Processes, 65*, 272–292.

Loomes, G. and Sugden, R. 1983. Regret theory: An alternative approach to rational choice under uncertainty. *Economic Journal*, 92, 805–824.

Luker, K. 1975. *Taking chances: Abortion and the decision not to contracept*. Berkeley: University of California Press.

Merz, J., Druzdzel, M. and Mazur, D. J. 1991. Verbal expressions of probability in informed consent litigation. *Medical Decision Making*, 11, 273–281.

Merz, J. F., Fischhoff, B., Mazur, D. J., and Fischbeck, P. S. 1993. A decision-analytic approach to developing standards of disclosure for medical informed consent. *Journal of Products and Toxics Liabilities*, 15, 191–215.

Mitchell, R. C and Carson, R. T. 1989. *Using surveys to value public goods: The contingent valuation method*. Washington, DC: Resources for the Future.

Morgan, M. G. 1993. Risk analysis and management. *Scientific American*, 269, 32–41.

Morgan, M. G. 1989. *Electric and magnetic fields from 60 Hz electric power: What do we know about possible health risks.* Department of Engineering and Public Policy, Carnegie Mellon University.

Morgan, M. 1986. Conflict and confusion: What rape prevention experts are telling women. *Sexual Coercion and Assault,* 1(5), 160–168.

Morgan, M. G., Fischhoff, B., Bostrom, A., Lave, L., and Atman, C. J. 1992. Communicating risk to the public. *Environmental Science and Technology,* 26, 2048–2056.

National Oceanographic and Atmospheric Administration 1994. *Federal Register,* 59(5), 1062–1091.

National Research Council. 1989. *Risk perception and communication.* Washington, DC: The Council.

Nisbett, R. and Ross, L. 1980. *Human inference: Strategies and short-comings of social judgment.* Englewood Cliffs, NJ: Prentice-Hall.

Nisbett, R. E. and Wilson, T. D. 1977. Telling more than we know: Verbal reports on mental processes. *Psychological Review,* 84, 231–259.

Pitz, G. F., Sachs, N. J., and Heerboth, J. 1980. Procedures of eliciting choices in the analysis of individual decision. *Organizational Behavior and Human Performance,* 26, 396–408.

Pommerehne, W. and Schwarz, N. (Eds.). 1995. *Determining the value of nonmarketed goods.* New York: Plenum.

Portney, P. R. 1994. The contingent valuation debate. *Journal of Economic Perspectives,* 8(4), 3–17.

Poulton, E. C. 1989. *Bias in quantifying judgment.* Hillsdale, NJ: Erlbaum.

Quadrel, M. J. 1990. *Elicitation of adolescents' risk perceptions: Qualitative and quantitative dimensions.* Ph.D. dissertation, Carnegie Mellon University.

Quadrel, M. J., Fischhoff, B., and Davis, W. 1993. Adolescent (in)vulnerability. *American Psychologist,* 48, 102–116.

Quadrel, M. J., Fischhoff, B., Fischhoff, M., and Halpern S. 1994. *Calibration of adolescents' and adults' confidence judgments.* Unpublished manuscript.

Quadrel, M. J., Fischhoff, B., and Palmgren, C. 1996. *Adolescents' definitions of risk behaviors.* Unpublished manuscript.

Schkade, D. and Payne, J. W. 1994. How people respond to contingent valuation questions: A verbal protocol analysis of willingness to pay for an environmental regulation. *Journal of Environmental Economics and Management,* 26, 88–109.

Shaklee, H. and Fischhoff, B. 1990. The psychology of contraceptive surprises: Judging the cumulative risk of contraceptive failure. *Journal of Applied Psychology,* 20, 385–403.

Slovic, P. 1987. Perception of risk. *Science,* 236, 280–285.

Thaler, R. 1980. Toward a positive theory of consumer choice. *Journal of Economic Behavior & Organization,* 1, 39–60.

Thaler, R. 1985. Mental accounting and consumer choice. *Marketing Science,* 4, 199–214.

Vlek, C. A. J. and Stallen, P. J. 1980. Rational and personal aspects of risk. *Acta Psychological,* 45, 273–300.

Wallsten, T., Budescu, D. V., Rapoport, A., Zwick, R., and Forsyth, B. 1986. Measuring the vague meanings of probability terms. *Journal of Experimental Psychology: General,* 115, 348–365.

16

ASSESSING ADOLESCENT DECISION-MAKING COMPETENCE

Baruch Fischhoff

DEPARTMENT OF SOCIAL AND DECISION SCIENCES AND DEPARTMENT OF ENGINEERING AND PUBLIC POLICY, CARNEGIE MELLON UNIVERSITY, PITTSBURGH, PA 15213–3890, USA

Abstract

Behavioral decision research offers a general approach to studying cognitive aspects of decision making, as well as a platform for studying their interplay with social and affective processes. Applied to any decision, behavioral decision research involves three interrelated tasks: (a) *normative* analysis, identifying the expected impacts of possible choices; (b) *descriptive* study, characterizing how individuals view the decision, in terms comparable to the normative analysis; and (c) *prescriptive* interventions, helping people to bridge critical gaps in their understanding. Applied to adolescents' decisions, behavioral decision research provides analytical and empirical procedures for clarifying the challenges that young people face and their success in addressing them. It recognizes that competence varies by individual and by decision, leading to domain-specific policies and interventions, affording teens as much autonomy as they can manage.

Introduction

High stakes ride on society's ability to assess adolescents' decision-making competence. If that competence is overestimated, then teens will face choices that are too difficult for them. If it is underestimated, then they will be kept from exercising warranted independence. If teens believe that the boundaries of their autonomy have been drawn wrongly, then they may feel unfairly restricted or unfairly left to fend for themselves.

Behavioral decision research offers a framework for studying decision-making competence, including methods for assessing it, theories for predicting it, and interventions for improving it (Hastie and Dawes, 2002; vonWinterfeldt and

Edwards, 1986; Yates, 1989). Behavioral decision research recognizes both individual and situational variability. A given decision might be harder for some people than for others. A given individual might find some decisions harder than others. That variability poses a challenge to teens, who must identify their personal "envelope of competence," circumscribing the decisions that they know how to make. It poses a challenge to adults, who must find the balance between affording teens too much freedom and too little.

Behavioral decision research cannot resolve such policy questions. It cannot say, for example, whether teens can make decisions about interpersonal violence well enough to be adjudicated as adults nor whether teens can make decisions about interpersonal intimacy well enough to assume control of various reproductive decisions. What the research can do is assess how likely teens are to make choices of varying soundness. Whether the benefits of autonomous decision making outweigh its risks is a political–ethical question, not a scientific one. What research can do is to clarify the expected costs and benefits of letting teens make various choices.

A normative standard

Behavioral decision research's starting point is a *normative* analysis, describing a decision precisely enough to identify the choice that a *rational* actor would make. In this usage, "normative" refers to the procedural norms embodied in the axioms of decision theory. A famous mathematical proof showed that following these seemingly simple rules (e.g., transitivity) leads to choosing the option with the highest expected utility, given an individual's beliefs and values (vonNeumann and Morgenstern, 1947). Plous (1993) provides a brief introduction; Hastie and Dawes (2002), vonWinterfeldt and Edwards (1986), and Yates (1989) provide fuller ones.

From this perspective, *social norms* are among the things that people might value. A rational choice might reflect just social norms, if people care solely about what other people value. A rational choice might also let social norms be overridden by other concerns or balance conflicting social norms (e.g., those of peers and parents).

Behavioral decision research does not assume that people are rational. Nor does it assume that people must follow (or even know) the axioms, in order to make rational choices. Some decisions are easy enough that even casual analysis leads to the rational choice; some good choices are learned by trial and error. Nor does behavioral decision research assume that people always want to make rational choices. People may prefer to follow their emotions or inviolate moral principles. Nor does behavioral decision research assume that rational choices will also be *optimal* ones, in the sense of making the greatest expected contribution to people's well-being. If people misunderstand their circumstances or themselves, rational choices may not bring the best outcomes.

Normative analysis plays several roles in behavioral decision research. One is facilitating a precise definition of each decision, against which people's performance can be compared. That means identifying people's goals, their options for achieving those goals, and the events that determine the chances of each goal being achieved by each option.

A second role of normative analysis is organizing evidence. Any issue that decision makers might consider must find its way into the normative analysis. It must affect either the definitions of the choice options, the valuation of potential outcomes, or the probabilities of experiencing those outcomes.

The third role of normative analysis is keeping researchers from focusing too narrowly. For example, researchers concerned about a risk behavior (e.g., unprotected sex) naturally see its link to health risks (e.g., sexually transmitted infections), but might neglect teens' other goals (e.g., showing trust in a partner) or obstacles (e.g., sexual coercion). Normative analysis requires a comprehensive view.

Sometimes, quantitative normative analyses are needed (e.g., for assessing the impact of underestimating condom effectiveness by 10%). Often, though, qualitative analysis is enough to clarify the decision's structure, identify its critical beliefs and values, and suggest potential problems or interventions (Fischhoff, 2005a; Fischhoff, Bruine de Bruin, Guvenc, Caruso, and Brilliant, 2006). Once completed, the normative analysis structures the next steps in behavioral decision research: *descriptive* research, empirically assessing people's competence to make the choice, and *prescriptive* interventions, attempting to increase that competence, by closing critical gaps between the normative ideal and the descriptive reality. The next sections consider research on three competencies: assessing beliefs, assessing values, and integrating beliefs and values.

Assessing beliefs

When outcomes are certain, decision making is just about values: identifying the most attractive set of outcomes. When outcomes are uncertain, decision makers must predict the outcomes of possible actions. In order to be meaningful, those predictions must be sufficiently precise to be evaluated in the light of experience. That requires a clear description of the event being predicted (Beyth-Marom and Fischhoff, 1997) and a numeric probability of its occurring.

Although people use verbal quantifiers (e.g., likely, rarely), in everyday speech, their meaning can vary widely across people and situations (Budescu and Wallsten, 1995; Schwarz, 1999). Thus, "likely drug side effect" may connote a different probability than "likely Stanley Cup winner." Although eliciting numeric values solves this problem, some researchers believe that it is too demanding for respondents. However, even imperfect measures can be useful, if their strengths and weaknesses are understood. Fortunately, the research on probability elicitation is very extensive, providing guidance on responsible usage (O'Hagan et al., 2006). Results include:

(a) Numeric probability judgments can be as reliable and acceptable as verbal ones. Woloshin, Schwartz, Byram, Fischhoff, and Welch (1998) found this pattern when comparing judgments of medical events, elicited with two verbal scales and two numeric scales, each of which had equally spaced options from 0% to 100%, with one expanding the 0–1% range with log values from 1/100 to 1/1,000,000 (under a cartoon magnifying glass).
(b) People often prefer to provide verbal judgments, but to receive quantitative ones. Receiving quantitative estimates provides useful information, while producing them requires greater effort and accountability (Erev and Cohen, 1990).
(c) Absolute values of numeric judgments can vary widely with response mode, while relative values are fairly invariant. For example, absolute values have been found to vary when using odds or probabilities, probabilities or relative frequencies, and individual items or grouped ones (Gilovich, Griffin, and Kahneman, 2003; Poulton, 1994).
(d) Some numeric values are treated specially. For example, people seldom use fractional values (Poulton, 1989), a pattern that motivated the log part of Woloshin et al.'s log-linear scale. When uncertain what to say, people sometimes say "50" in the sense of "50–50," a vague, non-zero value, rather than a specific numeric probability (Bruine de Bruin, Fischhoff, Halpern-Felsher, and Millstein, 2000).
(e) Probability judgments can be deliberately biased, unless honest responses are requested and rewarded. For example, Christensen-Szalanski and Bushyhead (1993) found physicians overestimating the probability of pneumonia, fearing that low-probability cases would be ignored. Forecasters sometimes overstate precipitation probabilities, in order to reduce users' chance of being caught in the rain (Lichtenstein, Fischhoff, and Phillips, 1982).
(f) Probability judgments for knowing the answer to a question are modestly correlated with the probability of being correct (Lichtenstein et al., 1982; Yates, 1989).
(g) People differ in their ability to use probabilities, an ability that correlates with performance on other tasks and with life events that might reflect decision-making competence (Parker and Fischhoff, 2005; Table 16.4).

Like much psychological research, these studies mostly involve people of at least college age. Jacobs and Klaczynski (2005) offer a collection of articles regarding developmental differences in decision-making processes, while Furby and Beyth-Marom (1992) and Reyna and Farley (2006) offer integrative essays. Perhaps a fair summary of the research is that, by mid adolescence, most individuals have approximately adults' imperfect cognitive skills. Of course, having skills is necessary, but not sufficient for relying on them. Nor does it guarantee having the domain-specific knowledge needed to make informed choices.

Fischhoff et al. (2000) analyze the construct validity of numeric probability judgments made by a nationally representative sample of 15 and 16 year olds, in the 1997 National Longitudinal Study of Youth (NLSY97). The analyses found, among other things, that teens use the full range of probability values, almost always give 100% for eating pizza in the next year (on a warm-up question), show no individual-difference tendency to give high or low values, and sometimes seem to say "50" in the "50–50" sense. These teens' probability judgments correlate sensibly with their answers to related questions on other NLSY97 modules (created by other researchers). For example, females who reported being sexually active also gave higher probabilities for getting pregnant in the next year and for having a baby by age 20; males showed similar correlations for making someone pregnant and for becoming a father. Teens reporting neighborhood gang activity gave higher probabilities for being arrested in the next year and for dying in the next year or by age 20. Thus, these teens seemed both sensitive to factors affecting their futures and able to express that knowledge in numeric probabilities.

The greater one's faith in a research method, the more seriously one can take the data it produces. For example, the log-linear scale that Woloshin et al. (1998) evaluated was first used by Quadrel, Fischhoff, and Davis (1993), in a study whose results challenged the conventional wisdom that adolescents have a unique sense of invulnerability. Recognizing the extra burden borne by methods producing unexpected results, Quadrel et al. (1993) examined some aspects of construct validity, finding sensible (and similar) usage by teens and adults. Confidence in their results is strengthened by Woloshin et al.'s study and by studies using other methods that also found no unique teen sense of invulnerability (Millstein and Halpern-Felsher, 2002).

That confidence is also strengthened by the NLSY97 finding that adolescents exaggerate their probability of dying soon. This extreme expression of vulnerability appears in the final two rows of Table 16.1, which shows the 12 NLSY97 questions whose predictive validity could be evaluated, with data from subsequent waves. Teens' median mortality judgments were 10%, for events with a tiny statistical probability (0.08%/year). Mean judgments were higher (about 20%), "inflated" by many 50s, some apparently expressing epistemic uncertainty, rather than numeric probabilities. Although non-numeric 50s are inappropriate responses, they still provide insight into teens' thinking, meaning perhaps, "I don't know if I'll live or die" or "I don't want to think about it."

Table 16.1's other rows show that teens' probability judgments are significantly correlated with these events' probabilities of occurring (e.g., teens who gave a higher probability for being in regular school a year hence were also more likely to be there). Teens' mean probability judgment sometimes approximates the observed rate (row 2), sometimes is optimistic (rows 1, 3, 4), sometimes is pessimistic (rows 9, 10), and sometimes has an unclear bias, depending on what teens value (rows 5–8). Because Table 16.1's occurrence rates reflect self-reports, teens' probability judgments are evaluated in terms of how they interpreted the events.

TABLE 16.1 Predictive validity and accuracy of adolescents' expectations in NLSY97

What is the percent chance that you will …		*N*	*Correlation (g) with outcome*	*Mean response (1997) (%)*	*Observed outcome rate (%)*
1.	Be a student in a regular school a year from now?	3160	.64***	92.5	79.6
2.	Have received a high school diploma by the time you turn 20?	3077	.60***	94.5	92.0
3.	If you are in school a year from now, … work for pay more than 20 hours a week?	2492	.29***	57.7	27.2
4.	If you are not in school a year from now, … work for pay more than 20 hours a week?	610	.31***	80.5	43.9
5.	Become pregnant within 1 year from now? (female)	844	.37***	8.9	20.1
6.	Get someone pregnant within the next year? (male)	1553	.35***	9.4	7.9
7.	Become the parent of a baby sometime between now and when you turn 20? (female)	1368	.38***	16.0	25.7
8.	Become the parent of a baby sometime between now and when you turn 20? (male)	1356	.27***	19.1	13.4
9.	Be arrested, whether rightly or wrongly, at least once in the next year?	3141	.41***	10.3	8.2
10.	Serve time in jail or prison between now and when you turn 20?	3300	.39***	5.4	2.8
11.	Die from any cause (crime, illness, accident, and so on) in the next year?	3165	ns	18.7	0.1
12.	Die from any cause (crime, illness, accident, and so on) between now and when you turn 20?	3169	ns	20.3	0.5

Source: Bruine de Bruin, Parker, and Fischhoff (2007a). *Note:* Complex skip patterns account for the different Ns.

*** $p < .001$; ns = not significant.

Table 16.1 evaluates teens' probability judgments in terms of *correspondence* tests, comparing them to *external* real-world events. Those judgments can be also subjected to *coherence* tests, examining their *internal* consistency, as defined by probability theory (Fischhoff and Beyth-Marom, 1983). Coherence tests include whether the probabilities assigned to an event and its complement sum to 1.0 and whether the probability for an event is strictly greater than that for a subset. Teens' probability judgments often show coherence (Parker and Fischhoff, 2005), but not with Table 16.1's mortality judgments. There, most teens gave the same probability for dying in the next year and by age 20 (a period of 4.5 years, on average). When the two values differed, one third of teens gave a *lower* probability to the longer period (Fischhoff et al., 2000). Thus, mortality judgments are anomalous in both correspondence and coherence terms, suggesting that many teens lack the competence here that they show for other probability judgments.

Assessing values

Behavioral decision research has two complementary approaches to assessing the values that individuals hope to realize when making decisions (Fischhoff, 1991, 2005a; Lichtenstein and Slovic, 2006). One approach, rooted in psychophysics, assumes that people know roughly what they want, regarding any question put to them. The second approach, rooted in decision analysis, allows that people may lack that knowledge, forcing them to derive specific preferences from potentially relevant basic values.

When people have well-articulated preferences, the two approaches converge. People know what they want and express those desires, whenever they are asked clear questions. The approaches diverge, when people lack stable preferences. Perhaps they have not given the topic much thought; perhaps they cannot reconcile conflicting values. In such cases, they don't know what they want, leaving them looking for clues and susceptible to manipulation by how questions are posed. As a result, their expressed values are *labile*, subject to *framing* or *context* effects (Fischhoff, 2005a; Kahneman and Tversky, 2000; Lichtenstein and Slovic, 2006; Schwarz, 1999).

As an extreme example, consider someone diagnosed with a brain tumor, who must choose between palliative care and aggressive treatment that is equally likely to lengthen or shorten her life. Over time, she might derive specific preferences from her basic values. Or, she might never know what tradeoffs to make between pain and hope. Or, she might shift between positions, depending on how she feels at the moment (Loewenstein, 1996). In such cases, the psychophysics approach to eliciting values makes little sense. Why ask people what they want, when they don't know?

The alternative value elicitation approach, that of decision analysis, seeks to help people reflect on what they might want (vonWinterfeldt and Edwards, 1986), by presenting a balanced set of potentially relevant perspectives, as identified in the normative analysis. When successful, decision analytic value elicitation deepens people's understanding. However, it can also bias their thinking, if the set of perspectives is not systematically developed and fairly presented (Fischhoff, 1991). Related procedures, facing similar risks, include motivational interviews (Rollnick and Miller, 1995) and deliberative polling (Fishkin, 1997). Both methods also have prescriptive goals: the former hopes that individual reflection will uncover a latent desire for healthy change; the latter hopes that group reflection will uncover latent agreement about public policies.

The values elicited by decision analysis approaches should predict the ones that people express in their lives, when those lives present similarly balanced perspectives. Psychophysics approaches should predict real-life behavior, when lives present the same incomplete set of perspectives as the research question. Comparing the perspectives presented by researchers and by life means looking for regularities, such as how young men shape the values salient to young women in sexual situations, how young men shape the values salient to other young men in conflict situations, and how advertisers shape the values salient to as many people as they can. Some regularities occur naturally; others come from training that makes specific values salient. Pledges (for abstinence, diet, non-violence, etc.) attempt to do that, by making internal cues stronger than external ones.

Value judgments are subject to the same two kinds of performance standard as probability judgments. *Correspondence* tests ask whether people accept externally prescribed values, such as social norms. *Coherence* tests ask whether people's values are internally consistent, across choices. If a value represents an unambiguous, inviolate principle, then the two tests converge. For example, people opposed to abortion should condemn it in all cases. In more complex situations, though, the tests require normative analyses, showing the roles that specific values play in specific choices. The importance of any value depends on how the decision options vary in that respect. For example, it is not inconsistent to pay attention to money when the options have different monetary outcomes, but to ignore money when the options have similar monetary outcomes. Conversely, ignoring money in the latter case would not represent anti-materialist virtue.

Asking people about specific values can present cognitively challenging tasks, just like asking about specific probabilities. Some of the most demanding tasks ask for values needed by regulatory analyses, such as pricing health services (Tengs and Wallace, 2000) or environmental damages (Mitchell and Carson, 1989). For example, "Would you be willing to pay $10/month for a special treatment that would relieve the breathing difficulty due to low level ozone from emissions?") The debate over these tasks raises essential questions regarding the nature of value elicitation (Fischhoff and Manski, 1999).

Although teens rarely participate in policy-related studies, they are often asked questions analogous to one policy concern: how much to discount future outcomes. Economists ask questions like, "How many dollars would you need one year from now, in order to forgo $100 today?" They would like to use the discount rates implied by the answers in cost–benefit analyses. Unfortunately, those answers vary widely.

Frederick, Loewenstein, and O'Donoghue (2002) argued that this variation reflects, not just differences in how people value future outcomes, but also differences in how they interpret the questions. Although superficially simple, such questions leave potentially critical details unstated. As a result, respondents must read between the lines in order to complete the task. If they make different inferences about missing details, then they are answering different questions.

Frederick et al. proposed that how people value a future outcome could depend on the seven factors in Table 16.2. The first is how much they care about the future per se (the usual notion of temporal discounting). Subsequent rows capture other possible reasons for caring less about future outcomes: (2) People are uncertain about getting those outcomes (e.g., because they will have died). (3) People expect the outcome to be worse for reasons not in the description (e.g., more poorly manufactured). (4) People expect to change in ways that diminish how much they will enjoy the outcome (e.g., losing their sweet tooth). (5) People see a psychological cost to waiting. (6) People will have less time to reminisce about the outcome after getting it. (7) People expect their other assets to change in ways that reduce the outcome's values (e.g., already having similar goods).

TABLE 16.2 Possible reasons for evaluating goods differently at different times

Model	*Corresponding description in words*
DU (time preference only)	Future utility should be discounted because we should care less about the latter parts of our life
DU + probability	Future utility should be weighted by the probability that the consequence that gives rise to the utility will actually occur
DU + changes in objective consequence	The objective properties of some coarsely defined consequence may depend on the time at which it occurs
DU + changes in utility function	The subjective utility associated with a particular objective consequence may change over time
DU + utility from anticipation	The utility at a given moment may be influenced by the anticipation of future utility
DU + utility from memory	The utility at a given moment may be influenced by the recollection of past utility
DU + opportunity cost	Utility depends on the current consumption level, and the potential consumption level depends on current income and past investment

Source: Frederick, Loewenstein and O'Donoghue (2002).

Analogous issues could arise when teens consider the relative value of, say, smoking today and enjoying better future health. Teens might care less about their future selves (row 1). However, they also might not expect to live that long, for reasons unrelated to smoking (row 2). Or, teens may expect a future world so degraded that all pleasures are diminished (row 3). And so on. Existing research provides a priori grounds for evaluating these options. For example, exaggerated mortality judgments (Table 16.1) lend credibility to row two concerns. However, without well-specified tasks, teens' values can be misunderstood, such as thinking that teens do not care about the future, when they actually do not expect to live to enjoy it.

Assessing decision-making processes

Experimental researchers have shown great ingenuity in creating tasks that distinguish among the various rules that people might use, when integrating their beliefs and values (e.g., Weber, 1994; Yates, 1989). Real-world decisions rarely allow the control possible with these artificial tasks. Fortuitously, for predicting behavior, such control is often unnecessary. Many different simple linear (weighted-sum) models can predict many behaviors, about equally well (Dawes and Corrigan, 1974; Goldberg, 1968).

The model-building process is straightforward: take variables that people consider when making a decision (or variables correlated with those variables), standardize them, give them the correct sign (indicating whether or not they favor a particular choice), and add. The resulting score will predict individuals' propensity to make the focal choice. Expect correlations around 0.3. Weighting the variables can sometimes improve predictions. However, robust weights are hard to find. Regression weights are often unstable, because of multicolinearity. Importance weights elicited with rating scales depend on people's limited powers of introspection (Ericsson and Simon, 1994). As a result, the best research strategy may be to use unit weights (± 1). In addition to its simplicity, that strategy keeps researchers from wasting their time, fruitlessly interpreting meaningless variations in weights (Camerer, 1980; Hastie and Dawes, 2002).

Because of these problems with weights, simple linear models provide limited insight into decision-making processes. As a result, when behavioral decision researchers want to explain (as opposed to predict) behavior, they typically focus on identifying choice rules that people find intuitively appealing. They are particularly interested in rules that violate the normative axioms. Such violations can be revealing about thought processes (which have survived despite being "irrational"), decision-making environments (which are forgiving of such violations), and interventions (which are needed for unforgiving environments).

As an example of such a violation, although decision theory requires evaluating all decision options, people often focus on just one. That leaves the expected outcomes of other options less clear than if they were examined equally well (Beyth-Marom, Austin, Fischhoff, Palmgren, and Quadrel, 1993). A corollary bias is insensitivity to the *opportunity costs* of choosing a focal option, namely, the other ways to invest those resources. A related corollary is undue commitment to *sunk costs*, resources already invested in an option. Normatively speaking, previous investments should be ignored, when evaluating future investments. However, people are reluctant just to look ahead, especially when that means realizing losses (Arkes and Blumer, 1985; Kahneman and Tversky, 2000).

Table 16.3 shows the types of options revealed, when teens described seven recent decisions about specific topics (school, free time, clothing, friendships, health, money, and parents), two recent hard decisions and one pending hard decision. Although the descriptions were often detailed, they typically focused on statements of resolve, like "eat more healthfully" and "stop blaming others." In effect, these decisions re-evaluate a single option, chosen previously, without stating any alternative options. Explicit alternatives were also absent from the next largest category, decisions about whether to do something (e.g., smoke cigarettes). Roughly equal numbers of decisions had two distinct options (e.g., whether to go to school or hang out with friends) or a set of identifiable options (e.g., which class to take, what to wear, with whom to

TABLE 16.3 Option structures in teens' open-ended decision descriptions

Decision	*Structure*					
	To do X	*Whether to do*	*Two choices*	*Finite choices*	*What to do about*	*Missing*[a]
Recent decisions						
School (%)	36.2	29.0	13.0	13.0	2.9	5.8
Free time	31.9	18.8	17.3	20.3	4.3	7.2
Clothing	30.4	10.1	11.5	31.9	2.9	13.0
Peers	47.8	24.6	10.1	5.8	4.3	7.2
Health	55.0	20.2	1.4	2.9	2.9	17.4
Money	52.2	11.6	10.1	4.3	10.1	11.6
Parents	30.4	23.2	18.9	4.3	8.7	14.5
Hard decisions						
First past	39.1	33.3	20.2	—	5.8	1.4
Second Past	20.2	44.9	18.8	5.8	4.3	5.8
Current	2.9	44.9	23.2	13.0	13.0	2.9

Source: Fischhoff (1996).

[a] Includes cases where respondents produced no answer or an uncodable one, or where the question was not asked due to a procedural error. These three cases constituted 27%, 51%, and 21% of missing responses, respectively.

have lunch). Few decisions involved "designing" options (e.g., how to spend free time, what to do about having fought with a friend).

Most of these decisions involved concrete, one-time choices. The main exceptions were the recent decisions about health and money, which often involved policies of some breadth (e.g., what kind of diet or spending pattern to adopt). It is not hard to imagine general policy choices in the other areas (e.g., how to spend free time, how to manage homework, how to keep parents happy). However, teens rarely described them. Although hard decisions provide an opportunity to reflect on big issues, these teens did not report seizing it.

These patterns were echoed in another study, asking young women to describe decisions about sexual activities. Their descriptions were so narrowly focused as to be barely decisions at all. In response, Downs, Murray, Bruine de Bruin, White, et al. (2004) developed an interactive DVD, hoping to reduce sexually transmitted infections (STIs) by helping young women to see that they had decisions to make. One element of that empowerment was affording cognitive mastery of the domain; by providing decision-relevant information in compact, comprehensible form. A second element was helping users identify decision-making opportunities, by interrupting video narratives in which young men pressed young women for sex, then asking users "What would you do?" A third element was eliciting cognitive rehearsal of refusal strategies, followed by an actress modeling refusal (Bandura, 2000). In a randomized control trial, the intervention outperformed an "ideal usual care" condition (with equal exposure to commercially available materials matched for topic), in terms of attitudes, knowledge, self-reported behavior, and sexual health.

Decision theory focuses on evaluating options and is mute regarding their creation. Indeed, it has been criticized for not empowering people to create (or demand) better options (Fischhoff, 1992; O'Brien, 2000). In order to create options, people need to understand the processes that affect important outcomes. Downs et al. (2004) used a *mental models* approach to teach teens about the factors affecting STI risks, focusing on commonly misunderstood facts relevant to formulating decision options (e.g., how risks mount up through repeated exposure, how hard it is for partners to self-diagnosis STIs). The mental models approach creates a formal model, summarizing the science regarding the processes shaping key outcomes. Semi-structured, open-ended interviews elicit beliefs on these topics, in their intuitive formulation. These serve as the basis for ecologically valid knowledge tests and interventions. The approach has been applied to informing decisions about many different risks (Fischhoff, 2005b; Morgan, Fischhoff, Bostrom, and Atman, 2001).

Cognition in decision making

Behavioral decision research attempts to treat cognitive aspects of decision making in a comprehensive, coherent way. Its normative analyses summarize what is known about the options, the probabilities of achieving valued outcomes, and those outcomes' relative importance. Descriptive analyses characterize decision makers' current beliefs in terms of deviations from the normative analysis. Prescriptive interventions try to bridge the gap between the normative ideal and the descriptive reality.

The norms in these normative analyses are typically those of Bayesian decision theory (Edwards, Lindman, and Savage, 1963; Fischhoff and Beyth-Marom, 1983). Studies showing non-normative behavior have prompted proposals for revising decision theory, so as to fit human intuitions better, as well as proposals that respect the decision theory norms, but study when less rigorous thinking will suffice (Lopes, 1987; Reyna and Farley, 2006; Shafer and Tversky, 1985; Simon, 1957). *Journal of Risk and Uncertainty* is one good place to follow this work.

Getting the cognitive part right is necessary, but not sufficient, for a full account of decision making, which also must accommodate social, emotional, and developmental factors (Fischhoff, Downs, and Bruine de Bruin, 1998). Behavioral decision research asks how these factors relate to the normative analyses that define sound decision making, the descriptive accounts that evaluate people's competence, or the prescriptive interventions that try to enhance it.

Social factors fit readily into normative accounts. If people care about social norms, then the expected costs and benefits of complying with them are treated like other outcomes. If social pressures affect how an option is implemented (e.g., if friends might not let friends drive drunk), then that becomes another source of uncertainty, when predicting outcomes. If social pressure becomes social coercion, then some actions might become impossible. Descriptive accounts can assess people's awareness of these social effects. Prescriptive interventions can seek to improve people's awareness or change the reality (e.g., Downs et al., 2004).

Emotional effects can also fit into normative analyses, as valued outcomes (e.g., if people want to be angry or sad or happy). They can be captured in descriptive accounts in terms of their effects on each aspect of decision making (defining options, predicting events, determining personal values, integrating beliefs, and values). Prescriptive interventions can help people manage their emotions, either by preselecting actions (in unemotional conditions) or choosing how to feel (e.g., anger management).

Cognitive appraisal theory (Lerner and Keltner, 2001) offers one such account, predicting the effects of specific emotions on specific judgments. For example, it predicts that anger encourages attributing problems to individuals (rather than situations) and increases the perceived probability of overcoming problems. Lerner, Gonzalez, Small, and Fischhoff (2003) and Small, Lerner, and Fischhoff (2006) demonstrate these effects with terror-related judgments, also finding that nationally representative samples of adults and adolescents responded similarly. Because the study elicited numeric probabilities for well-defined events, it was possible to evaluate the size of emotion effects (as well as their accuracy, given subsequent experience; Fischhoff, Gonzalez, Lerner, and

Small, 2005). Over eight events, respondents in an induced anger condition were about 6% more optimistic than respondents in an induced fear condition.

The impact of an effect of that size depends on the decision. A close decision might be sensitive to a 6% shift; a more clear-cut decision might not. Some observers have argued that the close decision to go to war in Iraq was tipped by anger for some people and by fear for others. The wisdom of emotional effects depends on the validity of the cues that emotions provide (e.g., do they overcome unwanted numbing? are they manipulated by others?) (Finucane, Alhakami, Slovic, and Johnson, 2000; Loewenstein, 1996; Loewenstein, Weber, Hsee, and Welch, 2001; Slovic, Peters, Finucane, and MacGregor, 2005; see also Rivers, Reyna, and Mills, 2008).

Cognitions can also affect emotions. For example, teens who see a 20% chance of dying in the next year (or think 50–50) might feel frustrated enough to act out or to disassociate themselves from long-term future outcomes (as in row 2, Table 16.2). The article reporting these exaggerated mortality judgments concludes by speculating that teens take "risks, in part, because they underestimate what is at stake, as a result of overestimating the risk of dying. That is, they take risks not just because of an exaggerated feeling that they are not going to die, but also because of an exaggerated feeling that they are not going to live" (Fischhoff et al., 2000, p. 200).

More generally, any cognitive process that undermines effective decision making may increase the roles of social and emotional factors. For example, Table 16.3 shows teens considering reduced sets of options. An overly narrow focus could keep teens from identifying good choices or from finding any acceptable choices. As a result, they may drift toward decision points—perhaps into situations where social and emotional concerns overwhelm cognitive ones. The cognitive rehearsal intervention used by Downs et al. (2004) sought to help young women make decisions prior to experiencing the passion and coercion of intimate encounters.

Thus, teens competent in cognitive aspects of decision making should make better decisions, not only because they can execute those cognitive elements better, but also because cognition gets them further, when making decisions. Table 16.4 presents results from a study of how cognitive decision-making competencies fit into young people's lives. It shows correlations with an individual-difference measure of decision-making competence (DMC) extracted from a factor analysis of performance on eight tasks, representing basic decision-making skills (e.g., assessing probabilities, applying decision rules). Respondents were 110 18–19 year old males, who, at age 10, had entered a longitudinal study at the Center for Education and Drug Abuse Research (Ralph Tarter, PI), returning every year or two for a day or two of testing, thereby creating an extensive battery of potentially related measures. DMC scores showed good test–retest reliability, as did scores on a version adapted for adults (Bruine de Bruin, Parker, and Fischhoff, 2007b).

TABLE 16.4 Correlations between decision-making competence (DMC) and other variables

DMC Correlated with	*Pearson r*	*Semi-partial correlation, controlling for*		
		Vocabulary	*ECF*	*Vocabulary and ECF*
Cognitive ability				
Vocabulary	.50	—	.28	—
ECF	.48	.26	—	—
Overall*	$p < .0001$	$p = .0009$	$p = .0008$	—
Cognitive style				
Polarized thinking	-.34	-.20	-.24	-.19
Self consciousness	.20	.14[b]	.05	.11
Self monitoring	.24	.29[b]	.30[b]	.32
Behavioral coping	.32	.27[a]	.28[a]	.26
Overall	$p < .0001$	$p < .0001$	$p < .0001$	$p < .0001$
Risk behavior				
Antisocial disorders	-.19	-.18[b]	-.05	-.09
Externalizing behavior	-.32	-.28[b]	-.18	-.20
Delinquency	-.29	-.28[b]	-.18	- .21
ln(lifetime # of drinks)	-.18	-.22[b]	-.15	-.18
ln(lifetime marijuana use)	-.25	-.30[b]	-.20	-.25
ln(# times had sex)	-.24	-.30[b]	-.21	-.27
ln(# sexual partners)	-.30	-.33[b]	-.29[a]	-.31
Overall	$p = .0004$	$p = .0002$	$p = .009$	$p = .002$
Social and family influences				
Risk status (HAR = 1; LAR = 0)	-.35	-.27	-.23	-.21
SES	.35	.20	.21	.15
Social support	-.30	-.21	-.23	-.19
Positive peer environment	.33	.35[b]	.32[a]	.35
Overall	$p = .0002$	$p = .002$	$p = .006$	$p = .007$

Abbreviations: ECF, executive cognitive function; SES, socio-economic status; HAR, high risk family; LAR, low risk family.

Source: Parker and Fischhoff (2005).

*Overall *p*-values were computed using Strube's (1985) method for combining significance levels from non-independent hypothesis tests. All reported *ps* are one-sided. A conservative Bonferroni correction on the 57 tests presented here and another table showing comparable correlations with the eight individual DMC tasks converts an individual $\alpha = .05$ into $\alpha = .0009$. Approximate cutoffs for individual zero-order correlations are $r = .16, p < .05$; $r = .22, p < .01$; $r = .29, p < .001$. For a semi-partial correlation, approximate cutoffs are $r = .18, p < .05$; $r = .25, p < .01$; $r = .32, p < .001$.

[a] Test A rejects the one-mediator null hypothesis.

[b] Test B rejects the one-mediator null hypothesis.

The first section shows that DMC scores were correlated with standard measures of verbal and fluid intelligence (Vocabulary and ECF, respectively). The second section shows positive correlations between DMC and "constructive" cognitive styles; these correlations generally remained after partialing out the two intelligence measures, indicating that DMC is independently related to these ways of thinking about the world. The third section shows that DMC is negatively related to several important risk behaviors; again, beyond correlations with intelligence. The fourth section shows that DMC is higher for teens coming from low-risk (LAR) families, higher SES families, and more positive peer environments. (The negative correlation with social support seemed to reflect low DMC teens' more frequent gang membership). As discussed by Parker and Fischhoff (2005), these results support the construct validity of DMC as a measure of decision-making skills that both cause and reflect important aspects of teens' lives. For example, teens with higher DMC seem to come from families that might both model and reward good decision making.

Conclusion

Cognition, like emotional and social processes, is, arguably, part of most decisions and all of none. Behavioral decision research provides an integrated approach to studying cognitive aspects of decision making that also accommodates research on non-cognitive aspects. It provides ways to analyze decisions, identify potential problems, and assess the importance of those threats. Its commitment to detail should reduce the risk of simplistic diagnoses. For example, when assessing teens' impulsivity, it encourages considering all the reasons in Table 16.2, lest one confuse inability to exercise control and deliberately choosing to discount future options.

A question that occupies many people concerned about teens' welfare is, "Does information work?", as a way to improve teens' decision making. From a behavioral decision research perspective, there can be no simple answer. In some situations, teens would not change their choices, whatever (truthful) information they received. In those cases, information has "worked," leading them to stable decisions (Reyna and Farley, 2006). Those choices might not please people who disapproved of the values that those decisions embodied; however, the problem would not be how teens had used the information. Stable choices might not even please the teens making them, if they wished that they had better options (e.g., those unable to stop smoking or escape abusive relationships).

Information interventions reveal nothing about their recipients' decision-making competence unless they address critical gaps between recipients' information priorities, as identified by normative analyses, and current beliefs, as identified by descriptive studies. Interventions also reveal little, unless performed to professional standards. That means keeping those critical facts from being buried in irrelevant information, including critical facts that recipients already know. That means taking advantage of research into how people process such information and conducting rigorous pretests.

Unless information interventions are tested fairly, their recipients may be blamed unfairly—for ignoring messages that deserved to be ignored, because their content was irrelevant, cluttered, incomprehensible, etc. Unfairly criticizing teens' competence can unfairly undermine their social standing. For example, a pundit recently chose to spin adolescent research as proving "We're perceivers first, not deciders" (Brooks, 2007). Any sweeping generalization diminishes the humanity of the individuals being depicted so formulaically. This particular generalization undermines any attempt to inform teens (e.g., sex education, over-the-counter labels on Plan B, driver education). Its acceptance would decrease the risk of holding teens responsible for decisions that they lack the competence to make, while increasing the risk of denying them choices that they could handle, were they properly informed.

If one succumbed to the temptation to make sweeping generalizations, but based them on detailed examination of specific decisions, one might conclude that teens do surprisingly well, given the difficulty of the decisions facing them (e.g., intimacy, friendship, drugs, careers, identity, money, appearance). These decisions often pose hard tradeoffs, have unpredictable effects, require mastery of unfamiliar facts, and lack trustworthy information sources. The number of poor decisions that teens make reflects not just their abilities, but also the number and nature of their challenges. Excellent third basemen still make a lot of errors, at the "hot corner," relative to other field positions.

Behavioral decision research's normative, descriptive, and prescriptive research provide an integrated structure for accomplishing tasks addressed by anyone concerned about teens: identifying the critical issues in teens' choices, assessing their current understanding, and helping them do better. It takes advantage of research into cognitive decision-making processes, while clarifying their interface with affective and social processes. It encourages the nuanced assessment of competence that teens deserve.

Acknowledgments

Paper preparation supported by National Science Foundation Grant SES 0433152. My thanks go to Wändi Bruine de Bruin, Julie Downs, Irene Janis, Tamar Krishnamurti, and Valerie Reyna.

References

Arkes, H. R. and Blumer, C. 1985. The psychology of sunk cost. *Organizational Behavior and Human Decision Processes*, 35, 124–140.

Bandura, A. 2000. Self-efficacy. In W. J. Perrig and A. Grob (Eds.), *Control of human behavior, mental processes, and consciousness* (pp. 17–33). Mahwah, NJ: Erlbaum.

Beyth-Marom, R., Austin, L., Fischhoff, B., Palmgren, C., and Quadrel, M. J. 1993. Perceived consequences of risky behaviors. *Developmental Psychology*, 29, 549–563.

Beyth-Marom, R. and Fischhoff, B. 1997. Adolescents' decisions about risk. In J. Schulenberg, J. L. Maggs, and K. Hurrelmann (Eds.), *Health risks and developmental transitions during adolescence* (pp. 110–135). New York: Cambridge University Press.

Brooks, D. 2007. When preaching flops. *New York Times,* June 22. Available from http://select.nytimes.com/2007/06/22/opinion/22brooks.html.

Bruine de Bruin, W., Fischhoff, B., Halpern-Felsher, B., and Millstein, S. 2000. Expressing epistemic uncertainty: It's a fifty–fifty chance. *Organizational Behavior and Human Decision Processes, 81*, 115–131.

Bruine de Bruin, W., Parker, A., and Fischhoff, B. 2007a. Can adolescents predict significant events in their lives? *Journal of Adolescent Health*, 41, 208–210.

Bruine de Bruin, W., Parker, A., and Fischhoff, B. 2007b. Individual differences in adult decision-making competence (A-DMC). *Journal of Personality and Social Psychology, 92*, 938–956.

Budescu, D. F. and Wallsten, T. S. 1995. Processing linguistic probabilities: General principles and empirical evidence. In J. R. Busemeyer, R. Hastie, and D. L. Medin (Eds.), *Decision making from the perspective of cognitive psychology* (pp. 275–318). New York: Academic Press.

Camerer, C. 1980. General conditions for bootstrapping methods. *Organizational Behavior and Human Performance*, 27, 411–422.

Christensen-Szalanski, J. and Bushyhead, J. 1993. Physicians' misunderstanding of medical findings. *Medical Decision Making, 3*, 169–175.

Dawes, R. M. and Corrigan, B. 1974. Linear models in decision making. *Psychological Bulletin, 81*, 95–106.

Downs, J. S., Murray, P. J., Bruine de Bruin, W., White, J. P., et al. 2004. An interactive video program to reduce adolescent females' STD risk: A randomized controlled trial. *Social Science and Medicine*, 59, 1561–1572.

Edwards, W., Lindman, H. R., and Savage, L. J. 1963. Bayesian statistical inference for psychological research. *Psychological Review, 70*, 193–242.

Erev, I. and Cohen, B. L. 1990. Verbal versus numerical probabilities: Efficiency, biases and the preference paradox. *Organizational Behavior and Human Decision Processes*, 45, 1–18.

Ericsson, A. and Simon, H. A. 1994. *Verbal reports as data*. Cambridge, MA: MIT Press.

Finucane, M. L., Alhakami, A., Slovic, P., and Johnson, S. M. 2000. The affect heuristic in judgments of risks and benefits. *Journal of Behavioral Decision Making*, 13, 1–17.

Fischhoff, B. 2005a. Cognitive processes in stated preference methods. In K.-G. Mäler and J. Vincent (Eds.), *Handbook of environmental economics* (pp. 937–968). Amsterdam: Elsevier.

Fischhoff, B. 2005b. Decision research strategies. *Health Psychology, 21*, S9–S–16.

Fischhoff, B. 1996. The real world: What good is it? *Organizational Behavior and Human Decision Processes, 65*, 232–248.

Fischhoff, B. 1991. Value elicitation: Is there anything in there? *American Psychologist, 46*, 835–847.

Fischhoff, B. 1992. Giving advice: Decision theory perspectives on sexual assault. *American Psychologist, 47*, 577–588.

Fischhoff, B., Bruine de Bruin, W., Guvenc, U., Caruso, D., and Brilliant, L. 2006. Analyzing disaster risks and plans: An avian flu example. *Journal of Risk and Uncertainty, 33*, 133–151.

Fischhoff, B., Downs, J., and Bruine de Bruin, W. 1998. Adolescent vulnerability: A framework for behavioral interventions. *Applied and Preventive Psychology,* 7, 77–94.

Fischhoff, B. and Beyth-Marom, R. 1983. Hypothesis evaluation from a Bayesian perspective. *Psychological Review*, 90, 239–260.

Fischhoff, B., Gonzalez, R., Lerner, J. S., and Small, D. A. 2005. Evolving judgments of terror risks: Foresight, hindsight, and emotion. *Journal of Experimental Psychology: Applied, 11*, 124–139.

Fischhoff, B. and Manski, C. (Eds.). 1999. *Preference elicitation*. Boston: Kluwer.

Fischhoff, B., Parker, A., Bruine de Bruin, W., Downs, J., Palmgren, C., et al. 2000. Teen expectations for significant life events. *Public Opinion Quarterly, 64*, 189–205.

Fishkin, J. S. 1997. *Deliberative democracy*. New Haven: Yale University Press.

Frederick, S., Loewenstein, G., and O'Donoghue, T. 2002. Time discounting and temporal preference. *Journal of Economic Literature, 40*, 331–401.

Furby, L. and Beyth-Marom, R. 1992. Risk taking in adolescence: A decision-making perspective. *Developmental Review, 12*, 1–44.

Gilovich, T., Griffin, D., and Kahneman, D. (Eds.). 2003. *Judgment under uncertainty II: Extensions and applications*. New York: Cambridge University Press.

Goldberg, L. R. 1968. Simple models or simple processes? Some research on clinical judgments. *American Psychologist, 23*, 483–496.

Hastie, R. and Dawes, R. M. 2002. *Rational choice in an uncertain world*. San Diego: Russell Sage.

Jacobs, J. E. and Klaczynski, P. (Eds.). 2005. *The development of judgment and decision making in children and adults*. Mahwah, NJ: Erlbaum.

Kahneman, D. and Tversky, A. (Eds.). 2000. *Choice, values, and frames*. New York: Cambridge University Press.

Lerner, J. S., Gonzalez, R., Small, D. A., and Fischhoff, B. 2003. Effects of fear and anger on perceived risks of terrorism: A national field experiment. *Psychological Science, 14*, 144–150.

Lerner, J. S. and Keltner, D. 2001. Fear, anger, and risk. *Journal of Personality and Social Psychology, 81*, 46–159.

Lichtenstein, S., Fischhoff, B., and Phillips, L. D. 1982. Calibration of probabilities. In D. Kahneman, P. Slovic, and A. Tversky (Eds.), *Judgment under uncertainty: Heuristics and biases* (pp. 306–339). New York: Cambridge University Press.

Lichtenstein, S. and Slovic, P. (Eds.). 2006. *Construction of preferences*. New York: Cambridge University Press.

Loewenstein, G. 1996. Out of control: Visceral influences on behavior. *Organizational Behavior and Human Decision Processes, 65*, 272–292.

Loewenstein, G., Weber, E., Hsee, C., and Welch, N. 2001. Risk as feelings. *Psychological Bulletin, 67*, 267–286.

Lopes, L. 1987. Between hope and fear: The psychology of risk. *Advances in Experimental Psychology, 20*, 255–295.

Millstein, S. G. and Halpern-Felsher, B. L. 2002. Judgments about risk and perceived invulnerability in adolescents and young adults. *Journal of Research on Adolescence, 12*, 399–422.

Mitchell, R. C. and Carson, R. T. 1989. *Using surveys to value public goods*. Washington, DC: Resources for the future.

Morgan, M. G., Fischhoff, B., Bostrom, A., and Atman, C. 2001. *Risk communication: The mental models approach*. New York: Cambridge University Press.

O'Brien, M. 2000. Risk analysis. Cambridge, MA: MIT Press.

O'Hagan, A., Buck, C. E., Daneshkhah, A., Eiser, J. E., et al. 2006. *Uncertain judgments: Eliciting expert probabilities*. Chichester: Wiley.

Parker, A. and Fischhoff, B. 2005. Decision-making competence: External validity through an individual-differences approach. *Journal of Behavioral Decision Making*, 18, 1–27.
Plous, S. 1993. *The psychology of decision making*. New York: McGraw-Hill.
Poulton, E. C. 1994. *Behavioral decision research*. Hillsdale, NJ: Erlbaum.
Quadrel, M. J., Fischhoff, B., and Davis, W. 1993. Adolescent (in)vulnerability. *American Psychologist, 48*, 102–116.
Reyna, V. F. and Farley, F. 2006. Risk and rationality in adolescent decision making: Implications for theory, practice, and public policy. *Psychology in the Public Interest,* 7, 1–44.
Rivers, S. E., Reyna, V. F., and Mills, B. 2008. Risk taking under the influence: A fuzzy-trace theory of emotion in adolescence. *Developmental Review, 28*, 107–144.
Rollnick, S. and Miller, W. R. 1995. What is motivational interviewing? *Behavioural and Cognitive Psychotherapy, 23*, 325–334.
Schwarz, N. 1999. Self reports. *American Psychologist, 54*, 93–105.
Shafer, G. and Tversky, A. 1985. Languages and designs for probability judgment. *Cognitive Science, 9*, 309–339.
Simon, H. A. 1957. *Models of man*. New York: Wiley.
Small, D. A., Lerner, J. S., and Fischhoff, B. 2006. Emotion priming and attributions for terrorism: Americans' reactions in a national field experiment. *Political Psychology, 27*, 289–298.
Slovic, P., Peters, E., Finucane, M. L., and MacGregor, D. 2005. Affect, risk and decision making. *Health Psychology, 24*, S35–S40.
Strube, M. J. 1985. Combining and comparing significance levels from nonindependent hypothesis tests. *Psychological Bulletin, 97*, 334–341.
Tengs, T. O. and Wallace, A. 2000. One thousand health-related quality-of-life estimates. *Medical Care, 38*, 583–637.
vonNeumann, J. and Morgenstern, O. 1947. *Theory of games and economic behavior*. Princeton: Princeton University Press.
vonWinterfeldt, D. and Edwards, W. 1986. *Decision analysis and behavioral research*. New York: Cambridge University Press.
Weber, E. U. 1994. From subjective probabilities to decision weights. *Psychological Bulletin, 115*, 228–242.
Woloshin, S., Schwartz, L. M., Byram, S., Fischhoff, B., and Welch, H. G. 1998. Scales for assessing perceptions of event probability: A validation study. *Medical Decision Making, 14*, 490–503.
Yates, J. F. 1989. *Judgment and decision making*. Englewood Cliffs, NJ: Prentice Hall.

17

QUESTIONS OF COMPETENCE

The duty to inform and the limits to choice

Baruch Fischhoff

DEPARTMENT OF SOCIAL AND DECISION SCIENCES, DEPARTMENT OF ENGINEERING AND PUBLIC POLICY, CARNEGIE MELLON UNIVERSITY

Sara Eggers

DECISION PARTNERS

Abstract

Many public policies embody assumptions about some individuals' competence to make some class of decisions. When that competence is underestimated, those individuals can be needlessly denied freedom of choice. When that competence is overestimated, those individuals can be denied needed protections. Questions of competence arise in policies as diverse as whether to make prescription drugs available over the counter, whether to declare martial law in an emergency, and whether to adjudicate adolescents as adults. We offer a general approach to assessing individuals' competence to make specific choices, under the conditions created by specific policies. Grounded in behavioral decision research, the approach characterizes policies by their ability to expand the envelope of competent individual decision making and reduce the need for paternalism.

Introduction

Many of our decisions are shaped by government policies that reflect policy makers' beliefs about our competence to make those choices. For example, policies establishing disclosure requirements for investments and pharmaceuticals reflect beliefs about our competence to recruit and comprehend the relevant evidence. Policies regulating claims made about consumer products and political candidates reflect beliefs about our competence to evaluate them. Policies governing living wills reflect beliefs about our competence to anticipate personally unprecedented circumstances.

The stakes riding on these beliefs are high. If our competence is overestimated, then we may be denied needed protections. If our competence is underestimated, then we may be wrongly denied the right to choose. If ways to enhance our competence are underestimated, then we may lose chances for empowerment. If those opportunities are overestimated, then we may receive useless help (e.g., overly complex financial or medical disclosures) and be held responsible for its failure.

Assessments of decision-making competence are subject to known biases (Gilovich et al., 2002). *Outcome bias* leads to judging decisions by the outcomes that follow them, rather than by the thinking that goes into them. As a result, people facing easy choices (e.g., among places to eat) seem more competent than people facing hard ones (e.g., among medical treatments). *Hindsight bias* leads to exaggerating the competence of people who experience good fortune and underestimating that of those who do not. *Defensive attribution* leads to deprecating the competence of people whom misery befalls, so that observers can feel less vulnerable to suffering a similar fate.

Beliefs about decision-making competence can also reflect motivated thinking (or even deliberate misrepresentation), when a policy's legitimacy depends on the perceived competence of those whose behavior it governs. For example, advocates of market-based policies want to see competent consumers (investors, patients, etc.), as do advocates of participatory policies (e.g., referenda, citizen advisory committees). Advocates of strong regulatory policies want to see incompetent consumers (investors, patients, etc.), needing that protection, as do advocates of policies that empower technocratic elites. Advocates of reproductive rights for adolescents make strong claims for teens' competence; opponents of adjudicating teens as adults make contrary claims (Scalia, 2005). Both cannot be right, at least without more discriminating accounts of the similarities and differences in these decisions and the teens making them (Parker and Fischhoff, 2005).

Sweeping generalizations about decision-making competence can stimulate useful public discourse, by encouraging partisans to assemble and defend examples supporting their positions. However, strategically chosen and interpreted examples are just that: things to consider, not systematic evidence. Sound policies require detailed analyses, able to capture the heterogeneity in both the demands that decisions make and the skills of those making them (Bruine de Bruin et al., 2007).

An approach to competency-based policy making

This chapter offers such a general approach to assessing and, where possible, improving, individuals' competence to make specific decisions, under the conditions created by specific policies. It illustrates the approach with risk-related decisions, in US policy contexts, chosen to suggest the variety of possible incentives and opportunities for implementing policies that enhance public decision-making competence. The examples vary in their topics (including drugs, pathogens, and contaminants), their policy-making locus (including regulators, courts, and emergency officials), and their decision makers (including teens, older men, hobbyists, and everyone). They are presented in roughly decreasing order of how explicit the decisions are, beginning with ones made at a clear point in time and ending with ones embedded in the flow of events.

Our approach follows the "traditional" behavioral decision research strategy (Edwards, 1954; vonWinterfeldt and Edwards, 1986):

(1) *Normative analysis*, identifying the best choices, using the available science to predict the outcomes of possible choices and decision makers' values to weight them.
(2) *Descriptive analysis*, predicting the choices that those individuals would actually make, under the conditions created by possible policies.
(3) *Prescriptive analysis*, characterizing the gap between the normative ideal and the descriptive reality, with each policy.

Determining the prescriptive implications of a normative-descriptive gap requires a value judgment. Among other things, policy makers must weigh the fates of different individuals. They might treat everyone equally or assign weights based on properties like age, health, pregnancy status, citizenship, or historical injustices (Fischhoff et al., 2007). Consider, for example, a choice between policies requiring just English or both English and Spanish on warning labels. Holding font size constant, the former allows more words, so that labels can address more problems or the same problems more thoroughly. However, that policy leaves Spanish-only speakers less protected. A third policy, reducing font size, could accommodate both languages, but lose users with limited vision (or aversion to fine print). A fourth policy, expanding warning label size, could allow more words or larger font, but crowd out benefit information. The prescriptive analysis makes these choices more explicit—which policy makers may or may not welcome.

The approach also allows clarifying the impacts of policies that honor procedural principles, like "freedom of choice," "consumer protection," "chances to learn from experience," or "full disclosure." For example, a First Amendment right to "commercial freedom of speech" has been invoked to expand the range of legal product claims. The result could be positive, if consumers can interpret the claims; negative, if not. "Full disclosure" has been advocated as a way to extract needed information from producers. The result could be positive, if consumers can extract the decision-relevant facts; negative, if the clutter overwhelms them. The approach characterizes such policies by their effects on consumers' ability to make the choices that they govern.

Applying the approach requires contributions from multiple disciplines. Identifying the optimal choice requires decision analysis, informed by behavioral research (capturing individuals' values) and subject matter expertise (regarding expected outcomes). Predicting individuals' choices requires behavioral research, into how individuals interpret the choices that emerge under different policies. Evaluating the gap between the normative ideal and the descriptive reality requires expertise in law and philosophy. Choosing policies requires political judgment, informed by scientific assessments of impacts on outcomes that matter to policy makers.

Although described as sequential, these steps are inherently interdependent. Without knowing individuals' values, analysts cannot identify evidence relevant to their choices. Without knowing policy makers' values, analysts cannot properly disaggregate outcomes (e.g., by age, gender). Without knowing the sources of poor choices (e.g., lack of skills, facts, or motivation), policy makers cannot understand their options. And so on.

The following case studies illustrate the approach with risk-related policies. Other applications include avian flu (Fischhoff et al., 2006), sexual assault (Fischhoff, 1992), nuclear energy sources in space (Maharik and Fischhoff, 1993), nuclear weapons (Florig and Fischhoff, 2007), and sexually transmitted infections (Downs et al., 2004).

The core of each application is a normative model informed by descriptive and prescriptive research. How deeply each component is pursued depends on the policy context. For some applications, rough normative models suffice; for others, quantitative solutions are needed. Some require dedicated behavioral research; others can rely on existing results, showing general tendencies. Some allow testing prescriptive interventions; others barely invite suggestions for change. Because each example responded to a perceived opportunity (and, sometimes, an actual invitation) to influence policy, they do not represent a well-defined universe of policy choices that depend on assessments of competence. Rather, they illustrate the variety of possible applications and their, sometimes surprising, policy implications.

Applications

Saw palmetto: consumers' competence to make decisions created by commercial freedom of speech

Policy context. In the US, dietary supplements are a $14 billion industry, with over half of Americans consuming at least one of the 29,000 products (FDA 2000a). Currently, the law treats supplements as "reasonably expected to be safe, unless adulterated." FDA bears the burden of proof for demonstrating harm and for ensuring that label information is "truthful, non-misleading, and sufficient to communicate any risk." Labels can make non-medical claims without FDA approval, but must be withdrawn, if FDA demonstrates that they are *potentially misleading* to a *reasonable consumer* (FDA 2002b) (emphasis added). Thus, the policy rests on whether consumer behavior meets a normative standard.

Some supplement manufacturers and consumer advocacy groups have argued that consumers are competent to evaluate such claims. In *Pearson v. Shalala* (1999), 164 F.3d 650 (D.C. Cir.), the US Court of Appeals accepted this argument, ruling that consumers might benefit from an unverified claim, if they knew that they had to evaluate it and could do so. One commentator praised the ruling as ending "FDA's paternalistic approach ... based on the counterintuitive notion [that] consumers lack the sophistication necessary to evaluate truthful and non-misleading health information" (Emord 2000, p. 140). If FDA's "counterintuitive notion" is accurate, though, the ruling denies consumers needed protection.

Eggers and Fischhoff (2004) analyzed this question for a supplement central to the litigation: saw palmetto (SP), a dietary supplement that might relieve lower urinary tract symptoms of benign prostatic hyperplasia (BPH), a chronic problem common among older men.

Step 1: Normative analysis. Figure 17.1 shows the decision facing men with those symptoms. The options (the square *choice* node) are (a) consume the recommended saw palmetto dose, (b) consult a physician (seek MD), and (c) do nothing. One uncertain *outcome* (triangular node) is the change in BPH-related symptoms. Others are the health effects of prostate cancer and other conditions with similar symptoms. The circular *chance* nodes capture the associated uncertainties. The probability of the other conditions should not depend on whether men self-treat with saw palmetto. However, it might take longer to discover those conditions, perhaps long enough to affect the chances of effective treatment (Coley et al., 1997).

The scientific evidence on saw palmetto (Marks et al., 2000; Schulz et al., 2002; Wilt et al., 1998) allowed creating a relatively simple decision tree. The tree omits side effects because they are too small to affect choices. It omits costs, because they, too, are small ($10/month). It has a single dose option (the recommended one) because taking more has no known (positive or negative) effects (Ernst, 2002). A "take less" option could be added; it would have less chance of symptom relief and, hence, greater chance of having other problems diagnosed in time. Eggers and Fischhoff (2004) summarize that evidence in decision-relevant terms, taking outcome probabilities from medical evidence and values from studies of utilities for health states (Tengs and Wallace, 2000). The message emerging from the normative analysis is "Saw palmetto might be worth a try, but don't neglect other possible sources of your symptoms." The behavioral analysis asks, in effect, how well men can extract this, easily understood, message from the information available to them.

Step 2: Descriptive analysis. The Court allowed any product claim, short of preventing or treating specific diseases, if accompanied by the disclaimer, "This statement has not been evaluated by the Food and Drug Administration. This product is not intended to diagnose, treat, cure or prevent any disease" (FDA, 2002a). Mason and Scammon (2000) found that people often ignore the disclaimer, hence might believe that FDA had approved the claim. Consumers who notice the disclaimer still might not understand its implications, perhaps assuming that FDA can regulate dietary supplements as stringently as food (GAO, 2000).

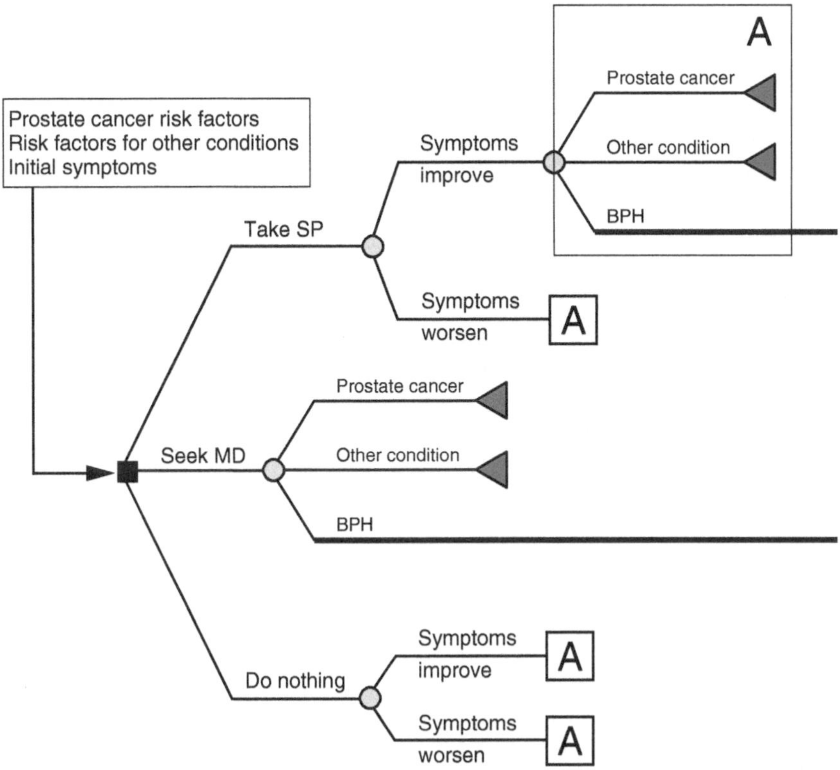

FIGURE 17.1 Decision tree for men considering saw palmetto (SP) as a treatment for lower urinary track problems. Other decision options (square node) are getting medical help (Seek MD) and doing nothing. The primary uncertainty (circular node) for Take SP and Do Nothing is how symptoms will respond. For Seek MD, it is whether the diagnosis is benign prostatic hyperplasia (BPH), prostate cancer, or some other conditions. The outcomes (triangles) depend on the underlying condition and how it responds to treatments, which will depend on how soon it is diagnosed. Box A reflects outcomes with delayed diagnosis.

Source: Eggers and Fischhoff (2004)

Whether such problems are severe enough to undermine men's competency to make these choices is an empirical question. Eggers and Fischhoff (2004) addressed it by asking older men to think aloud as they read, in turn, four labels with increasing detail: (a) *no health claim*, beyond what was inferred from the product name on a green background; (b) an *unqualified claim* of "improving prostate health"; (c) the same health claim, *qualified* by the Court-mandated disclaimer; and (d) *full information*, summarizing decision-relevant evidence in a drug fact box (Schwartz et al., 2009; Woloshin et al., 2008). After reading each label, respondents said whether they would use the product, if they had BPH symptoms, and for how long, if symptoms persisted. Given respondents' extensive study of the materials, we treated their choices after reading the fact box as the ones that they should make. For 55%, that choice was to take saw palmetto.

The interview transcripts provide rich qualitative detail regarding respondents' beliefs about supplements, labeling, and regulation. Most saw the health claim as advertising and the disclaimer as perfunctory (e.g., "for liability protection"). Many interpreted "prostate health" as referring to prostate cancer or sexual function. The (unqualified) health claim prompted many to offer higher estimates of positive effects than with full information. Adding the disclaimer decreased the product's perceived efficacy for some respondents, while increasing it for others, who gave explanations like, "FDA doesn't believe in alternative medicine.' The full information label sometimes increased judged side effects, sometimes reduced them. Almost all respondents said that they would check with their physicians, should symptoms persist. Gades et al. (2005) found that about 50% of men with lower urinary tract symptoms seek medical treatment.

Step 3: Prescriptive analysis. Table 17.1 compares whether respondents *should use* saw palmetto with whether they said that they *would use* it, after reading each label. With No Claim, all those for whom the optimal choice is taking saw palmetto would miss its potential benefits (row 2). With the Unqualified Claim, most of them would use it (44% in row 1 vs. 11% in row 2), as would some for whom it was inappropriate (row 3). Adding the Disclaimer to the Unqualified Claim reduces appropriate choices, by discouraging some men who should try saw palmetto (row 1).

Policy analysis. Table 17.1 predicts the distributions of outcomes with three policies, each embodied by a different label. The policy choice should depend on the weight that regulators assign to each cell. If they weight all cells equally, then they should prefer the Unqualified Claim, which produces the most appropriate choices (row 1 + row 4). They should prefer No Claim, if mistakenly taking the product (row 3) is much worse than missing its potential benefits (row 2). They should never add the Disclaimer, which is worse than the (somewhat misleading) Unqualified Claim.

Consumers are competent to make a decision, if the distribution of outcomes, under a policy, falls within regulators' tolerances. Here, although the Unqualified Claim label dominated the Claim+Disclaimer label, its outcomes would be acceptable only if the 77% correct choices (row 1 + row 4) outweighed the 22% incorrect ones (row 2 + row 3). Having 11% take saw palmetto inappropriately seems of little consequence, given that almost all respondents said that they would see a doctor, if symptoms persisted, reducing the opportunity costs of trying saw palmetto. Given its limited efficacy, not much is lost by having another 11% fail to give it a warranted try. By the same logic, even the confusing disclaimer might allow competent choices, with this benign product and such moderately engaged, informed, and skeptical consumers.

TABLE 17.1 Predicted optimality of consumer choices with alternative saw palmetto labels

Normative Decision	Predicted Decision	Label		
Should Use	Would Use	No Claim	Unqualified Claim	Claim+Disclaimer
Yes	**Yes**	**0%**	**44%**	**27%**
Yes	No	55	11	27
No	Yes	0	11	13
No	**No**	**45**	**33**	**33**

Note. The optimal decision (should consume) reflects the choices of respondents who studied the full disclosure label. Bold indicates appropriate decisions. See Eggers and Fischhoff (2004) for details.

A product that might produce similar conclusions is black cohosh, a supplement that might reduce menopausal symptoms. Its decision tree resembles that of Figure 17.1, with the only major risks arising from delayed treatment. Behavioral research might reveal similar consumer competence, perhaps reflecting heuristics like "Supplements are worth a try, but not for too long, if they're not working' and "Tell your doctors what you're taking, even if you expect them to be skeptical." With other products, though, the disclaimer might exact a high price, if (as seen here) consumers ignore it, brush it off as perfunctory, or have it remind them about the possibilities of alternative medicine. With most dietary supplements, the risks and benefits are unknown. Saw palmetto and black cohosh are unusual in having been studied enough to be able to estimate their effects.

Schwartz et al. (2009) found that most people can understand fact boxes, summarizing the risks and benefits of possible treatments, like that used in the Full Information label. Consumers are better served by policies that require providing the facts needed to make sound choices, rather than just claims and disclaimers that "work well enough" in forgiving situations. Developing such messages requires analytical and empirical research. It should not be left to intuition, even that of well-intended jurists.

Plan B morning-after pill: adolescents' competence to make reproductive decisions

Policy context. Plan B is an emergency contraceptive (EC) pill that reduces the probability of pregnancy, if two doses are taken within 72–120 hours of unprotected sex (von Hertzen et al., 2002). In August 2000, US FDA approved

OTC sale of EC to women aged 18 and over, in outlets that have accredited pharmacies and avoid selling it to younger women. In 2003, the drug's manufacturer, Barr Pharmaceuticals, petitioned FDA to approve OTC status for all women, in less restricted outlets. Such approval requires demonstrating the drug's safety and efficacy without physician supervision (Pub. L. No. 82–215, 65 Stat. 648, 1951). The producer submitted clinical trial data, along with behavioral evidence regarding label comprehension and use, under simulated OTC conditions. An advisory panel (Sherman, 2004) recommended OTC status, by a vote of 23 to 4 (FDA, 2003). However, FDA's Center for Drug Evaluation and Research (CDER), which governs prescription-to-OTC switches, denied approval (FDA, 2004). In his explanatory note, CDER's acting director objected to extrapolating behavioral data from older adolescents to ones under 16, who might exhibit "impulsive behavior, without the cognitive ability to understand the etiology of their behavior (GAO, 2005, p. 46)." He also expressed concern about "the potential impacts that the OTC marketing of Plan B would have on the propensity for younger adolescents to engage in unsafe sexual behaviors due to their possible lack of cognitive maturing compared to older adolescents (GAO, 2005, p. 5)."

The parties to this rulemaking invoked opposing theories of teen decision-making competence. Critics of OTC status predicted that making Plan B more available would encourage unprotected intercourse, with teens knowing that they can still prevent pregnancy – so that Plan B becomes Plan A. Sexually transmitted infections (STIs) would then increase, even if unplanned pregnancies did not. Some critics worried about increased intercourse among unmarried teens per se, an outcome with no status under the law (FDA, 2003). In contrast, proponents of OTC status argued that teens were sufficiently competent as decision makers that they would not only maintain their current contraceptive practices, but make better choices given the additional option. These advocates also predicted that reducing unintended pregnancies would reduce abortions.

Step 1: Normative analysis. The deliberations over Plan B addressed two decisions within FDA's statutory public health mission: how adolescents choose to protect themselves against (a) unwanted pregnancy and (b) STIs. Some panel members seemingly considered decisions outside the statutes: whether adolescents have sex and abortions. Figure 17.2 analyzes one decision whose options and outcomes clearly fall within FDA's mandate: whether women use Plan B after suspected contraceptive failure. Krishnamurti et al. (2007) provide additional details on this choice and two others: whether women have sex and what protection they use, if they do.

Predicting the outcomes in these normative analyses is generally straightforward, as the health outcomes are often well studied. Evaluating the outcomes is more difficult. Some outcomes are valued similarly by most women, hence might be assessed with standard utility measures; others depend on the woman facing them. Some women, in some life situations, find unplanned pregnancies

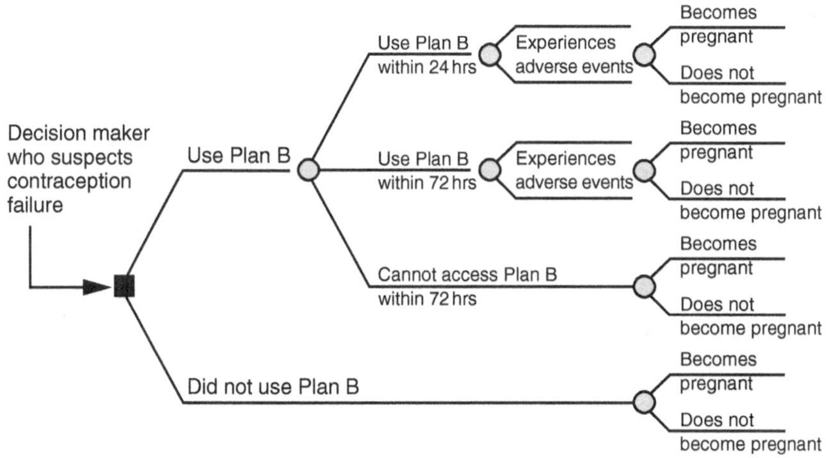

FIGURE 17.2 Decision tree for women considering the use of emergency contraception, contingent on suspecting failure of contraceptive method.

Source: Krishnamurti, Eggers and Fischhoff (2008)

enormously disruptive; others take them in stride. Some women view pregnancies avoided by Plan B as abortion; others do not. Teen (or unmarried) sex is acceptable to some people, sinful to others. These normative analyses used the values of the women making them. Decisions optimal for them might be unacceptable to people with different values (e.g., regarding the morality of extramarital sex or various birth control options).

Step 2: Descriptive analysis. Krishnamurti et al. conducted semi-structured, open-ended interviews with 30 adolescents from high-risk populations, whose ages straddled 16 years old (critical to FDA's decision). The interviews followed a mental-models protocol (Morgan et al., 2001), which directs respondents increasingly toward issues in the normative analysis. Such interviews have the statistical power to reveal large group differences. Their results guide the design of structured surveys suited to larger samples (which confirmed the results reported here).

As with any other study, the results of these interviews should be interpreted in the context of general scientific knowledge. In this case, a broad summary (Fischhoff, 2008; Reyna and Farley, 2006) might be that, by the mid-teen years, most adolescents have roughly the cognitive decision-making skills of adults. GAO (2005) noted that FDA's Plan B decision departed from earlier rulings that had found it "scientifically appropriate to extrapolate data from older to younger adolescents" (p.5), consistent with the research.

Whatever their skills, teens' knowledge of a topic depends on their opportunities to learn about it, through exposure to information (e.g., instruction, media, word-of-mouth) and experience. Krishnamurti et al. found that their respondents generally said that they (a) knew about Plan B; (b) would not use it as their primary birth-control method (usually citing discomfort or cost); (c) would consider it, if they suspected an unplanned pregnancy; (d) saw barriers to access that expanded OTC status would reduce; and (e) had considered whether Plan B constituted abortion. Based on these accounts, Krishnamurti et al. concluded that expanded OTC status would not affect sexual behavior or the choice of primary birth control method; however, it would increase Plan B use when needed. They also found considerable confusion about the timing of usage, with more teens underestimating the effective period. The only observed difference between teens over and under 16 was that younger teens saw greater costs and less benefits to Plan B, hence were less likely to use it.

Step 3: Prescriptive analysis. Based on these interviews, expanded OTC availability would affect only one decision: Plan B use after unprotected sex. There, it would allow some young women to achieve a desired end that is currently infeasible. Although the teens' decision-making processes were thoughtful, they were not always well-informed, particularly regarding Plan B's effective period. Women who underestimate effectiveness will not receive full benefit from greater availability. Those who overestimate it will not receive the benefits that they expect.

Policy analysis. In terms of the health outcomes in FDA's regulatory mandate, these results suggest that expanded OTC availability dominated limited availability: There should be fewer unplanned pregnancies, with no changes in other health effects. The largest gap in teens' knowledge, the period of EC effectiveness, should not be hard to close, with a suitable communication channel. Emotion might, in principle, undermine the cognitive competencies seen here (and in other studies). In practice, though, it seems that OTC availability should reduce emotional pressure in the one decision where EC seems to play a role: dealing with unplanned unprotected sex.

Going beyond FDA's mandate allows considering abortion-related issues. In the study, most respondents who were opposed to abortion thought that Plan B was used too soon after intercourse to constitute abortion. For them, expanded availability would allow more optimal choices. That conclusion should be rejected by individuals who oppose abortion and place conception at intercourse. They should fault such teens' beliefs (about how Plan B works), but not their cognitive abilities.

Carotid endarterechtomy: medical informed consent

Policy context. In the US, roughly half of the states hold physicians to a *materiality* standard, when securing informed consent. That is, they must ensure that patients understand all facts material to their decisions. (The other states have a *professional* standard, requiring adherence to common practice.) That standard implies a normative analysis, ordering information by its materiality, signaling when physicians can stop communicating, because they have conveyed all that patients need to know. Having that standard should provide some protection against unfair legal judgments, when things go wrong and patients claim that they were not adequately informed. Merz et al. (1993) offer an approach to formalizing the materiality standard, which they illustrate with a common surgical procedure whose large risks and benefits have engendered extensive research on expected outcomes.

Carotid endarterechtomy, scraping out the artery to the brain, can reduce stroke risk for patients with atherosclerosis. However, many things can go wrong, ranging from death and iatrogenic strokes to headaches and broken teeth. Some risks are specific to the procedure; others are part of major surgery. The full list is a lot to consider, when facing difficult tradeoffs between these risks and the expected benefits of treating a life-threatening illness. "Informed consent" is a vague policy without specifying what patients need to know.

Step 1: Normative analysis. The method creates hypothetical patients, for whom surgery would be the optimal choice were there no risks (and were money no object). These "patients" vary in their physical condition, represented by probability distributions over possible outcomes, and in their personal values, represented by utility distributions over the outcomes. Patients are created by sampling values from these distributions (with the simplifying assumption of independence, in this application). The expected utility of surgery is calculated for each such patient, ignoring all risks. It is positive, given how because the population is created. Expected utility is then recalculated, incorporating knowledge of each possible side effect. A side effect's materiality is defined as how often it gives the surgery negative expected utility, so that it is no longer recommended.

Merz et al. found that only three side effects should matter to many of these simulated patients: About 15% should decline surgery, if told the risk of immediate death; another 5%, if told the risk of iatrogenic stroke; and 3% more, if told the risk of facial paralysis. Learning about other risks would tip the scales for very few patients.

Step 2: Descriptive analysis. Because few candidates for this surgery have faced it before, they cannot be expected to have authoritative risk information. As a result, Merz et al. (1993) conducted no behavioral research, assuming that all information would have to be conveyed. This strategy would be flawed, if patients held strong beliefs about other risks that communications about the three focal risks would ignore (Downs et al., 2008). In cases where there are many material facts, assessing current beliefs allows identifying those that go without saying—and can be skipped in favor of new ones.

Step 3: Prescriptive analysis. Thus, effectively conveying the three major side effects would keep 15–20% of these potential patients from undergoing surgery that is suboptimal for them (because its risks outweigh its benefits). Based on risk-communication research (Fischhoff, 2009; Woloshin et al., 2008), that task seems manageable. Each critical side effect involves a fairly familiar event (death, stroke, facial paralysis), with a fairly large probability (not, say, 1/46,000). Thus, it should be possible to afford patients the knowledge needed for competent decision making. A greater prescriptive challenge may be helping patients to be rational enough to take advantage of that information. The empirical-analytical question is how sensitive these choices are to imperfect information integration (vonWinterfeldt and Edwards, 1986).

Policy analysis. If the materiality standard is implemented in this way, the duty to inform is fulfilled once the three critical side effects are communicated. Doing so absolves physicians of responsibility for suboptimal choices that reflect poor information integration, rather than poor understanding. Focusing communication on the most material facts should improve patients' information integration by protecting them from "full disclosure" practices that drown them in immaterial facts. Full disclosure is needed to perform the analyses that identify the critical risks (and to ensure that nothing is hidden). However, forcing patients to sort through all the details can undermine their competence.

Materiality analysis can also help to implement research-funding policies, by assessing the value of potential results for patient decision making. Those analyses could help patients decide whether to wait for clinical trial outcomes.

Methylene chloride-based paint stripper: consumer competence to make decisions created by voluntary self-regulation

Policy context. In the mid-1990s, the International Agency for Research on Cancer declared the solvent methylene chloride a probable carcinogen. Methylene chloride also gives off carbon monoxide, which can cause heart attacks in confined spaces. One use of the solvent with no clear substitute is stripping paint. Although industrial users might be required to take protective measures (e.g., ventilating hoods, respirators), home users must protect themselves. If they can, then it is safe to leave paint stripper on the market. That depends on their competence to make decisions about how to use the product and about whether it has acceptable risks relative to its benefits.

Step 1: Normative analysis. Riley et al. (2001) characterized consumers' health risks with an analysis sensitive to their usage patterns. It predicted consumer exposures based on physical principles (e.g., air circulation, chemistry), whose interactions had been calibrated under laboratory conditions, and on the effects of actions that users might decide to take (e.g., open windows, wait in corner while stripper is curing). The analysis created a *supply curve*, ordering the actions by decreasing marginal effectiveness for reducing exposure. That order is the logical one for adopting risk-control measures (assuming similar costs). The analysis found that, for many jobs (varying in room size, duration, etc.), consumers could greatly reduce their risks by deciding to implement two simple actions: opening a window and having a fan blow outward.

Step 2: Descriptive analysis. Interviews with paint stripper users found that they (a) were motivated to use control measures, (b) could easily understand the two key measures, and (c) seemed realistic about their ability to execute them (e.g., "I'm not going to open a window in the winter, however bad the fumes"). However, those actions were sufficiently unintuitive that consumers needed to have them explained. For example, without instruction, many would have fans blow inward, in order to feel the airflow, not realizing that internal circulation left concentrations unchanged. Many also reported choosing actions whose ineffectiveness could be easily explained (e.g., gloves get ruined because the solvent dissolves them; fumes diffuse, so don't bother crossing the room while the solvent cures). Thus, users appeared competent to make sound choices, if provided the relevant information.

Step 3: Prescriptive analysis. In lieu of in-home observation of consumers' usage decisions, Riley et al. (2001) estimated exposures for users who understood and conscientiously followed everything they read, but had different labels and reading patterns (e.g., just the instructions, just the warnings, just the bolded material, just the first five items). The analyses found wide variation in exposures, across labels and reading patterns. Some labels provided useful information, whatever the reading strategy; some had no information on reducing exposures anywhere. Thus, consumers are competent to make sound usage decisions, if they receive relevant information about how to use the product. Without that information, they cannot know what risks they are taking or whether to use the product. Predictions of actual label-reading behavior, and resultant risk levels, could use direct observation or general patterns (Wogalter, 2006).

Policy analysis. Flawed communications limit otherwise competent consumers' ability to make effective choices. As a result, they may purchase unduly risky products and use them in needlessly risky ways. Or, consumers may forego useful products and take needless precautions. When a product's risks and benefit depend on how it is used, its label is as much a part of the product as are its physical constituents. Regulators should want to know what consumers take away from labels, so that they can ensure proper protections. Producers should want the same knowledge, so that they can help consumers get the greatest value from their products—and defend themselves against charges of failing to fulfill their duty to inform. Individual firms produced the labels in this study. The great variation in the value of the information that they provided (and the associated risks) suggests that such voluntary self-regulation was inadequate.

Cryptosporidium: consumer competence to cope with contamination emergency

Policy context. Cryptosporidium is a common protozoan parasite, with mammalian hosts, that can infect public water supplies, typically through uncontrolled sewage discharges and fecally contaminated runoff after heavy storms (e.g., from feedlots, deer). Typical water treatment systems cannot fully remove or deactivate it. Symptoms, appearing after 1–7 days, include nausea, vomiting, diarrhea, cramps, and low fever. Although cryptosporidiosis has no medical cure, most infected individuals recover, many without exhibiting symptoms. However, the disease, which attacks the liver, can be fatal to immunocompromised individuals (e.g., those with AIDS). Water and public health authorities have a duty to inform consumers so that they can make competent water-usage decisions, namely when and how to use boiled or bottled water.

Step 1: Normative analysis. Casman et al. (2000) created a model predicting the health effects of a *Cryptosporidium* intrusion. The model includes inputs from microbiology (dose-response relationships), civil engineering (filtration, testing), ecology (upstream land use), communication research (dissemination of "boil water" notices), and psychology (perceived risk, actual response). It allows assessing when consumers are competent to make water-usage decisions, under the conditions created by different intrusion scenarios. Among other things, it examines how quickly they receive messages, how adequately they boil water, and how much they rely on personal testing (which has little value).

Step 2: Descriptive analysis. The model uses estimates from observational studies of consumers' water-use decisions in past intrusions, which have found that people often use improperly treated water. New interviews found that the main sources of these poor decisions were ignorance (about how to boil water well enough to destroy the parasite) and suspicion (about how seriously to take warnings). Both the procedures and the context are simple enough that it should be possible to explain them well enough to allow most people to make competent choices. As a result, the competence of consumer choices depends on whether they get sound messages in time to act.

Step 3: Prescriptive analysis. The model was used to predict health effects under the conditions created by plausible intrusion scenarios—varying, for example, in how long it took to detect and repair problems. One set of simulations examined the effects of ensuring that all consumers received sound messages, as soon as an intrusion was established. Such communication was found to have no effect, a result that was traced to *Cryptosporidium* testing being too slow for messages to arrive in time to prevent exposures. As a result, relying on "boil water" notices ensures suboptimal choices, by individuals who could be competent, given the right information.

Policy analysis. In this case, the analysis revealed an inherent flaw in a standard policy, relying on consumer decision making to manage risks—a faith that deflected attention from other possible policies. One other possibility is improving the speed of detection, so that officials can provide timely warning. Another possibility is reducing the risk, through better land use or water purification. A third possibility is routinely providing highly vulnerable populations with safe water. Repeated with contaminants allowing rapid detection (e.g., some *E. coli* strains), the analysis might reveal that timely, comprehensible warnings would allow effective consumer decision making. Consumers should not be blamed for negative outcomes, when they cannot help themselves.

Emergency evacuation: citizen competence to make voluntary decisions

Policy context. Mass emergencies can be defining moments for a society (Boin et al., 2005), in which leaders' decision-making competence will be examined intensely. One aspect of that examination will be how their actions affected their citizens' decision making. Were citizens afforded the information needed to make effective choices? Did they have the resources needed to act on that information? Were they treated like competent adults? Was martial law imposed when they could have managed without it?

Emergency plans must make some assumptions about the public's decision-making competence under stressful conditions. One such class of hazards is terror attacks with contaminating materials, like radioactive dispersion devices (RDDs), or "dirty bombs." RDDs use ordinary explosives to spread radioactive materials, causing immediate casualties from the blast, long-term casualties from radiation poisoning, and potentially great social and economic costs. The extent of that disruption will depend on their leaders' perceived competence. That will depend, in turn, on how well those leaders assess their public's competence. If they expect too much, then the public will be denied needed protection. If they expect too little, the public will be denied deserved freedom.

Step 1: Normative analysis. Dombroski et al. (2006a, b) developed a general model for predicting the health effects of RDD attacks, with one predictor being citizens' decisions about evacuating or sheltering in place. The model incorporates features from research into explosive impacts, aerosol dispersion, traffic flows, dose–response relations, and so on. Choosing values for model parameters (e.g., location, time of day, explosive force, contaminant, weather) produces attack scenarios specific enough to predict morbidity and mortality (from which economic and social effects might be predicted). For public health officials, the optimal response minimizes those health effects. One key decision is whether to recommend evacuation or sheltering in place. For individual citizens, the optimal response minimizes those risks, subject to other concerns (e.g., protecting family members, helping co-workers, demonstrating resilience). For them, one key decision is whether to follow that recommendation.

Step 2: Descriptive analysis. Estimates for most model parameters were taken from the research literature. However, although there are many studies of emergency behavior, they are rarely in model-ready form. Therefore, judgments of the model's behavioral parameters were elicited from ten social science experts and 36 local disaster specialists. These experts predicted behavior for variants of a scenario involving a 10-kilo Cs-137 RDD, exploded at Pittsburgh's USX Tower, at 10am on a summer weekday. Their judgments included the percentages of citizens complying with instructions to evacuate or shelter in place, when at home and at work. These experts generally agreed about citizens' decisions. Consistent with historical experience, although not with popular myth (Tierney, 2005; Wessely, 2005), the experts expected no panic. Rather, they expected most people to follow instructions, with higher rates for sheltering at home and evacuating from work.

Step 3: Prescriptive analysis. Incorporating the experts' judgments in the model revealed that, for this scenario, the predicted rates of compliance with official instructions (60–80%) are good enough to minimize health effects. Enough people would shelter in place to keep the roads open enough to allow first responders to treat those injured in the blast, while not trapping evacuees in a radioactive cloud. Little would be gained by making the recommendation compulsory.

Policy analysis. These results suggest that citizens are competent enough to make choices that achieve generally optimal outcomes. Thus, voluntary compliance should satisfy a consequentialist regulatory philosophy. It should have additional procedural value for demonstrating faith in the public, compared to compulsory policies, like martial law. That value underlies the commitment to, "Keep the public fully informed—tell what we know, tell what we don't know, and tell it often. … Maintain credibility and public trust, by providing accurate, science-based information" (DHHS, 2006). Producing and disseminating useful information is one way to earn trust.

Strategies for competence assessment

Reprise

More or less the same kinds of (ordinary) people, with more or less the same general decision-making skills, were involved in each of these examples. Yet, their competence emerges differently in each, depending on the difficulty of the choice and the adequacy of others' attempts to fulfill a duty to inform. Normative, descriptive, and prescriptive analyses allowed evaluating proposed policies, in terms of how well they fit the competence that individuals bring to them and to design better ones.

According to these analyses:

With saw palmetto, the policy of allowing any non-health claim, if accompanied by a court-mandated disclaimer, created decisions that consumers were competent to make. That occurred despite the disclaimer's flaws, as a result of the product being benign and consumers being skeptical. Taking saw palmetto should not hurt them or lead them to delay medical care too long. The disclaimer's main impact is leading some consumers to forgo a product that might help them. With other products and consumers, the policy might have much worse effects.

With Plan B, young women should be competent to make the decisions created by the policy of expanded OTC availability. It should help them to avoid unwanted pregnancies, without violating their abortion views.

Expanded availability should not affect their decisions about sexual behavior or contraceptives.

With carotid endarterechtomy, most patients should be competent to make the decisions created by a policy of focusing patient briefings on the most material facts. In states with a materiality standard, conveying those few, simple facts might allow physicians to claim to have secured informed consent.

With methylene chloride paint stripper, most consumers would not be competent to make the decisions created by the current policy of allowing producers to design their own labels. However, the critical facts are simple: decide to use the product only if you can have a fan blow outward through an open window. Thus, a policy that mandated labels with that information should allow competent choices.

With *Cryptosporidium*, consumers are not competent to make water usage decisions, under the conditions created by a policy of relying on them to protect themselves—because critical information will not arrive in time. In effect, such a policy asks consumers to do the impossible, whatever their decision-making skills. It might allow them to make competent choices about contaminants with faster testing.

With the radioactive dispersion device scenario, citizens are competent to decide whether to obey recommendations to evacuate or shelter in place. Trusting that competence should enhance citizens' trust in their authorities. Thus, a policy of making clear recommendations should dominate draconian policies, like martial law.

Organizing for assessing (and improving) decision-making competence

Sweeping claims about decision-making competence cannot do justice to the diversity of decisions and decision makers. Blanket claims of competence leave some people without needed protections. Blanket claims of incompetence deprive some of deserved freedoms. Blanket claims create the temptation of working backward from desired policies to behavioral assumptions that justify them. An incompetent public suits those who favor strong regulations and technocratic management. An incompetent public suits those who favor free markets and participatory processes. A disciplined approach, combining empirical and analytical research, is needed to determine the legitimacy of such policies.

Executing this approach to assessing (and improving) decision-making competence requires an interdisciplinary team, with

a. *Subject-matter experts*, able to identify the decision options and characterize the processes determining their effects;
b. *Decision analysts*, able to estimate the risks and benefits of those options, showing the facts most relevant to decision making—and competence;

c. *Social scientists*, able to assess decision makers' beliefs and values, guide attempts to enhance competence, and evaluate their success; and
d. *Designers*, able to implement measures that achieve theoretically possible competence.

Assembling such a team requires leadership. Often, policy-making organizations are dominated by experts from one field, with little interest in collaborating with others. They may want to avoid sharing scarce resources. They may not recognize the limits to their own expertise. When such organizations expand, they face the challenge of evaluating unfamiliar expertise. If they cannot tell "what good is" and select poor representatives of another field, they may both get poor advice and devalue that field.

Once present, these experts must be coordinated. A strong team will accept ideas from anyone. However, it must assign responsibility for each task to the appropriate experts. Subject-matter experts can best predict the outcomes that concern decision makers, but social scientists are needed to assess what those outcomes are. Subject-matter experts will know which facts are central to their professional community, but decision analysts are needed to determine their relevance. System designers will know how to get messages out, but social scientists are needed to determine how well their content is absorbed. Social scientists will know how difficult tasks are, but decision analysts are needed to assess how sensitive decisions are to those effects.

Once assembled and coordinated, the competence-assessment team must be integrated into the policy-making process. Figure 17.3 shows an organizational model with that goal. Taken from the quasi-governmental Canadian Standards Association (1997), it is consistent with recommendations from HM Treasury (2005), National Research Council (1996), and the Presidential/Congressional Commission on Risk (1997), among other bodies. The center of the figure depicts a standard policy-making process, unusual only in evaluating each stage, before proceeding to the next (with the implicit possibility of never finishing). Notably, each stage requires two-way risk communication. For example, the Initiation stage entails experts learning which issues matter to those whom a policy affects and telling them how those issues will be addressed. These communications might be direct or indirect, with social researchers soliciting views and conveying results.

Such a process allows considering decision-making competence early enough to shape the design of policies and mid-course corrections. It invites recruiting the kinds of expertise needed to make the work *behaviorally realistic*, in its assumptions about individuals' ability to secure, comprehend, and use information, and *analytically sound*, in its sensitivity to the heterogeneity in people's abilities and decisions. Without such integrated expertise, it is impossible to do justice to individuals' needs and limitations, and create policies that afford them as much autonomy as they want and can handle.

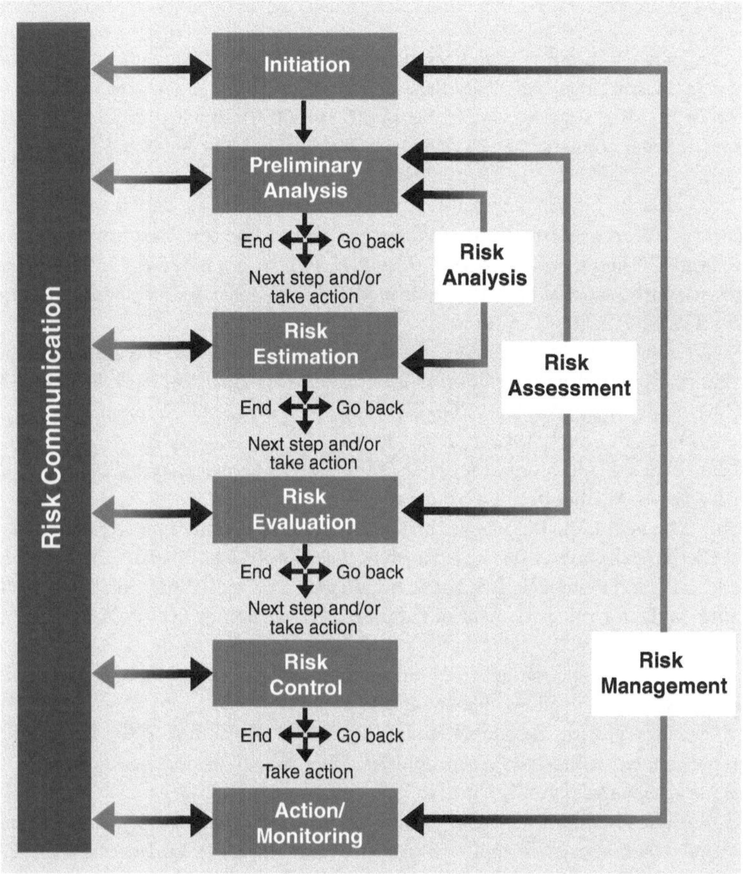

FIGURE 17.3 Steps in the Q850 risk management decision-making process—simple model.

Source: Canadian Standards Association (1997)

Note: Risk communication with stakeholders is an important part of each step in the decision process.

Acknowledgements

Chapter preparation supported by US National Science Foundation Grant SES 0433152 and the Veterans Administration Center for Health Equity Research and Policy. We thank Wändi Bruine de Bruin, Julie Downs, Irene Janis, Valerie Reyna, and two anonymous reviewers. The views expressed are those of the authors.

References

Boin, A., 't Hart, P., Stern, E., and Sundelius, B. 2005. *The politics of crisis management.* Cambridge: Cambridge University Press.

Bruine de Bruin, W., Parker, A., and Fischhoff, B. 2007 Individual differences in adult decision-making competence (A-DMC). *Journal of Personality and Social Psychology. 92,* 938–956.

Camerer, C., Issacharoff, S., Loewenstein, G., O'Donoghue, T., and Rabin, M. 2003. Regulation for conservatives: Behavioral economics and the case for "asymmetric paternalism." *University of Pennsylvania Law Review, 151,* 1211–1254.

Canadian Standards Association. 1997; reaffirmed 2002. *Risk management guidelines for decision makers.* (CSA-50). Ottawa: Author.

Casman, E., Fischhoff, B., Palmgren, C., Small, M., and Wu, F. 2000. Integrated risk model of a drinking waterborne *Cryptosporidiosis* outbreak. *Risk Analysis, 20,* 493–509.

Coley, C.M., Barry, M.J., Mulley, A.G., Thibault, G.E., et al. 1997. Screening for prostate cancer. *Annals of Internal Medicine, 126,* 480–84

Department of Health and Human Services. 2006. *DHHS Communication Plan for First Case of H5N1 in US.* Washington, DC: Author

Dombroski, M., Fischhoff, B., and Fischbeck, P. 2006. Predicting emergency evacuation and sheltering behavior: A structured analytical approach. *Risk Analysis, 26,* 501–514.

Dombroski, M.J. and Fischbeck, P.S. 2006. An integrated physical dispersion and behavioral response model for risk assessment of radiological dispersion device (RDD) events. *Risk Analysis, 26,* 501–514.

Downs, J. S., Bruine de Bruin, W., and Fischhoff, B. 2008. Patients' vaccination comprehension and decisions, *Vaccine. 26,* 1595–1607.

Downs, J. S. Murray, P. J., Bruine de Bruin, W., White, J. P., et al. 2004. An interactive video program to reduce adolescent females' STD risk: A randomized controlled trial. *Social Science and Medicine, 59,* 1561–1572.

Eggers, S. L. and Fischhoff, B. 2004. Setting policies for consumer communications: A behavioral decision research approach. *Journal of Public Policy & Marketing, 23,* 14–27.

Emord, J.W. 2000. Pearson v. Shalala: The beginning of the end for FDA speech suppression. *Journal of Public Policy and Marketing, 19*(1): 139–43.

Ernst, E. 2002. The risk–benefit profile of commonly used herbal therapies: Ginkgo, St. John's Wort, ginseng, echinacea, saw palmetto, and kava. *Annals of Internal Medicine, 136* (1), 42–53.

FDA. 2000a. *Dietary supplement claim for saw palmetto extract and Benign Prostatic Hyperplasia: Denied.* (accessed May 2002), [available at http://www.cfsan.fda.gov/~dms/dspltr01.html].

FDA 2002a. "Regulations on Statements Made for Dietary Supplements Concerning the Effect of the Product on the Structure or Function of the Body," Docket No. 98N–0044, Federal Register, 65 (4), (January 6).

FDA 2002b. "Guidance for Industry: Structure/Function Claims Small Entity Compliance Guide," (accessed January 2003), [available at http://www.fda.gov].

FDA. 2003. *Nonprescription Drugs Advisory Committee in joint session with the Advisory Committee for Reproductive Health Drugs, Meeting Transcript.* http://www.fda.gov/ohrms/dockets/ac/03/transcripts/4015T1.htm

FDA. 2004. *FDA's decision regarding Plan B: Questions and answers.* Accessed January 2006 at http://www.fda.gov/cder/drug/infopage/planB/planBQandA.htm.

Fischhoff, B. 1992. Giving advice: Decision theory perspectives on sexual assault. *American Psychologist, 47*, 577–588.

Fischhoff, B. 2008. Assessing adolescent decision-making competence, *Developmental Review, 28*, 12–28.

Fischhoff, B., Atran, S., and Fischhoff, N. 2007. Counting casualties: A framework for respectful, useful records. *Journal of Risk and Uncertainty, 34*, 1–19.

Fischhoff, B., Bruine de Bruin, W., Guvenc, U., Caruso, D., and Brilliant, L. 2006. Analyzing disaster risks and plans: An avian flu example. *Journal of Risk and Uncertainty. 33*, 133–151.

Florig, K. and Fischhoff, B. 2007. Individuals' decisions affecting radiation exposure after a nuclear event. *Health Physics, 92*, 475–483.

Gades, N.M., Jacobson, D.J., Girman C.J., Roberts, R.O., et al. 2005. Prevalence of conditions potentially associated with lower urinary tract symptoms in men. *BJU International, 95*, 549–53.

GAO 2000. *Food safety: Improvements needed in overseeing the safety of dietary supplements and "functional foods."* Washington, DC: Author.

GAO 2005. *Decision process to deny initial application for OTC Marketing of the emergency contraceptive drug Plan B was unusual.* GAO-06–109. Washington, DC: Author.

Gilovich, T., Griffin, D., and Kahneman, D. (Eds.). 2002. *The psychology of judgment: Heuristics and biases.* New York: Cambridge University Press.

Her Majesty's Treasury. 2005. *Managing risks to the public.* London: Author.

Krishnamurti, T.P., Eggers, S.L., and Fischhoff, B. 2008. The effects of OTC availability of Plan B on teens' contraceptive decision-making. *Social Science and Medicine, 67*, 618–627.

Maharik, M. and Fischhoff, B. 1993. Public views of using nuclear energy sources in space missions. *Space Policy, 9*, 99–108.

Marks, L.S., Partin, A.W., Epstein, J.I., Tyler, V.E., et al. 2000. Effects of a saw palmetto herbal blend in men with symptomatic Benign Prostatic Hyperplasia. *Journal of Urology, 163*, 1451–1456.

Mason M.J. and Scammon, D.L. 2000. Health claims and disclaimers: Extended boundaries and research opportunities in consumer interpretation. *Journal of Public Policy and Marketing, 19*, 144–50

Merz, J., Fischhoff, B., Mazur, D.J. Fischbeck. P.S. 1993. Decision-analytic approach to developing standards of disclosure for medical informed consent. *Journal of Toxics and Liability, 15*, 191–215.

Morgan, M., Fischhoff, B., Bostrom, A., and Atman, C. 2001. *Risk communication: The mental models approach.* New York: Cambridge University Press

National Research Council. 1996. *Understanding Risk.* Washington, DC: National Academy Press.

Parker, A. and Fischhoff, B. 2005. Decision-making competence: External validity through an individual-differences approach. *Journal of Behavioral Decision Making, 18*, 1–27.

Presidential/Congressional Commission on Risk. 1997. *Risk management.* Washington, DC: Author.

Reyna, V.F. and Farley, F. 2006. Risk and rationality in adolescent decision making. *Psychological Science in the Public Interest, 7(1)*, 1–44.

Riley, D.M., Fischhoff, B., Small, M., and Fischbeck, P. 2001. Evaluating the effectiveness of risk-reduction strategies for consumer chemical products. *Risk Analysis, 21*, 357–369.

Scalia, J. 2005. Dissent. Roper v. Simmons. 543 US 03–633. Supreme Court of the United States.

Schulz, M.W., Chen, J., Woo, H.H., Keech, M., et al. 2002. A comparison of techniques in patients with Benign Prostatic Hyperplasia. *Journal of Urology, 168* (July), 155–59.

Schwartz, L.M., Woloshin, S., and Welch, H.C.G. 2009. Using a drug facts box to communicate drug benefits and harms. *Annals of Internal Medicine, 150,* 516–527.

Sherman, L.A. 2004. Looking through a window of the FDA. *Preclinica, 2(2),* 99–102.

Tengs T.O., Wallace A. 2000. One thousand health-related quality-of-life estimates. *Medical Care* 38: 583–637

Tierney, K.J., Lindell, M., and Perry, R.W. 2001. *Facing the unexpected: Disaster preparedness and response in the US.* Washington, DC: National Academy Press.

Von Hertzen, H., Piaggio, G., Ding, J., Chen, J., et al. 2002. Low dose mifepristone and two regimens of levonorgestrel for emergency contraception: A WHO multicentre randomised trial. *Lancet, 360,* 1803–10.

Wessely, S. 2005. Don't panic! Short- and long-term psychological reactions to the new terrorism. *Journal of Mental Health, 14*(1), 106.

Wilt, T., Ishani, A., Stark, G., MacDonald, R., et al. 1998. Saw palmetto extracts for treatment of Benign Prostatic Hyperplasia. *Journal of the American Medical Association, 280,* 1604–09.

Wogalter, M. (ed.) 2006. *The handbook of warnings.* Hillsdale, NJ: Lawrence Erlbaum Associates.

Woloshin S., Schwartz L,. and Welch G.H. (2009). *Know your chances.* Berkeley, CA: University of California Press.

CONCLUSION

Baruch Fischhoff

I have been privileged to be an early member of an intellectually exciting and practically useful field that also touches on fundamental issues, such as individuals' competence to manage their own affairs, the balance of power between experts and laypeople in society, and even our view of human nature. The field's results have been replicated, elaborated, extended, qualified, and sometimes contradicted in companion fields such as behavioral finance, behavioral accounting, behavioral economics, and behavioral law.

It would be presumptuous to prescribe where the field should go next. A guarded prediction for where it will go next might be found in an article that I came across, when working as a graduate research assistant for Danny Kahneman. At a time when decision research had no dedicated journals (although Jim Naylor had already the vision to encourage the submissions that led to *Organizational Behavior and Human Performance* eventually becoming *Organizational Behavior and Human Decision Processes*), relevant research was widely scattered.

One place where I looked was the *Journal of Applied Psychology*. There, I found a retrospective by John G. Darley (1968). Honoring the journal's fiftieth anniversary, he read every issue of each fifth volume. He observed that there were relatively few new big ideas. Rather, researchers revived old ideas, when they had slipped from memory or when new methods arose for studying them. Darley's clearest exceptions to this rule were the "tests of yesteryear ... that have by now disappeared into deserved oblivion as the psychotechnology and standards of test construction have become more sophisticated and rigorous." The most sustained topics were formal "concepts [such as] cross-validation, moderator variables, and question format"(p. 9). Darley also lauded the interplay of experimental research and naturalistic observation.

Analogous trends in behavioral decision research might entail heightened interest in:

a. *Formal decision theory*. As Darley observed, powerful concepts endure. The core concepts of decision theory have emerged over centuries, often refining behavioral observations. Revisiting them from new perspectives will continue to bear fruit. Although performing formal analyses is not to everyone's tastes, they can, for example, ensure the claims of "bias" are supported, that the similarities and differences among biases are understood, and that the size of biases is assessed. Small biases can be theoretically revealing, robust, and important if their size makes them hard to detect and correct—before, one day, they are large enough to matter.
b. *Sustained applications*. Seeing projects through to their full implementation requires a multiple-method approach, as Darley admired. In behavioral decision research, that requires cycling through the normative-descriptive-prescriptive loop, seeing new issues emerge as one gets deeper into a problem. It also involves collaboration with subject-matter experts who can identify the boundary conditions for the theories being tested. It may even reveal value assumptions unwittingly shared by researchers who have been isolated within their own community (e.g., about others' competence or objectives).
c. *Measures of individual differences*. Measures of decision-making style have been, in Darley's term, the "tests of yesteryear." Indeed, their results were so meager that Huber (1983) suggested abandoning them altogether, arguing that anything that had yet to be found wasn't worth finding. However, as Goldberg (1974) observed, sound tests require clear concepts. Although "style" concepts are often vague, the core concepts of decision theory are very clear. Moreover, researchers have found ways to convey them to people in experimental tasks. Measures based on those tasks have begun to show stable individual differences that correlate with decision-making behavior (Bruine de Bruin et al., 2007; Stanovich, 2011; Weber, 2011). More could follow.

A likely byproduct of these trends is the evolution of a decision science that is both more ambitious and more modest. Fulfilling Edwards's vision, that science would look at all decisions from normative, descriptive, and prescriptive perspectives. Doing so would mean facing the complexity of people, decisions, and situations. As a result, it would provide an antidote to psychology's natural temptation to make sweeping generalizations about human behavior. The result would be a maturing science – perhaps a little less fun, but also more useful, challenging, and interesting.

References

Bruine de Bruin, W., Parker, A., and Fischhoff, B. 2007. Individual differences in adult decision-making competence (A-DMC). *Journal of Personality and Social Psychology. 92*, 938–956.

Darley, J.G. 1968. 1917: A journal is born. *Journal of Applied Psychology, 52*(1), 1–9.

Goldberg, L.R. 1974. Objective diagnostic tests and measurements. *Annual Review of Psychology, 25*, 343–366.

Huber, G.P. 1983. Cognitive style as a basis for MIS and DSS designs: Much ADO about nothing? *Management Science, 29*, 567–579.

Stanovich, K. 2011. *Rationality and the reflective mind*. New York: Oxford University Press.

Weber, E. 2010. Risk attitude and preference. *Wiley Interdisciplinary Reviews: Cognitive Science,* 1(1), 79–88.

INDEX